Computers and Medicine

Helmuth F. Orthner, *Editor*

Computers and Medicine

Gilad J. Kuperman   Reed M. Gardner
T. Allan Pryor

HELP: A Dynamic Hospital Information System

With 110 Illustrations

Springer-Verlag
New York Berlin Heidelberg London Paris
Tokyo Hong Kong Barcelona Budapest

Gilad J. Kuperman, M.D.
National Library of Medicine Fellow in
  Medical Informatics
LDS Hospital/University of Utah
Salt Lake City, UT 84143, USA

T. Allan Pryor
Professor, Department of Medical
  Informatics
University of Utah;
Co-chairman, Department of Medical
  Informatics LDS Hospital;
Salt Lake City, UT 84143, USA

Reed M. Gardner, Ph.D.
Professor, Department of Medical
  Informatics
University of Utah;
Co-chairman, Department of Medical
  Informatics LDS Hospital;
Salt Lake City, UT 84143, USA

Series Editor:
Helmuth F. Orthner
Professor of Computer Medicine
The George Washington University Medical Center
Washington, DC 20037, USA

Library of Congress Cataloging-in-Publication Data
Kuperman, Gilad J.
  HELP: A dynamic hospital information system/ Gilad J. Kuperman,
Reed M. Gardner, T. Allan Pryor.
    p. cm. — (Computers and medicine)
  Includes bibliographical references.
  ISBN-13: 978-1-4612-7785-9   e-ISBN-13: 978-1-4612-3070-0
  DOI: 10.1007/978-1-4612-3070-0
  1. HELP (Information retrieval system)   2. Medical care — Data
processing.   3. Hospital care — Data processing.   4. Hospitals —
Administration — Data processing.   I. Gardner, Reed M.   II. Pryor,
T. Allan (Thomas Allan)   III. Title.   IV. Series: Computers and
medicine (New York, N.Y.)
  [DNLM: 1. Decision Making, Computers-Assisted.   2. Hospital
Information Systems.   3. Medical Records.   WX 26.5 K96h]
R858.K86   1991
610′.285 — dc20
DNLM/DLC
for Library of Congress   90-10349

Printed on acid-free paper.

Typeset by Bytheway Typesetting Services, Norwich, NY.
Printed and bound by Edwards Brothers, Inc., Ann Arbor, MI.
Softcover reprint of the hardcover 1st edition 1991

9 8 7 6 5 4 3 2 1

ISBN-13: 978-1-4612-7785-9 Springer-Verlag New York Berlin Heidelberg

*To my parents and brother and close friends
in Utah, California, and Hawaii.*

—G.J.K.

To Dr. Homer R. Warner, our mentor and department chairman.

—R.M.G.
—T.A.P.

Note to the Reader

Series Preface

This monograph series is intended to provide medical information scientists, health care administrators, health care providers, and computer science professionals with successful examples and experiences of computer applications in health care settings. Through the exposition of these computer applications, we attempt to show what is effective and efficient and hopefully provide some guidance on the acquisition or design of information systems so that costly mistakes can be avoided.

The health care industry is currently being pushed and pulled from all directions – from the clinical side to increase quality of care, from the business side to improve financial stability, from the legal and regulatory sides to provide more detailed documentation, and, in a university environment, to provide more data for research and improved opportunities for education. Medical information systems sit in the middle of all these demands. They are not only asked to provide more, better, and more timely information but also to interact with and monitor the process of health care itself by providing clinical reminders, warnings about adverse drug interactions, alerts to questionable treatment, alarms for security breaches, mail messages, workload schedules, etc. Clearly, medical information systems are functionally very rich and demand quick response time and a high level of security. They can be classified as very complex systems and, from a developer's perspective, as 'risky' systems.

Information technology is advancing at an accelerated pace. Instead of waiting five years for a new generation of computer hardware, we are now confronted with new computing hardware every 18 months. Similarly, the forthcoming changes in the telecommunications industry will be revolutionary. Within the next five years, certainly within the next decade, new digital communications technologies, such as the integrated services digital network (ISDN), will not only change the architecture of information systems but also the way we work and manage health care institutions.

The software industry is trying to provide tools and productive development environments for the design, implementation, and maintenance of information systems. Still, the development of information systems in medicine is to a large extent an art and the tools we use are often self-made and crude. One area that needs desperate attention is the interaction of health care providers with the computer. The user interface needs improvement and the emerging graphical user interfaces may form the basis for such improvements. Eventually, multi-media information must be incorporated into the workstations used by the health care providers.

To develop an effective clinical system requires an understanding of what is to be done, and how to do it, and an understanding on how to integrate information systems into an operational health care environment. Such knowledge is rarely found in any one individual; all systems described in this monograph series are the work of teams. The core of these teams is usually small but dedicated to working together over periods sometimes spanning decades. Clinical information systems are dynamic systems; the functionality is constantly changing because of external pressures and administrative changes in the health care institution. This dynamic functionality is often underestimated when systems are acquired from and maintained by vendors. Good clinical information systems will and should change the operational mode of providing care which, in turn, affects the functional requirements of the information systems. This interplay requires that medical information systems are flexible on the one hand and willingness by the organization to adjust and, most of all, provide end-user education. This interplay may take some time, perhaps a year. Although medical information systems should be functionally integrated, these systems must be modular so that upgrades, additions, and deletions of functional modules can be done incrementally in order to match the pattern of capital resources and investments available to an institution.

We seem to build medical information systems just as automobiles were built early in this century (1910s), i.e., in an ad-hoc manner disregarding standards even if they exist. Technical standards addressing computer and communications technologies are necessary but not sufficient. We need to develop conventions, agreements, standards, and perhaps even a few regulations that address the core of medicine and the principal use of medical information in computer and communication systems. I presume, if the building industry would not have developed its conventions, agreements, standards, and regulations, most of us would be living in tents or wooden shacks since we could not afford a house built with only custom parts. Standardization allows the mass production of low cost parts, which can be used to build more complex structures. What are those parts in medical information systems? We need to identify them, classify them, describe them, publish specifications, and perhaps even standardize them.

Clinical research, health services research, and medical education will benefit greatly when controlled vocabularies are used more widely in the practice of medicine. For practical reasons, the medical profession has developed numerous classifications, nomenclatures, dictionary codes and thesaurus terms (e.g., ICD, CPT, DSM-III, SNOMED, COSTAR dictionary codes, BAIK thesaurus terms, MESH, etc.). The collection of these terms represents a considerable amount of clinical activities, the large portion of the health care business, and a focus for clinical research. These terms and codes form a sort of 'glue' linking the practice of medicine with the business of medicine or the practice of medicine and the literature of medicine. Since information systems are more efficient and effective in retrieving information when controlled vocabularies are used in large databases, the attempt to unify and/or build bridges between these coding systems is a great example of unifying the field of medicine with informatics tools as is done by the unified medical language system (UMLS) projects performed and coordinated by the National Library of Medicine, NIH, in Bethesda, Maryland.

The purpose of this series is to capture the experience of medical informatics teams who have successfully implemented medical information systems. We hope the individual books in this series will contribute to the evolution of medical informatics towards a recognized professional discipline. We are at the threshold where there is not only the need but also the momentum and interest in the health care and computer science communities to identify and recognize Medical Informatics.

Washington, DC   HELMUTH F. ORTHNER

Preface

The HELP computerized hospital information system has been under development here at LDS Hospital in Salt Lake City for more than 20 years, first under the direction of Dr. Homer R. Warner and more recently under the direction of Dr. T. Allan Pryor, Dr. Reed M. Gardner, and others. Throughout its history, the system has served as a vehicle for the study of the use of computers in medical care. Research areas have included signal processing, automated data collection, medical decision making (including reasoning under uncertainty), and human–machine interfaces. As the title of this work implies, the system has been constantly under development, with new capabilities continually being added and outdated functions being either enhanced or discontinued.

Adherence to an integrated computerized patient database has remained a constant feature of the system. The integrated database ensures that all applications will have access to all patient data stored in the system regardless of what application initially stored the data. For example, a program designed to review medication orders for potential adverse reactions has access to information from the clinical laboratory even though the laboratory information was stored by a completely different program (and even by a different computer).

LDS Hospital, the integrated computerized patient database, and computerized medical decision-making techniques have combined to provide the HELP system designers with an environment in which to develop many diverse and innovative clinical computing applications. New computerized applications have been incorporated into clinical practice and have been tested and evaluated. Although the number of clinical applications has grown, accurate descriptions of these applications have been scattered throughout many journals and conference proceedings. Interested parties could only be introduced to the system on an individual basis and describing the entire system became overwhelming. Even some people working closely with certain aspects of the HELP system would be unaware of details of other parts of the system. There was obviously a need for a description of the system for all those expressing interest in it.

This book attempts to fill that need. The task of concisely and completely describing the HELP system began as an assignment from one of the authors (RMG) to another (GJK) when the latter began graduate work in Medical Informatics at LDS Hospital and the University of Utah in 1988. The result was a manual describing each of the HELP system's applications. The manual was distributed to many of the visitors coming to view the HELP system at LDS Hospital and it received favorable responses. Because the manual was well received, we felt a comprehensive description of the HELP system may have broad appeal. The manual was given a positive review by the Computers and Medicine series editor and was revised for this publication.

Part I of the book has been written to give those not familiar with the HELP system (or perhaps not familiar with hospital information systems in general) an overview. Chapter 1 is a brief history of the HELP system. Chapters 2 and 3 describe the current hardware and software structures used by the HELP system. Since computers must interact with people (and other computers) to be useful, Chapter 4 has been included as an overview of the various interfaces used by HELP. Much of the material in Chapter 4 is included piecemeal in other chapters but a comprehensive overview of the topic was felt to be worthwhile. Chapter 5 discusses the many ways in which the HELP system provides decision support in the hospital setting and Chapter 6 discusses issues related to the impact of the system on users and includes a look toward the future by the system designers.

Each chapter in parts II through V describes a different HELP system application as it is in current clinical use at LDS Hospital. Where relevant, the chapter describes how use is made of the integrated patient database and where computerized medical decision-making techniques have been applied. Examples of computerized clinical reports generated by the HELP system are included as figures. Results of evaluation studies are presented. The chapters that describe the applications have been collected as follows: administrative and quality-assurance related functions (Part II), applications that figure prominently in the intensive care units (Part III), other clinical applications in active use (Part IV), and applications that are experimental or for some other reason are not in current clinical use (Part V). Parts of the book or chapters may be read individually or sequentially depending on the reader's preference. References to related chapters (a book on an integrated system necessarily contains many) are included in the text.

This book has been written to describe the system from many different perspectives. It is intended to convey technical details and some of the system's specifications as well as to give a feel for what end-users experience when using the system in their routine hospital tasks. It has been written to be read by one who is either slightly familiar with the HELP system or by someone who is completely unfamiliar with it. It has been designed to be read by physicians, nurses, health-care administrators, medical students, or anyone with an interest in the diverse field of medical informatics.

The applications have been described as they currently (Summer 1990) function. Applications under development or scheduled to be implemented are indicated as such. As the title of this book implies, however, new applications not described here will soon become functional and applications described here as functional may, for one reason or another, fall into disuse. Although change in any computerized information system is inevitable, this description of the HELP system should be useful as a basis to understand future developments.

Many people must be acknowledged in a work of this size. Dr. Helmuth Orthner, the Computers and Medicine series editor, has been encouraging and provided valuable comments on the first draft. The editors at Springer-Verlag have been helpful in arranging the text and the figures. The authors also wish to acknowledge the staff of LDS Hospital who use the HELP system daily, and the programmers at LDS Hospital and Intermountain Health Care Information Systems Division who have created the system, both of whom gave generously of their time to describe the HELP system in intense detail to one of the authors (GJK) and allowed most of the information in this book to be gathered.

GILAD J. KUPERMAN

Contents

II. Administrative Functions on the HELP System

IV. Other Clinical Applications on the HELP System

I
Overview of the HELP System

1
History of the HELP System

The HELP (for *H*ealth *E*valuation through *L*ogical *P*rocessing) system is the integrated computerized hospital information system in current clinical use at LDS Hospital in Salt Lake City, Utah. In addition to assisting with administrative and financial hospital functions (as many hospital information systems do), the HELP system distinguishes itself by additionally storing a wide variety of clinical patient data. Having access to clinical patient data enables the HELP system to assist with hospital functions such as the management of clinical patient data, order communications, results review, and the production of comprehensive clinical reports. The HELP system is also able to interpret the clinical data for the purpose of computer-assisted decision making.

The system, as it exists today, is the result of a long evolution. From unlikely beginnings, the system developed incrementally, providing a setting for valuable lessons to be learned and unavoidable mistakes to be made. The HELP system developed in parallel with the emerging discipline of medical informatics and many of the issues faced by the system developers were general medical informatics research questions. This chapter describes the development of the HELP system from its start as a tool to acquire and display physiological data, to the inclusion of other forms of clinical data, and, finally, to the development of the system into a routine clinical and decision-making tool at LDS Hospital.

1.1 The Cardiovascular Period: 1954 to 1967

1.1.1 Introduction

The HELP system had its origins in the LDS Hospital[1] Cardiovascular Laboratory, which was established by Dr. Homer R. Warner in 1954. Dr.

---

[1]The hospital was owned by the Church of Jesus Christ of Latter-day Saints (Mormon Church) and was officially known as Latter-day Saints Hospital until 1975. At that time, the Hospital came under the management of Intermountain Health Care, Inc. (with headquarters located

Warner had been involved in research in cardiovascular physiology at the Mayo Clinic in Rochester, Minnesota before coming to Utah. In 1956, in some of the earliest work done using computers in medicine, Dr. Warner used analog computers to study the propagation of pulse waves through the arterial bed. More powerful digital computers were first used when simulations of blood flow through the liver, kidney, and other vital organs needed to be performed. This work was successful and resulted in some of the earliest published work of computers in medicine [1].

1.1.2 1959: Bayesean Diagnosis Using Digital Computers

In the late 1950s, congenital heart disease was an object of study in the Cardiovascular Laboratory. Cardiac catheterization and new surgical techniques made accurate diagnosis and treatment of these diseases practicable. Diagnosis previously had been made strictly on clinical grounds. There now was an opportunity to compare diagnoses based purely on clinical findings to a "gold standard"—the diagnosis based on the catheterization results. Dr. Warner, along with fellow researchers Dr. Alan F. Toronto and Dr. L. George Veasy, noted that the presumptive diagnoses of the physicians referring the patients for catheterization was often incorrect when compared with the final catheterization results. These observations, along with awareness of early work by Ledley and Lusted [2] who had theorized that the likelihood of disease could be inferred by applying a statistical (Bayesean) approach, inspired Dr. Warner and his colleagues to see how well the computer could be programmed to make diagnoses based on clinical findings. The resulting computer program took patient findings and conditional probabilities as input and listed the diseases that could account for the findings in order of greatest to smallest likelihood. The program performed well [3]. The ability of the program to identify the actual diagnosis (based on the catheterization results) as the most likely was equal to, or better than, each of three experienced cardiologists who were asked to predict the diagnosis based on the clinical findings. The computer also generated a more complete differential diagnosis. Heart catheterizations became a routine procedure prior to surgery and the computerized diagnostic methods were never used for routine clinical practice. However, the statistical techniques that were developed are still considered important in the realm of computerized medical diagnosis.

At the time, an IBM 1620 computer was used and data entry was performed with a card punch and card reader. Programs were written in Fortran and intermediate code was output on punched cards as "memory."

---

in Salt Lake City), and the name was officially changed to LDS Hospital. Throughout this chapter and the rest of the book, only the name LDS Hospital will be used.

These punched cards were run through the card reader, which generated machine-executable code. The output was then generated and printed. Data entry required days to do what could be done in seconds on a modern personal computer. The IBM computer had a memory capacity of 8000 to 12,000 bytes. Random access memory and disk drives became available only in the early 1960s. The first disk drive used in the laboratory was half the size of a home refrigerator and stored a grand total of 5 megabytes of data.

1.1.3 1963: Digital Signal Processing

Experiments in the catheterization laboratory produced a variety of physiological signals (blood pressures from the heart chambers and vessels, dye dilution curves, the ECG, and nerve potentials) from experimental animals. Drs. Homer R. Warner, W. Sanford Topham, and one of the authors (RMG)[2] were interested in processing these signals to aid their understanding of physiological control systems. At first the signals were processed manually but the laboratory staff soon realized the computer could be used to acquire, process, and report the signals. The signals were being digitized by hand sampling and the results keypunched into the computer. When analog-to-digital converters became available, the signals could be automatically captured by the computer and converted into digital form for further processing and reporting. The first analog–digital converter (or ADC) filled a 6-foot high rack, cost $80,000, and had a (what was at the time thought to be very fast) sampling rate of 50,000 samples per second. A modern ADC with a sampling rate of more than 1 million samples per second can be purchased for $200 to $300. Much of the important early research in the field of physiological data acquisition and processing took place at LDS Hospital and contributed greatly to the field of what is now called patient monitoring. Patient monitoring has played an important role in medical care since the 1960s permitting, to a great extent, the advent of critical care medicine and intensive care units.

In 1964, the computer hardware was changed from an IBM 1620 to a Control Data Corporation 3200 (CDC 3200) computer. The CDC 3200 had serial number 3! This machine had a stupendous amount of main memory — 32K bytes. Each clinical program was designed to run in a 2000-byte partition, tiny by today's standards. A time-sharing system was developed

---

[2]Dr. Reed M. Gardner first came to LDS Hospital during the winter of 1960 as a National Science Foundation research fellow in Electrical Engineering doing work on analog computer simulations. He returned to the Laboratory in the fall of 1962 as a graduate student in Electrical Engineering doing work on acquisition and analog-digital conversion of physiological signals. He received a Ph.D. in Biophysics and Bioengineering from the University of Utah in 1968.

that would permit 15 users to be active simultaneously and would time share and swap users if they became inactive.[3]

In mid-1964 the Department of Biophysics and Bioengineering was established as an academic department at the University of Utah. The department later changed its name and is now known as the Department of Medical Informatics. Dr. Warner was the first chairman and Drs. Topham and Toronto were the first faculty members. Students were first graduated in 1965 and since then more than 60 students have received doctorates and more than 70 students have received masters from the department.

1.1.4 1964: Heart Catheterization Laboratory Computerization

After the success in obtaining and processing physiological signals (i.e., ECG and blood pressure) in the laboratory, it was decided to try to obtain the same results from patients undergoing heart catheterization. Drs. Warner, Toronto, Topham, Gardner, and W. Clinton Day were involved in the project.

Automation of the heart catheterization laboratory required development of a computer terminal. At the time storage cathode-ray tubes (CRTs) were just becoming available. (Storage CRTs write the information only once to the screen as opposed to the constant refresh technology used today.) The first computer terminals were custom-built and cost about $4,000 each. Sterile "wands" were developed to permit physicians to operate a keyboard and maintain sterile conditions.

The automatic acquisition of heart catheterization data became part of the clinical routine at LDS Hospital [4]. The system was more accurate than hand measurements, gave real-time results, and allowed physicians to have automatically printed reports within minutes of finishing the case. At that time, most of the heart diseases treated were caused by rheumatic fever or were congenital defects. Acquisition of the physiological data was crucial to planning the surgical treatment.

1.1.5 1966: Intensive Care Unit Computerization

After experiencing the benefits of monitoring in the heart catheterization laboratory, heart surgeons realized that the computer could help them to monitor their patients during and after the surgical process as well. The methods developed in the catheterization laboratory were extended to two open-heart surgical suites and to a six-bed intensive care unit for post-

---

[3]It was at this time that T. Allan Pryor came to work at the Cardiovascular Laboratory. He had previously worked at Rockwell International in Cape Canaveral, Florida programming a similar Control Data Corporation computer for projects involved in the early space program. He eventually earned an M.S. in Mathematics and a Ph.D. in Computer Science from the University of Utah.

open-heart surgical patients. The computer installed in the open-heart unit used vacuum tube and analog computer technology since it was impractical to have huge digital devices in a clinical setting. Connecting the computers to the bedside monitors required special integrated circuits, switching circuits, and digital multiplexors, all of which had to be custom-built.

1.1.6 1967: Regional Medical Program Outreach

In 1967 the Regional Medical Outreach Program granted LDS Hospital $2 million to be used over 3 years to study the feasibility of using computers to communicate with geographically distant hospitals. The purpose of the program was to develop techniques for bringing state of the art medical techniques and medical knowledge to rural hospitals. The grant allowed for expansion of the LDS Hospital computer system (this included an upgrade to a CDC 3300) and for the development of communications techniques. Four other hospitals were all equipped to be able to perform computerized intensive care unit (ICU) monitoring, electrocardiography (ECG), pulmonary function testing, and cardiac catheterization. Signals would be transmitted to the LDS Hospital computer system that would provide interpretation functions. Communications were via analog modems over telephone lines. With the help of AT&T, special analog modems were developed that allowed the storage tube CRTs to be used remotely. Although the concepts and their operation were a success, problems such as noise on the communications lines and repeated failures of the devices pointed out the importance of reliability in a system's routine clinical functioning.

Computers and Biomedical Research was established in 1967, with Dr. Warner as editor, to provide a forum for researchers in computers and biomedicine. The journal continues to be one of the leading publications in medical informatics.

1.2 Expansion of the Database: 1967 to 1972

In 1965 the sequential multichannel chemistry analyzer became available. This system (the Technicon SMA-12) could analyze 12 chemical values from a single serum sample. John D. Morgan was assigned to interface the SMA-12 to the computer system so that laboratory data could be reviewed in the catheterization laboratory and ICUs along with the hemodynamic data. The benefits of having many forms of clinical data available concurrently quickly became evident and projects that captured other types of patient data were soon begun.

Drs. Warner, Gardner, and Pryor determined that the best way to store and retrieve clinical information required some kind of medical terminology coding scheme. Work began on the coding scheme known as pointer-to-text, or PTXT (pronounced "P-text"). This scheme is still used and is described in detail in Chapter 3. Coding of medical terms was a concept that would later permit the computer to perform automated medical decision making. At the same time, John Morgan, who had interfaced the SMA-12 machine to the central computer, began to explore the possibility of using the SMA-12 results to infer diagnoses. He soon found that no facile computerized diagnostic coding scheme existed so he encoded the Systematic Nomenclature of Pathology (SNOP) into a computer-usable form [5] and medical records diagnostic data were now entered into the computer using SNOP codes. Soon thereafter, admit-discharge-transfer (ADT) data from the admitting office were added to the database. (An extension of Dr. Morgan's coding effort was later sold to 3M Corporation and became the Code 3 product line, which helps medical records departments find appropriate DRG, HCPCS, ICD-9-CM, and CPT-4 codes for hospitalized patients at discharge.)

1.2.1 Clinical Laboratory Automation

The direct interfacing of the laboratory analyzers to the central computer in 1966 allowed laboratory data to be collected automatically and transmitted to the ICU and used in the clinical care of patients. This system functioned until 1972, at which point a special purpose, stand-alone laboratory system was installed. The new laboratory system, functioning on a CDC 1700 computer, interfaced with the laboratory instruments (chemistry analyzers, Coulter counter, manual tests, etc.) and managed the laboratory workload. After the laboratory computer and the central computer were interfaced, the laboratory computer could receive data (e.g., demographics) from the centralized hospital computer and could transmit laboratory results back to the central computer. Implementation of the computerized laboratory system was crucial to many subsequent computerization projects and was the beginning of an integrated patient database.

In 1970 a local company was formed for the purpose of marketing the combined ICU-laboratory system running on the CDC 3300 and 1700 machines. The company (called Medlab) did not prosper initially and after about 2 years was acquired by CDC. Control Data Corporation did not have much success either marketing the combined ICU-laboratory system; however, they eventually were able to be fairly successful with the laboratory system alone, installing more than 150 systems. The ICU module did have limited success at George Washington University in Washington, D.C., and at the VA Hospital in Minneapolis, Minnesota.

1.2.2 Computerized Electrocardiogram Interpretation

The year 1968 brought automated interpretation of the electrocardiogram (ECG) to LDS Hospital. The work in this area was done primarily by Dr. Allan Pryor with the assistance of Drs. Allan E. Lindsey and Homer Warner and other cardiologists who participated in development of the interpretive computer algorithms. The basic scheme for the computerized ECG interpretation used the Frank Lead system (the X, Y, and Z vector electrocardiogram). Analog ECG signals were acquired in the patients' rooms and were transmitted via telephone lines to the computer's ADC. The central computer performed the interpretation.

This system operated for more than 15 years as the primary source of computerized ECG interpretation at LDS Hospital. It was only in 1987 that the system was replaced with the Marquette MUSE System. The Marquette ECG interpretation system, discussed in Chapter 28, uses the conventional 12-lead interpretation rather than the vector ECG. The vector ECG interpretation project provided experience with computerized medical data interpretation and was a forerunner to more general medical decision-making strategies.

1.3 Growth of Clinical Applications/Addition of Decision-Making Capability: 1972 to 1990

1.3.1 Growth of Clinical Applications

With several years' experience gained in the ICUs, catheterization laboratories, ECG analysis, and laboratory data reporting, it became apparent that a clinical hospital information system could do more than just acquire and present data—the data could be interpreted as well. In 1972 data automatically gathered by the computer system included chemistry, ECG, blood gas, and spirometry results. Demographics, medical records data, and certain clinical observations were also entered. The two first LDS Hospital computer decision-making applications to become clinically operational were interpretations of the ECG and blood gas results. The basics of decision making with, what was now called HELP, were described in 1972 [6]. The name "HELP" was chosen because the system would "help" the physician. It was about 2 years later before the phrase "*H*ealth *E*valuation through *L*ogical *P*rocessing" was chosen as the phrase that HELP should stand for.

A key factor of the system was that computerized logic could be activated when new patient data was added to the database (i.e., the system could be "data-driven"). Thus, programs functioned automatically and did not have to be manually activated. Another important component of the HELP computer program was the fact that the "knowledge base" (i.e., the computerized medical logic that the computer used to make its decisions) was distinct from the routines that governed the procedural functioning of the program. This allowed the "medical knowledge" in the program to be manipulated as a distinct entity.

In the early 1970s a language called HCOM (for *HELP com*piler) was developed to facilitate the development of computerized logic modules, or sectors as they were called. With HCOM a user could define the desired database variables and related time constraints and use if–then or Bayesean rules to develop logic modules that would evaluate patient data. The Bayesean strategies were developed to derive diagnoses from self-administered patient histories [7]. The computerized history-taking program was not successful clinically because the ability to put the data to the best use had not been developed yet and because the computational power to make the program run smoothly was lacking. The technology still has not yet matured to the point where acquiring complete patient histories from patients sitting at terminals is an easy task. However, the Bayesean strategies developed in the 1960s and 1970s have been used elsewhere, including in ILIAD [8], a computerized medical education and medical testing tool in which Dr. Warner has been intimately involved.

Just after 1972 several new clinical applications were developed and more nursing divisions at LDS Hospital were provided with computerized services. Medication ordering, with an automated medication monitoring system, was developed and installed. Development of a computerized nurse charting program was initiated. A computerized general order entry program was established. Services were expanded into other ICUs and some general nursing floors. System tools such as the General Questionnaire Asking Program (GQAP, described in Chapter 4) were developed to facilitate application development.

1.3.2 Conversion to the Tandem Computer System

In the late 1970s the limitations of providing clinical services on an outdated computer were painfully apparent. Reliability and rapid response time were determined to be the two features of a computerized hospital information system crucial to achieve user acceptance and functionality. To achieve high reliability, a second (research) CDC 3300 computer was purchased. If there were failures in the clinical system, the research system could be activated as a backup, but several cables needed to be moved. The dual CDC 3300 was available about 97% of the time. For clinical applications, however, greater than 99% availability was felt to be necessary.

A search was begun for a more reliable and faster computer system. The decision was made to convert HELP from CDC hardware to hardware supplied by Tandem Computer, Inc., of Cupertino, California. The complete transition took from 1980 to 1982 and involved rewriting system programs and creating new development tools. The 2 years of transition was a frustrating and difficult experience as many functioning programs had to be completely rewritten. However, the final result was the desired clinically reliable system.

During the conversion to the Tandem System, many terminals and printers were installed throughout the hospital. Each nursing division was given at least one terminal and one printer. Charles River Data Systems (CRDS) unix-based supermicrocomputers were implemented (and continue to be used) as multiplexors and data acquisition devices. The current hardware configuration of the HELP system is discussed in detail in Chapter 2.

1.3.3 Expansion of Decision-Making Capability

In the 1980s several clinical programs were developed that made use of HELP's decision-making abilities. Between 1984 and 1986, an initial version of a high-level database query language (PTXT Application Language, or PAL) was developed to assist in application development, report generation, and research projects. Previously, all applications programs had been written in the Tandem system language, Transaction Application Language (TAL). (PAL, TAL, and the HELP frame language — the current system languages — are discussed in Chapter 3.) Computerized clinical applications such as respiratory care charting, infectious disease monitoring, blood ordering, ventilator management protocols, and laboratory alerting were all developed in the 1980s and made use of the HELP system's decision-making ability. These computerized clinical applications are discussed in detail in later chapters. Several other computerized clinical applications (e.g., computerized surgery scheduling, nurse charting, x-ray scheduling, APACHE scoring, etc.), which capture patient data that are used by computerized decision-making applications, were developed or expanded at the same time. All of these applications have become part of the routine day-to-day functioning of LDS Hospital. An overview of information flow within the HELP system is shown in Figure 1.1.

1.3.4 Intermountain Health Care Net

In late 1987 Intermountain Health Care (IHC), Inc., the parent corporation of LDS Hospital, decided to install the HELP System in 10 of its larger hospitals and link the systems to allow interhospital communications. The project is underway and is called IHC Net. When completed, nearly identical HELP systems will function at each of the 10 hospitals and will be able to communicate with each other. For example, if a patient is admitted to McKay-Dee Hospital (an IHC hospital located in Ogden, Utah, 40 miles north of Salt Lake City) and subsequently transferred to LDS Hospital (e.g., for a special procedure), the patient data can be rapidly transferred to LDS Hospital's HELP system electronically. The proposed data-sharing scheme has raised several interesting issues, such as the standardization of procedure nomenclatures, data structures, and classification schemes (e.g., of physicians, supplies, etc.) across institutions that previously had no reason to be concerned with how the other institutions carried out their work.

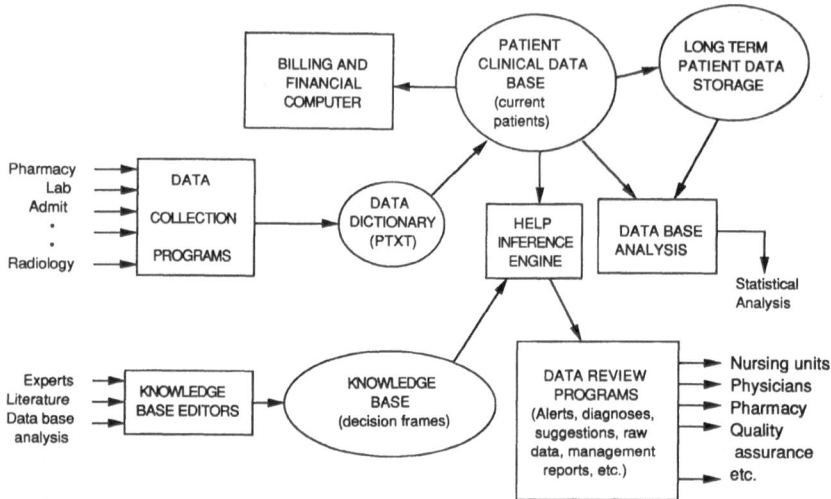

Figure 1.1. HELP system data flow diagram.

Solving these problems is challenging but must be done in order to obtain the potential benefits.

1.3.5 The 3M/Intermountain Health Care/ Medical Informatics Joint Venture

In late 1980, CDC acquired the rights to market the HELP system on Tandem hardware from the Department of Medical Informatics. The department received funding that allowed for system expansion and additional research.

In the mid-1980s 3M Corporation's Health Information Systems Division (located in Salt Lake City) purchased the rights to both the HELP System and the Medlab product from CDC. The Medlab system was rewritten and now functions on Data General hardware. In 1988, 3M, IHC, and the Department of Medical Informatics at LDS Hospital entered into a three-way joint venture for the purpose of developing and marketing the HELP system. Intermountain Health Care and 3M are contributing the development resources and the Department of Medical Informatics at LDS Hospital is providing the direction. The goals of the joint venture are for 3M Corporation to market a high quality computerized hospital information system, for IHC to have the HELP system installed in each of its major hospitals, and for the Department of Medical Informatics to have the opportunity to experiment with current theories of medical information science in a functioning system.

1.4 Conclusion: The HELP System Today

The research and development that have taken place at LDS Hospital in physiological monitoring, systems integration, database development, man–machine interface, and medical decision making makes the LDS Hospital experience in the field of computerized hospital information systems unparalleled. Additionally, in a research field where 1975 is often considered ancient history, the HELP system has benefited from an early start, from having the opportunities to experiment over a long period of time with developing technologies, and from the practical need to provide real-world solutions for an active and mature clinical setting.

Today HELP is part of the routine at LDS Hospital. Functioning without the computer would be unimaginable to most of the hospital personnel. The system, however, is far from static. Research into the unsolved question of how best to use computers in the hospital setting continues all the time and development of new projects that test research ideas or fill practical needs in the hospital are also always ongoing. Future research directions for the HELP system are discussed in section 4 of Chapter 6. The dynamic nature of the HELP system makes it an exciting system to observe but a difficult one to describe precisely. The goal of the rest of this book is to describe the state of the HELP system at a single point in time.

References

[1] Warner HR. A study of the mechanism of pressure wave distortion by arterial walls using an electric analog. *Circ Res* 1957;5:79.

[2] Ledley RS, Lusted LB. Reasoning foundations of medical diagnosis. *Science* 1959;130:9–21.

[3] Warner HR, Toronto AF, Veasey LG, Stephenson R. A mathematical approach to medical diagnosis: application to congenital heart disease. *JAMA* 1961;177:177–183.

[4] Warner HR, Gardner RM, Pryor TA, Stauffer WM. Computer automated heart catheterization laboratory. UCLA Forum in Medical Science #10. Berkeley: University of California Press, 1968.

[5] Morgan JD. *A Computerized "Conversational" Technique to Form Numerically Coded Medical Diagnoses*. Salt Lake City, UT: University of Utah; 1971. Dissertation.

[6] Warner HR, Olmsted CM, Rutherford BD. HELP—A program for medical decision making. *Comput Biomed Res* 1972;5:65–74.

[7] Warner HR, Rutherford BD, Houtchens B. A sequential bayesean approach to history taking and diagnosis. *Comput Biomed Res* 1972;5:256–262.

[8] Warner HR, Haug PJ, Bouhaddou O, et al. ILIAD as an expert consultant to teach differential diagnosis. *Symposium for computer applications in medical care (SCAMC)* 1988;12:371–376.

2
HELP System Hardware

HELP has always been a mainframe-based computerized hospital information system. As discussed in Chapter 1, the system initially operated on
CDC 3300 computers and by the late 1970s was providing patient monitoring, signal processing, and laboratory interpretation services. As the demands on the system grew, this hardware started to become inadequate in
terms of response time and expandability and in 1979 the decision was
made to convert the HELP system to operate on an expandable multiprocessor Tandem computer system. The Tandem computer became operative
in 1980 and continues to be the heart of the HELP system.

During the period that the HELP system ran on the CDC computers,
and for the first few years that the Tandem computer was used, the central
computer provided the vast majority of the system's total processing power.
Increasingly, however, specialized computing services are being provided
by other computers that are being connected to the Tandem computer. The
use of other computer systems yields two benefits. First, it permits exter-
nally developed applications to be used by the HELP system and, second,
it relieves the central Tandem computer of some of the total processing
burden. Figure 2.1 shows the many computer systems that are linked to
the Tandem computer to compose the current configuration of the HELP
system.

2.1 Tandem Computer

The heart of HELP is a system of 10 (expandable to a maximum of 16)
Tandem TXP processors, each with 8 megabytes of main memory. Each
TXP processor has a capacity of approximately 2 MIPS (millions of instructions per second). Multiple simultaneous processes may be distributed
among the multiple processors; however, any single process is constrained
to a single processor. The average response time for the Tandem varies

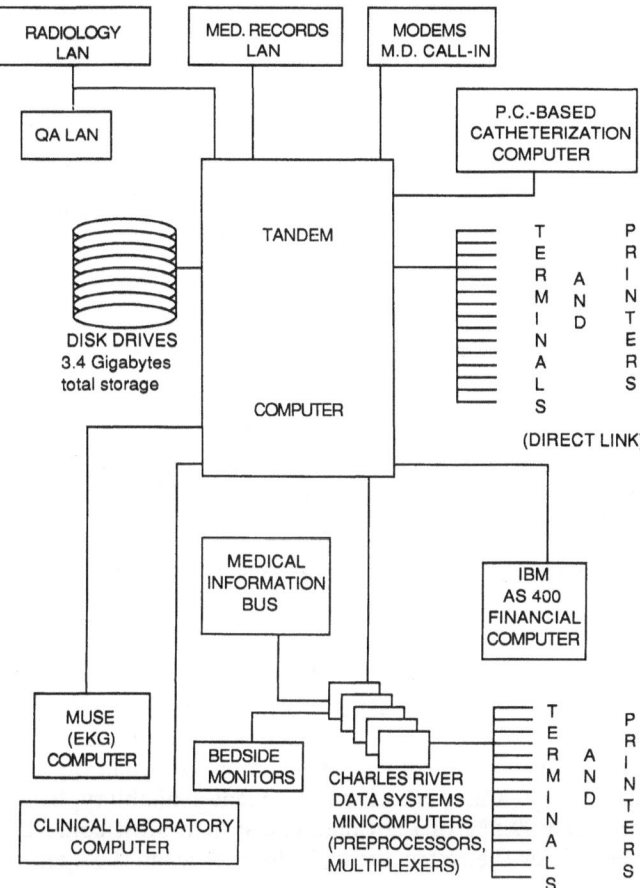

Figure 2.1. HELP system hardware configuration.

from 2 to 4 seconds depending on the time of day. Average response times over a 24-hour period are shown in Figure 2.2. The response time depends on the system load with maximum use coming in the middle of the day. At about 5 AM and at 6 PM the system load is increased due to the generation of the computerized nursing shift reports (discussed in Chapter 13), a processor-intensive task. The later evening peak corresponds to the time of creation of tape backups. Whereas users of any computer system have an insatiable desire for faster and faster response times, LDS Hospital personnel seem to be generally satisfied with the current system performance.

At the time of the 1979 conversion, Tandem computers were chosen because of their central processor and disk hardware redundancy and intelligent system software that results in "fault-tolerance." Fault-tolerance implies that no one single hardware failure will halt the system — an important

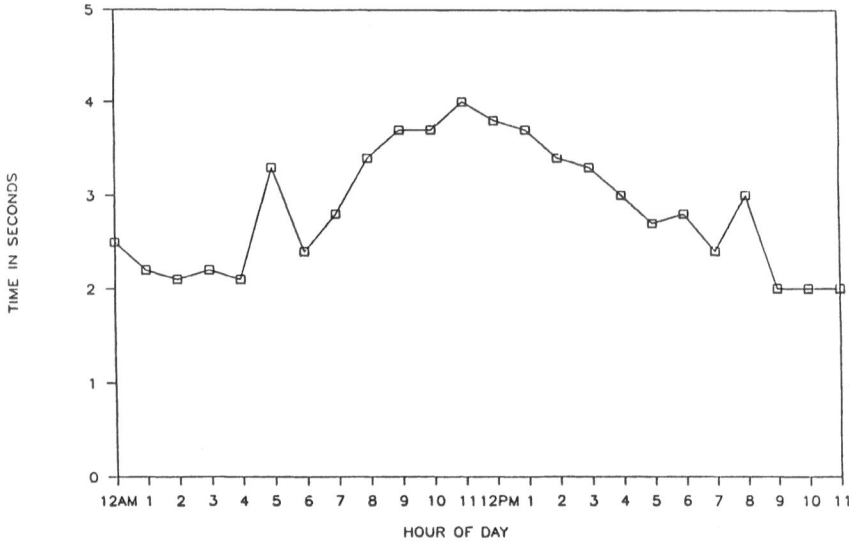

Figure 2.2. Average Tandem computer response times at LDS Hospital.

feature in a system that is used 24 hours a day, 7 days a week, as the primary means of providing clinical information to hospital personnel. The system incorporates backup central processors, data buses, and power supplies that activate automatically in the event of failure in the primary unit. As a result of this feature, HELP has excellent availability, being clinically active more than 99.5% of the time. The system must come down when, for example, major file restructuring is done or when the operating system is upgraded.

The Tandem computer operates under the proprietary Guardian Operating System. Although Tandem supports standard computer programming languages such as Fortran, C, and Pascal, most of the HELP system's file utility programs have been written in the low-level Transaction Application Language, TAL, supplied by Tandem. Higher level languages have been written for data query and decision-making purposes and are discussed in Chapter 3.

Programs files and patient data are stored on 14 disk drives. The eight drives that are currently used for clinical purposes store 2.6 gigabytes of data, while the remaining six (used for research purposes) hold 0.8 gigabytes. Another example of redundancy is that the clinical drives are mirrored (i.e., two drives hold the same data) to reduce the possibility of data loss or unavailability in the event of a disk failure. When data are accessed, the system retrieves the data from the drive that has the read/write head closest to the desired data, thus minimizing data retrieval time.

Eight hundred fifty terminals and 100 line and laser printers are currently active throughout LDS Hospital. About half of these are connected directly to the Tandem, the other half are connected via unix-based CRDS mini-computers serving as multiplexors. All of the nursing units have terminals located at the nursing station and at various hallway sites. The four ICUs (60 beds) and three 48-bed acute care units also have terminals at each bedside. A terminal is also located in each of the 24 operating rooms.

Two recent developments are changing the number and nature of terminals throughout LDS Hospital. First, bedside terminals are being installed throughout the hospital. This will allow functions such as data review and computerized nurse charting to be performed at the bedside. Second, terminals are being converted to 80286-based personal computers as opposed to the simple ASCII terminals used previously. The processing power of the personal computers is used to generate a more sophisticated, window-based, user interface and for a small amount of intermediate data handling. The use of bedside terminals for computerized nurse charting is discussed in Chapter 13.

2.2 Other Computer Systems Connected to the Tandem

Many other computer systems are networked to the Tandem to provide specialized capabilities and to add processing power. Computer systems that feature prominently in clinical applications are discussed in the chapters that discuss the particular application.

Eighteen CRDS unix-based minicomputers are interfaced to the Tandem. The CRDS machines serve as multiplexors and preprocessors for terminals on the nursing divisions, and in the surgery, medical records, pulmonary, and medical informatics departments. For example, the CRDS machines preprocess the large volume of physiologic patient data generated by bedside monitors in the ICUs. The CRDS machines decide which data are relevant and then transfer those data to the Tandem for permanent storage. Copies of frequently accessed data files (e.g., patient census data) are kept on the CRDS machines to decrease access to the Tandem computer. Using the CRDS machines as multiplexors also decreases the total number of Tandem input-output ports that are needed for comprehensive intrahospital communication. The cost of Tandem ports is higher than that of CRDS ports.

The Medical Information Bus (MIB), Marquette ECG System, and the LabForce clinical laboratory system are examples of computer systems that provide specialized functions for HELP. The MIB (Chapter 20) provides a standard hardware and software environment for data communications between a host computer and patient monitoring and therapeutic devices frequently found in ICU settings. Examples of devices communicating with the HELP system through the MIB are intravenous pumps, ventilators, hemodynamic monitors, and oximeters. The MIB data are collected and processed by a CRDS computer before being transmitted to the Tandem. The Marquette ECG computer system, MUSE (Chapter 28), is used for signal processing of ECGs and is used by cardiologists to review and finalize ECG interpretations. The final ECG interpretations (rather than the waveforms themselves) are communicated in batch mode to the Tandem computer. The laboratory system (running on Prime hardware with LabForce software) manages workflow within the laboratory, interacts with computerized measuring devices, and transmits results to the Tandem computer. The laboratory system is described in detail in Chapter 16.

Many of the specialized computer systems connected to the HELP system maintain a local database. However, a copy of all relevant clinical patient data is sent to the Tandem. All patient data are stored in the computerized patient files (discussed in Chapter 3) residing on the Tandem's disk drives so that the Tandem is able to manage the patient data (e.g., for results review or decision-making purposes) without having to access data residing on other computer systems.

Several personal computer (PC) applications have been developed that interact with the Tandem computer. HELP-related PC applications have been created both for stand-alone PCs as well as for local-area networks (LANs). A PC-based program has been written to help analyze and record data from patients undergoing cardiac catheterization (Chapter 27). The analyzed information is transmitted to the Tandem Computer after completion of the case. Physicians admitting patients to LDS Hospital may use modems with PCs (Chapter 31) to remotely access (e.g., from their homes or offices) administrative and clinical patient data stored on the Tandem computer.

Various departments within the hospital have installed PC LANs to assist with local data management and analysis. Communication protocols have been developed to transfer data between the LANs and the Tandem computer. For example, the Radiology Department (Chapter 29) has installed a PC LAN to facilitate transcription of radiology reports. The reports are entered via a standard word processor into the LAN but are transmitted to the Tandem for printing, natural language processing, and on-line results review. The quality assurance (Chapter 9) and the medical records (Chapter 8) departments both use PC-based database programs to analyze data downloaded from the HELP system as part of their normal work routine.

The HELP system keeps track of tests performed and services provided throughout a patient's hospital stay, but the Tandem does not directly generate the patient bills. Rather, billing functions are performed by a free-standing IBM System AS 400 Computer. The HELP system keeps track of the billable items for each patient in a transaction file (discussed further in Chapter 3). A copy of the transaction file is transmitted from the

Tandem to the AS 400 on a nightly basis and is used to generate the patient bills.

2.3 Summary

The multiprocessor, fault-tolerant, Tandem computer is the core of the HELP system. Whereas many other computers are linked to the Tandem to provide specialized functions, all data in the integrated database are stored on the Tandem disk drives. Users interact with the HELP system through terminals and printers located throughout the hospital. The next chapter describes the data structures, file structures, and medical terminology representation schemes employed by the HELP system.

3
The Data Dictionary and Data
and File Structures

The fundamental software structures of the HELP system have been designed to maximize efficiency and flexibility. A numerical coding scheme for medical terms was chosen in the early 1970s so computerized interpretation and decision making could be performed along with data review. The HELP system's data and file structures have been designed so that data retrieval can be performed rapidly enough to function satisfactorily in a demanding clinical setting. Much of the HELP software has been locally developed to maximize efficiency and also because standardized tools were not widely available through much of the system's development.

This chapter discusses the data dictionary (medical term representation scheme), data structures, and file structures used by the HELP system. Where relevant, the rationale for specific design decisions is given as well as the limitations that have resulted from these choices. The chapter ends with an introduction to the methods available for querying the HELP database. Approaches to computer-assisted decision making with the HELP system are discussed in Chapter 5.

3.1 Medical Terminology Representation Scheme

3.1.1 Hierarchical Structure

HELP uses a hierarchical, numerically based coding scheme to represent medical terms (e.g., names of drugs and laboratory tests, diagnoses, physical findings, nursing tasks, etc.) internally (see Figure 3.1). A numerically based scheme allows each medical term to be represented uniquely. As shown in Figure 3.1, the medical term "systolic blood pressure" is represented by the numerical code "7 1 2 1." A hierarchical scheme allows properties of a general term to be "inherited" by a more specific term. For example, it can be determined that systolic blood pressure is a physical finding since the first digit of the code is a 7. Furthermore, if we retrieve all

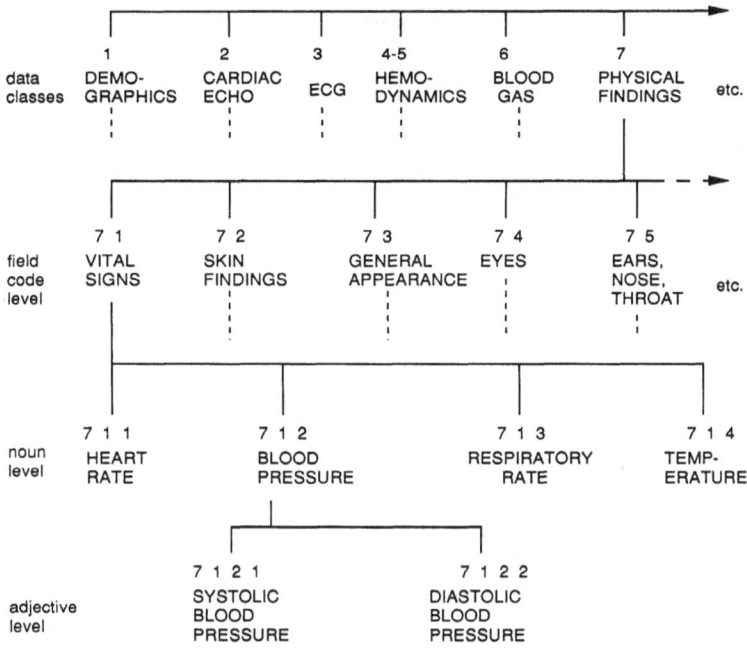

Figure 3.1. Hierarchical representation of medical terms in the HELP system.

of the physical findings for a patient, systolic blood pressures will automatically be retrieved.

3.1.2 Pointer-to-Text

The HELP system uses an eight-integer (or 8-byte) coding scheme to represent medical terms, although only 5 of the bytes are involved in the hierarchy. The 8 bytes are known as the term's PTXT (pronounced "P-text," for *Pointer-to-text*) code. There are currently more than 120,000 medical terms in the PTXT file and management of this file (e.g., restricting the number of new entries, ensuring that existing codes are used appropriately, etc.) is an ongoing effort. The eight bytes are known respectively as the code's:

1. data class
2. data type
3. field code
4. level

5. noun
6. adjective
7. adverb
8. modifier

Only bytes 1, 3, 5, 6, and 7 are involved in the hierarchy. The remaining bytes convey other information that is explained below. The depth of the hierarchy limits the precision to which the medical terms can be subdivided for the purpose of categorization. The five levels of hierarchy supported by the HELP system have been adequate for most medical descriptions but occasionally the fixed depth has imposed some restrictions. Another problem with the hierarchical representation scheme is the inability to impart more than one concept to a single medical term. Giving a different meaning to a medical term depending on the context is known as aliasing. As an example, at different times one would wish to consider hepatitis B as a disease of the gastrointestinal system and at other times as a viral disease. In a pure hierarchical scheme, one solution is to enter the term redundantly (i.e., at more than one point in the hierarchy); however, this increases the total number of terms and still requires the two codes representing the same term somehow to be linked. Solutions to the problems of fixed depth and the need for aliasing are currently being sought. The grammatical designations of bytes 5, 6, and 7 (i.e., noun, adjective, and adverb) name the levels of the hierarchy and do not correspond precisely to grammatical equivalents. There is a loose correspondence, however, in that a term at the adjective level "modifies" (i.e., is more specific than) the corresponding term at the noun level. Similarly, an adverb "modifies" an adjective.

The medical term's data class is represented by the first byte in the PTXT code. Data classes (of which there are currently slightly fewer than 60) represent broad categories of patient data. In addition to being the first level of the term's hierarchical representation, data classes are important in the manner in which data is physically stored on the disk. Data storage schemes are discussed later in the chapter. A list of the HELP system data classes is shown in Table 3.1.

Field codes represent the next level down in the hierarchy. For example, radiology data are contained in a single data class; within that data class the field codes symbolize different kinds of radiological examinations (e.g., CT, MRI, plain films, etc.). In the pharmacy data class, field codes refer to classes of drugs (e.g., antibiotics, antipyretics, etc.). The field code is also important because it is the "event level" in the HELP system hierarchy and specifies which data items will be physically stored together on the disk. Nouns, adjectives, and adverbs, are used to subcategorize further the medical terms in the hierarchy.

The data type, level, and modifier (bytes 2, 4, and 8 of the PTXT code) are the three bytes of the PTXT code not involved in the hierarchy. Data type was introduced in the HELP system to accommodate fixed length

Table 3.1. HELP system data classes

1. Patient identification data	29. Electron microscopy
2. Cardiac echo	30. Infectious diseases
3. ECG	32. Cardiovascular catheterization
4–5. Monitoring/physiology (ICU hemodynamics)	laboratory
6. Pulmonary function/blood gas	33. Central service
7. History and physical	34. Blood bank
8. Pharmacy/medications/drug allergies	35. Dietary
9. Medical records (SNOMED)	36. Respiratory therapy
10. Medical records (protocols)	37. Physical therapy
11. Decision logic	38. Surgery procedures
12. Triage (patient protocols)	39. Radiation
13. Laboratory (chemistry)	40. Room charges
14. Laboratory (chemistry)	41. EEG/EMG
15. Laboratory (hematology, urinalysis, coagulation)	42. Cardiovascular monitoring
16. Laboratory (microbiology)	43. Peripheral vascular laboratory
17. Laboratory (drug levels in body fluids)	44. Emergency room physician RVS codes
18. Family member data	45. Emergency room supplies
19. Kidney dialysis	46. Operating room supplies
20. X ray	47. Endoscopy
21. X ray (research)	48. Given drug strings
23. Echocardiograms (research)	49. Administration
24. Medical records (ICD-9-CM codes)	50. Shortstay surgery
27–28. Nursing care plans/actions/orders	51. Anesthesiology
	52. Miscellaneous service charges
	54. Trauma registry
	55. Nurse staffing

(called type 0) and variable length (called type 1) data strings. Fixed length data strings are being phased out. The structure of a variable length string (type 1) is discussed later in this chapter. For simplicity, attempts are being made to represent all data in the HELP system as type 1. The level refers to the number of relevant bytes in the PTXT code of that item (see Table 3.2). The level is important in data storage and in identification of modifiers. A modifier on the HELP system is a medical term that can modify other

Table 3.2. Meaning of level byte of a PTXT code

level = 0: field code is the deepest level for this PTXT code
level = 1: noun level is the deepest
level = 2: adjective level is the deepest
level = 3: adverb level is the deepest
level = 4: the term is a data class modifier
level = 5: the term is a field code modifier

medical terms in the same field code or data class. For example, a sodium value in a patient's record may have been obtained from a sample of serum, urine, or cerebrospinal fluid. The source of the bodily fluid will be designated by a modifier appended to the sodium value. Similarly, a modifier will be used to indicate if a measured blood pressure was obtained in the supine or standing position. If a modifier's PTXT level byte is equal to 4, then the term can modify any term in the data class and is called a data class modifier. If the level is 5 then it is a field code modifier and can only modify terms belonging to that particular field code. An example of the use of modifiers is given below.

Some examples of medical terms and their respective PTXT codes are shown in Table 3.3. Codes shown are for complete blood count (CBC, a common hematological profile) and for some of the specific tests that help to make up the CBC. (WBC refers to white blood cell count and RBC refers to red blood cell count.) Hemoglobin, WBC, RBC, and hematocrit are below CBC in the hierarchy. All of the terms belong to data class 15, one of the data classes for laboratory data. The data type is type 1; that is, the data are stored in a variable length string. The terms all belong to field code 62. The level for the term CBC is 0 since the field code is the lowest relevant byte in the hierarchy for CBC; however, the level for the other terms is 1 since they have another relevant byte.

3.1.3 The PTXT File

The medical terms, their PTXT codes, and other information relevant to the terms is contained in the PTXT file. As mentioned, there are currently about 120,000 terms in the PTXT file. The PTXT file is a key sequenced data file consisting of eight fields as follows:

1. the 8-byte numerical code representing the medical term
2. the medical term: that is, the text associated with the code (up to 474 characters)
3. keywords (used for cross referencing, up to 32 in total)

Table 3.3. Examples of medical terms and their PTXT codes*

	DC	type	FC	lv	no	adj	adv	mod
CBC	15	1	62	0	0	0	0	0
WBC	15	1	62	1	1	0	0	0
RBC	15	1	62	1	2	0	0	0
Hemoglobin	15	1	62	1	3	0	0	0
Hematocrit	15	1	62	1	4	0	0	0

*DC = data class, FC = field code, lv = level, no = noun, adj = adjective, adv = adverb, mod = modifier.

4. a flag field (e.g., active/inactive, male/female only, stat/not stat, etc.)
5. transaction code (an alternate key for billable items, used by the billing office)
6. charges associated with the terms (up to 10 fields, e.g., inpatient, outpatient, ER, etc.)
7. fees (up to 10 fields)
8. buffers (an all purpose data area, application specific).

System level utilities exist that allow the PTXT file to be searched by code, text, or keyword. Similarly, utilities exist that allow the file to be modified. The PTXT file is sometimes called the "data dictionary" of the HELP system since it is a listing of all the possible data elements in the system. The PTXT file is accessed when information about the code (e.g., the term's text, cost information, etc.) is required. The values contained in these various fields are used in many diverse administrative and clinical hospital functions.

3.2 HELP System Data Structures

3.2.1 Introduction

HELP uses magnetic disks for long-term storage of patient data. The data storage schemes have been designed in a way that optimizes data retrieval. Rapid data retrieval is essential since much of the data review at LDS Hospital is done using the HELP system. Indeed, data review is the most commonly used feature of the HELP system. Two important facts about retrieving data from a hard disk (called a disk access) have helped to determine the manner in which patient data are stored on the HELP system. First, a disk access takes a relatively long period of time to perform (milliseconds as opposed to the microseconds required by other computer operations). Although milliseconds seem like a small period of time, many accesses may be required for a routine function (such as data review) and system performance may become unacceptably slow if too many disk accesses are required. Second, disk accesses retrieve entire *blocks* of data for processing, rather than single data items. (A block of data on the HELP system is 2048 bytes.)

To minimize the number of disk accesses for data retrieval, data elements that are likely to be retrieved together are stored in physical proximity (i.e., within the same HELP system block on the disk). Therefore, the manner in which data are retrieved has been considered in design of the storage schemes. For example, if the temperatures of all the patients on a given ward were the routine type of data retrieval, all these data items would have been stored together in a single block and available for retrieval in a single access. In contrast, at LDS Hospital retrieval of one kind of data (e.g., chemistry values) for a single patient is the most commonly performed function. Therefore, all the data for a single kind of data (i.e., a single data class) for a single patient are stored in a single block (or consecutive blocks if a single block is filled). A query across patients (e.g., all medications for all patients on a given ward) takes longer to complete; however, this type of search is a less frequent event.

3.2.2 Structure of the Patient Data File

Computerized clinical patient data for all patients at LDS Hospital is stored in the patient data file. As part of the admission process, patients at LDS Hospital are assigned a patient number. (The patient number is discussed in detail in Chapter 7.) The patient number is used to create logical areas for data storage for each patient in the patient data file (see Figure 3.2). For each patient a separate data block is created for every data class; however, the block is not physically created until data from that data class

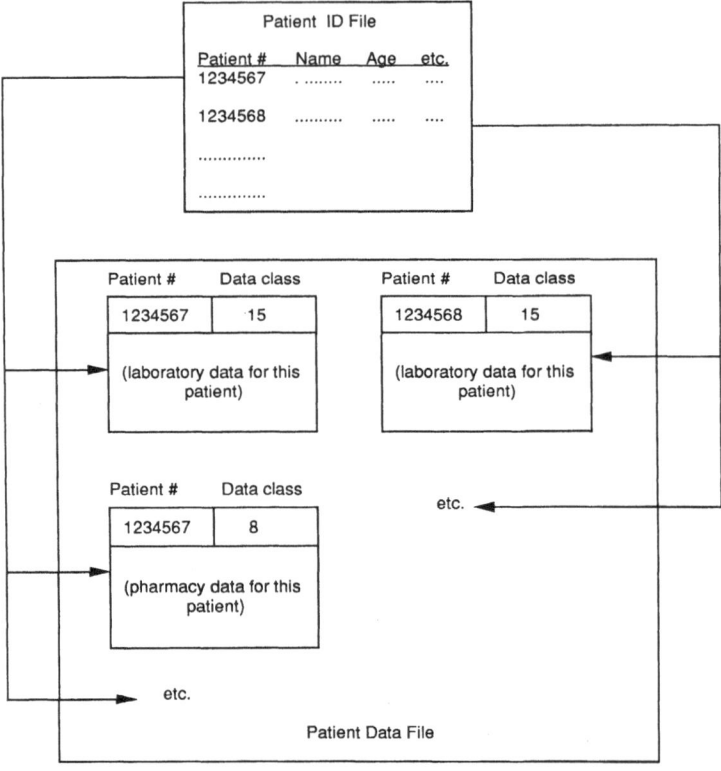

Figure 3.2. Structure and relationship of patient identification and patient data files (see text for details).

are actually stored. For example, a patient with number 1234567 would have a data block created in the patient data file the first time hematology data are stored and another created the first time pharmacy data are stored. The hematology block would be uniquely identified by the patient number (1234567) and the data class for hematology, 15. When pharmacy data are first stored the new block would be identified by the patient number (1234567) and the number representing the pharmacy data class, 8. In computer science parlance, the patient number and the data class are the key into the patient data file. If a block for one data class fills up, a new block is created.

Data Blocks. An example of a data block is shown in Figure 3.3. The block consists of a header and the data. The header consists of the patient ID

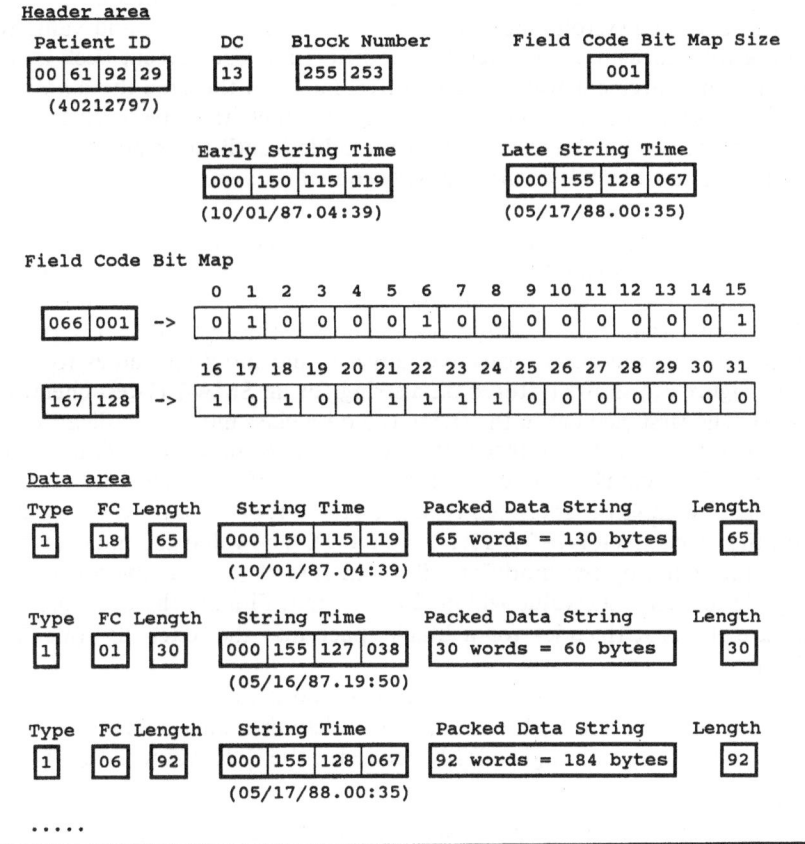

Figure 3.3. Structure of data block within patient data file (see text for details). DC = data class, FC = field code.

number, the data class, the block number (indicating the number of blocks used for this patient for this data class), a byte indicating the size of a field code bit map (discussed shortly), and an early string time and a late string time (i.e., the earliest and latest times of data stored in the block). The field code bit map follows and indicates which field codes have been represented by data stored within this data class. The early and late string times and the field code bit map facilitate data review because data retrieval programs can determine if data from a certain time period or with a certain field code exist within the block without having to scan the entire block.

Data are stored in the block in the form of data strings. A single string corresponds (as often as feasible) to a single "event" such as a single laboratory test, a single physiological measurement, or the administration of a single drug. Each of these events corresponds to a single field code and may contain more than one PTXT code. For example, a blood pressure measurement would contain the PTXT code for systolic blood pressure, the value for the systolic measurement, and the PTXT code and value for the diastolic measurement. Certain modifiers may be present such as the time the measurement was made (if different from the time the data were stored), and/or the anatomical location at which the measurement was made. Once the block exists, strings are added until the block is full. At that time a new block is created for subsequent data storage.

Data Strings. A data string also has a header and data area. As can be seen in Figure 3.4, the string header contains the data type, the field code of the string (since one string represents one field code and one event),[1] the length of the string, and a time stamp (i.e., the time the data was stored). The string contains the noun, adjective, adverb, and modifier values for each code. Figure 3.4 shows a stored data string for an SMA-7 (7-test chemistry panel). The first element in the string is a data class modifier indicating the specimen type. The first integer (250) is a data class modifier delimiter (i.e., indicates this item is a data class modifier). The integer 1 specifies that this modifier indicates the specimen type (i.e., the PTXT code for the data class modifier, "specimen type" is 13 1 1 4 0 0 0 1). The value 231 indicates that the value taken by the modifier will be in ASCII (alphanumeric) format. The 002 indicates the value will be 2 bytes long. Finally, the bytes 032 and 066 are the ASCII codes for a blank space and the letter "B" (meaning blood).

The next item (to the right) stores the blood urea nitrogen (BUN) value from the SMA-7. The 253 is a noun level delimiter and indicates the next value (5) will be a noun. [The PTXT code for the BUN value is 13 1 1 1 5 0

---

[1]Although all efforts are made to limit "events" to one field code, some strings must be stored that consist of PTXT codes from more than one field code. In this case the string header contains the field code value 254 and the field code values of the PTXT code items must be stored in the string data area.

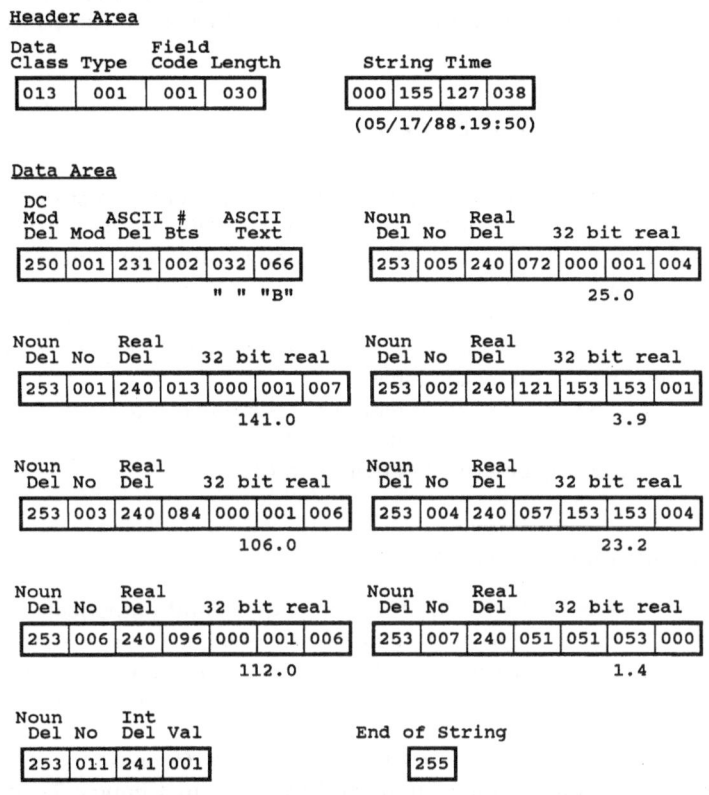

Figure 3.4. Structure of data string within data block (see text for details). DC = data class, mod = modifier, no = noun, int = integer, bts = bytes, del = delimiter, val = value.

0 0 0. The first three bytes (data class, type, and field code) are indicated in the header. If the term had values at the adjective and/or adverb level, those data would also appear at this point.] The value of 240 indicates the result will be a 32-bit real value number. The next 4 bytes are the octal value for the BUN value of 25.0.

The rest of the string contains the information to indicate the PTXT codes and values for sodium, potassium, chloride, bicarbonate, glucose, and creatinine, respectively. The penultimate entry (with the noun of 11 and the integer value of 1) indicates the test is an SMA-7. An end-of-string (255) character terminates the data area. Integers, 32-bit and 64-bit real numbers, time values, and alphanumeric characters (free text) may be stored as values in strings.

Data strings are thus inserted into blocks sequentially until the block fills, at which point a new block is created. Gradually the patient data file is built up. Next, the other HELP system patient files will be described.

3.3 HELP System File Structures

Computerized data from current LDS Hospital patients are stored in the computerized current patient file, the structure of which is described in the previous section. A variety of other computerized files containing patient-related data also exist. For example, data from discharged patients are kept in archived files that are accessible but are distinct from the current patient file. Also, billing information is kept in a file that may be easily transferred to the hospital's IBM financial system.

Patient data on the HELP system are maintained in a variety of long- and short-term files. The patient files consist of the patient identifying (ID) file and the previously described patient data file.

3.3.1 Patient Identification File

One entry is made in the ID file for all inpatients and outpatients at LDS Hospital. The only exception to the one-visit one-entry rule is outpatients who are receiving an ongoing course of therapy (e.g., radiotherapy). Such patients are identified as "series" patient and are given a single entry in the patient file.

The ID file is a key-sequenced file containing information such as the patient number and radiology number (discussed in Chapter 7), patient name, date of birth, physician identifier, area where the service was provided, room number, and other financial and demographic data. The ID file is keyed (indexed) by the patient number. The radiology number is also assigned to patients at the time of admission. Although a single patient will have a different patient number from visit to visit, the radiology number remains constant.

3.3.2 Patient Data File

Once a record has been created in the ID file, data can be stored in the current patient data file. As mentioned, a data block is created for each patient for each data class. Data are kept in the current patient file for at least 10 days after discharge, at which time the data are copied electronically onto another disk file. After 3 to 6 months the data are moved onto a removable disk pack for archival purposes. The archived data are available for research purposes but a request must be made to the system manager for the disk pack to be loaded.

3.3.3 Free-Text File

The text of dictated medical reports can be stored on the HELP system. Currently, only radiology reports are being stored (see Chapter 29 for further discussion). However, plans call for operative, pathology, and histology reports, and history and physical and discharge summaries to be included. These reports will then also be available for remote review and natural language processing (see Chapter 29). These large free-text items are stored in a separate free-text file.

3.3.4 Billing and the Transaction File

Charges for billable items are stored in the PTXT file. When a billable item is ordered through the HELP system (e.g., ECG, medical supplies, etc.), an electronic copy of the transaction is sent to a transaction file.

Every night a program automatically checks the HELP computer's "bedsfile" and adds a 1-day's room charge and a daily supply charge to the transaction file for every patient occupying a bed for that day. Charges for nursing services (calculated from computerized nurse charting or mark sense acuity cards, see Chapter 13) are similarly transferred.

The contents of the transaction file are automatically transferred to LDS Hospital's IBM AS 400 financial computer nightly. Demographic data (name, address, insurance company, etc.) are also sent at the time of admission and discharge. Bills for services are sent to the responsible parties 3 days after the patient's discharge.

3.3.5 Long Term Files

Shortly after an inpatient is discharged, a computerized abstract of the hospital stay is automatically generated by the HELP system. The abstract includes diagnosis and procedure codes entered by the Medical Records Department. A paper copy of the abstract (see Chapter 8) is entered into the medical record and an electronic copy is entered into a long-term abstract file. The abstract file is available on-line for research and utilization review purposes.

The long-term patient file, or Master Patient Index (MPI), is structured in a similar way to the patient data files. The MPI ID file contains a 32-bit "unit number" as an indexing term. Other data in the MPI ID file include the social security number, medical record number, and radiology number. Data such as addresses, insurance carrier information, dates of admissions and discharges, and attending and referring physicians identifications are stored in the MPI data file. The MPI data file is a record of all admission dates, past and present insurance information, past and present addresses, etc. The information in the MPI is available on-line at any time. The MPI is discussed further in Chapter 7.

The management corporation for LDS Hospital and 24 other hospitals in the Utah–Idaho–Wyoming region (IHC) is planning to link electronically 10 of its hospitals through a plan known as IHC Net (discussed in Chapter 1). With the advent of IHC Net, centralized MPI ID and data files will be maintained at IHC headquarters in downtown Salt Lake City for all IHC hospitals. The centralized files will contain data for all patients who have been admitted to any of the IHC hospitals and the data will be available electronically to any hospital in the Net.

```
! SOURCE FILE: $RALPH.RUSSELL.SZAPREC
! OBJECT FILE: $BILL.ERLOG.ZAPREC
! APPLICATION: DELETES ARRIVAL & DISPOSITION STRINGS FOR ED LOG FROM
  PATIENT'S RECORD
! AUTHOR: JOEL E. RUSSELL

SECTION ZAP^RECORD;
BEGIN

  VARIABLE ARR^TIME,
           DSP^TIME,
           PAT^NUM,
           NAME   CHAR[40],
           KEY    CHAR[4];

    VARIABLE LNM            CHAR[20],
             FNM            CHAR[15],
             PC             CHAR,
             I,
             LI;

  SECTION GET^DATA SUBROUTINE;
  BEGIN

    WHILE RECORDIN $FILES( 7 ) DO   !RECORDIN $BILL.ERLOG.ZAPSTRNG
    BEGIN

      EXTRACT PAT^NUM FIRST 0 LENGTH 4,
              ARR^TIME FIRST 4 LENGTH 4,
              DSP^TIME FIRST 8 LENGTH 4;

      KEY := WRITE PAT^NUM FORMAT( B4 );

      KEYON $FILES( 0 ) USING KEY GENERIC;

      WHILE RECORDIN $FILES( 0 ) DO
      BEGIN

        EXTRACT $PATIENT FIRST  0 LENGTH  4,
                NAME       FIRST 56 LENGTH 40;

        WRITE PAT^NUM,NAME FORMAT( ERASE,4/,F10,4X,A40 );

        !******* PARSE NAME *******

        I := 1;
        WHILE PC <> "," AND I < 21 DO
        BEGIN
          PC := $MID( NAME, I, 1 );
          IF PC <> "," THEN LNM := LNM & PC;
          I := I + 1;
        END;   !FAILS WHEN COMMA ENCOUNTERED, OR CHARACTER COUNT > 20
```

Figure 3.5. Example of PAL programming code.

3.4 Database Query Languages

As mentioned in Chapter 2, a low-level compiled language for file maintenance, TAL, is provided with the Tandem computer. Programs to retrieve
data may be written in TAL. Because of the low-level nature of TAL, a
higher level query language was designed in the 1980s. PTXT Application
Language (PAL) was originally designed as a report generation language.
PAL permits a user to define a local database that is a subset of the stored
data for a single patient. The locally defined database may then be used to
create displays or reports. Today PAL is also used to create interactive data
entry programs. PTXT Application Language is a structured, compiled,
programming language. A sample of source code is shown in Figure 3.5.
The compiled output of PAL source code, called PCODE, is not executed
directly but rather serves as the input to an interpreter written in TAL.

Report generation programs may also be written with the HELP Frame
Language. The HELP Frame Language (discussed further in Chapter 5) is
used to represent medical knowledge in a modular fashion. A compiler
exists to translate HELP frames into PAL source code.

3.5 Summary

The data and file structures used by the HELP system have been developed
locally over the past 20 years. The early framework was designed by one of
the authors (TAP) working with Dr. Homer R. Warner. An approach was
chosen that allowed expandability and flexibility. The flexibility of the
system is evident in that its use continues to be modified (e.g., phasing out
of type 0 data). The success of the approach is evident in that the system is
still being used.

4
HELP System Interfaces

In general, a system is a collection of components organized to perform specified functions. Large systems are frequently composed of smaller, integrated subsystems that are known as modules. Each system or module has a boundary that defines its limits (see Figure 4.1). The function of HELP is to expedite and simplify the flow of data and information throughout the hospital. This general task can be subdivided into three distinct phases: (1) data acquisition, (2) data manipulation and analysis, and (3) data reporting. Of these three phases, only data analysis and manipulation takes place entirely within the boundaries of the system. Data acquisition and data reporting require the system to interact with entities (e.g., other computer systems, human users) beyond the system boundary.

All system applications that interact with entities beyond the system boundary require an interface between the HELP system and the external object. Interfaces between the Tandem computer and other computer systems require a physical (hardware) interface and software communications protocols. Examples of such machine–machine interfaces occur repeatedly within the HELP system and are described elsewhere in this book (e.g., Chapter 20, Medical Information Bus; Chapter 16, Laboratory Information System; Chapter 28, MUSE ECG System, etc.). Human–computer interfaces, on the other hand, may be so simple as to not be immediately apparent. In the case of printed reports, the piece of paper (which will be read by the human user) is the interface. With video terminals, the keyboard and the terminal screen are the interfaces. Other examples of human–computer interfaces exist (e.g., mice, joysticks, light pens, voice input, etc.); however, none of these is currently being used on the HELP system. Newer, easier to use, man–machine interfaces for computer systems are continually being sought.

Interfaces are indeed "where the rubber meets the road." They permit systems to communicate with each other and with the outside world. In

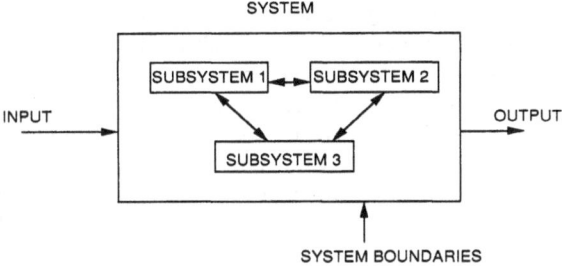

Figure 4.1. General system diagram.

hospital information systems they permit raw data to be entered and processed data to be transmitted. Interfaces are also necessary to convey the
results of automated medical decision support programs (e.g., alerts, interpretations of test results) to the appropriate personnel. This chapter discusses the manner in which data are entered into the HELP system and
how results are presented to users. The format of the some common HELP
system reports will also be discussed.

4.1 Data Acquisition

Patient data enter the HELP system through one of two primary modes:
either (1) electronically from other computer systems, or (2) manually with
human users using interactive data entry programs.

4.1.1 Data Capture from Other Computer Systems

Examples of data sent to the central patient database from other computer
systems include (1) physiologic patient data captured directly (in digital
form) from patient monitoring devices, (2) narrative reports entered into
PC-based word processors, (3) laboratory data from the laboratory computer system, and (4) ECG findings transmitted from the ECG computer
system.

Patient Monitoring [1,2]. Patient monitoring refers to the process of measuring and displaying the parameters that describe a patient's current physiologic state. Data from patient monitoring are used extensively for the
purposes of medical decision making in the intensive care setting [3] and
has accounted to a great extent for the growth of the field of critical care
medicine. Examples of parameters that are commonly measured include
vital signs (temperature, heart rate, respiratory rate, and blood pressure),
oxygen saturation, urine flow, and the electrocardiogram signal. In a com-

plicated intensive care setting, patient monitoring is extended to included the status of critical therapeutic devices such as microprocessor controlled intravenous pumps and mechanical ventilators.

The flow of information in patient monitoring is shown in Figure 4.2. An analog signal is obtained from the patient (e.g., blood pressure waveform via a pressure transducer). The signal is amplified as necessary, converted to digital form through an ADC (analog-to-digital converter), and analyzed. Part of the analysis consists of feature extraction such as determination of the systolic and diastolic readings in the blood pressure waveform, or determining if an ECG waveform is normal or not. The signal processing just described used to require large computers but today is performed by the monitors themselves. Bedside monitors are now able to extract, display, and transmit the essential features of the acquired physiological signals. It is the analyzed results of patient monitoring and not the waveforms themselves that are transmitted from the monitors to the HELP system. The MIB (Chapter 20) is being developed to standardize and simplify communications between a central computer and the wide variety of electronic devices likely to be present in an ICU and other acute care settings.

The ability to analyze and capture easily a patient's physiologic data has led to another question in computer-assisted patient monitoring: which of the data to commit to long-term storage. Quick calculations show that if all the data acquired via patient monitoring are stored, many megabytes of data can accumulate in the course of a day and quickly choke any computer system [4]. Whereas the general problem of how much monitored patient data should be stored permanently is still unsolved, the approach on the HELP system has been to store (and thus, make available for decision making) the value of monitored parameters every 15 minutes. The choice of 15 minutes is an ad hoc approach and improved data collection schemes are being studied.

The greatest use of patient monitoring is for patients in the ICUs and

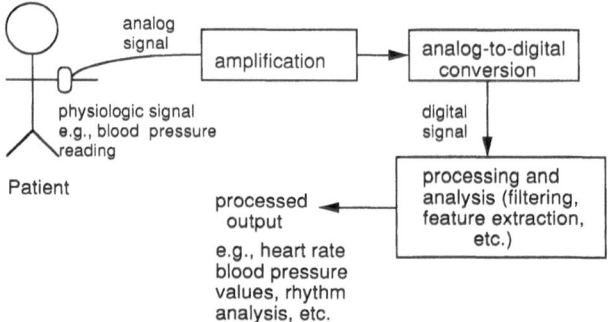

Figure 4.2. Flow of information in patient monitoring.

for patients undergoing open-heart surgery. The results of monitoring are stored in the central data base and are accessible remotely. Physicians may follow a patient's status from any HELP system terminal in the hospital, or, with the use of remote physician call-in services (Chapter 31), from their homes or offices. Also, nurses in the thoracic ICU awaiting the return of an open-heart patient from surgery can be aware of the patient's physiologic status and which surgical procedures have been performed before his/her return. Further uses of monitored data are discussed in the chapters on Hemodynamic Monitoring (Chapter 21) and Surgery Monitoring (Chapter 24).

Transfer of Data from Ancillary Service Computer Systems. As mentioned in Chapter 2, HELP's central Tandem computer is supplemented by many secondary computers providing specialized services. Whereas these secondary systems sometimes maintain local databases for certain functions, copies of the data must be transferred to HELP's centralized database for use in decision making and for the production of comprehensive reports.

Subsystems transferring data to the HELP system include the clinical laboratory system (Chapter 16), the Marquette MUSE ECG system (Chapter 28), and the PC used in the cardiac catheterization laboratory for data capture and analysis (Chapter 27). Various data communications protocols are used by each system and are discussed in the respective chapters.

Capture of Transcribed Textual Data from Word Processors. HELP system programs have been written that allow transcribed narrative reports that have been entered into PC word processing programs to be transferred to the Tandem computer through a Tandem-LAN gateway product. This project has been successfully piloted for radiology reports and will be extended to history and physical reports, discharge summaries, and operative, histology, and pathology reports. Text thus stored on the Tandem is available for review on video terminals and analysis by natural language processing programs (see Chapter 29 for further discussion).

4.1.2 Interactive Data Entry Programs

The major mode of data acquisition on the HELP system is manual data entry through interactive programs (besides from other computer systems). As mentioned, light pens, automated pointing devices such as mice, and voice input are not used on the HELP system. Rather, the keyboard is the primary means of manual data entry.

QSTN and the General Questionnaire Asking Program. A HELP system program (called QSTN) allows interactive questionnaires to be designed. Another program, called GQAP, allows the QSTN questionnaires to be executed on the Tandem computer to acquire data from HELP system users. Questionnaires have been designed for the entry of nurse charting data, respiratory therapy data, x-ray ordering, general order entry, and a number of other purposes.

QSTN allows (1) individual data entry screens to be designed, (2) a hierarchical sequence of screen presentations (based on responses to previous screens) to be structured, (3) the values for the entered data to be linked with the correct PTXT codes (see Chapter 3 for description of PTXT), and (4) the time associated with the value to be entered (this may not be the time of the actual data entry, as in the case of entering events occurring earlier). When executed by GQAP, the questionnaire leads the user through a series of screens designed to capture the desired patient data. The data are then stored in the patient data base.

QSTN allows for four basic types of data entry screens: (1) a multiple choice data entry screen. For example, a respiratory therapist entering data on a ventilator patient will be presented with a list of ventilator modes (e.g., intermittent mandatory ventilation, assist mode, etc.) and asked to enter which applies to the given patient; (2) a data entry screen for numerical values (e.g., patient age, units of medication, etc.); (3) a screen permitting free text to be entered (since free text serves no purpose in computerized medical decision making, the amount of free text entered into the HELP system is kept to a minimum); and (4) time entry screens. For example, a ward clerk indicating that a patient has been discharged or transferred may indicate that this event occurred at the current time (the default value) or at a specified earlier time. QSTN supports range checking so that nonsensical values (e.g., age < 0, a negative blood pressure, etc.) will not be accepted.

Transaction Application Language and PTXT Application Language Programs. The TAL and PAL procedural programming languages are described in Chapter 3. Many special purpose data entry programs that do not lend themselves to being designed with GQAP have been written with these two languages for the HELP system. Examples include the pharmacy ordering program (Chapter 15) written in TAL and the blood ordering program (Chapter 26) written in PAL.

4.2 Outputs

Once present in the system, computerized patient data are available for processing and output by the HELP system. Patient data are displayed in a variety of informative formats for use by health-care workers. The results of automatic and continuous computerized data surveillance are conveyed to the appropriate personnel and large clinical databases are constructed and analyzed retrospectively for research purposes.

4.2.1 Reports

HELP system reports may contain any data present in the patient database, without concern as to what module originally captured the data. The reports serve a variety of purposes. Some reports (e.g., computerized nurse charting notes) contain a record of patient data and become part of the permanent medical record. Other reports, such as the ICU Rounds Report described below, assist physicians by displaying a comprehensive record of the patient's most recent data but are temporary and are discarded after use.

Certain clinical reports are packed with patient data and can be confusing to the uninitiated viewer. Because these reports represent a significant component of the HELP system output and because they are relevant to many of the HELP system modules described later, the format of two important reports, the ICU Rounds Report and the 7-day Report, are described in detail here. Although these reports can be produced for any patient on a nursing unit using the computerized nurse charting program (Chapter 13), they are most useful in the intensive care settings where the complexity of the patients and amount of pertinent data are higher than average. Many other HELP system clinical, administrative, and management reports exist and are described where relevant throughout the book. Almost all reports can be displayed on HELP system video terminal screens or printed on conveniently located laser printers.

Routine Laboratory Reports. Routine laboratory data display on the HELP system allows the results of one or a series of laboratory panels to be displayed in reverse time sequential order. An example is shown in Figure 4.3.

Intensive Care Unit Rounds Report (Figure 4.4). The ICU Rounds Report contains data for a single patient for the previous 24 hours. It is intended to give a comprehensive overview of the patient's current status and is primarily used by the Shock–Trauma–Respiratory Intensive Care Unit (STRICU) team at morning rounds. In practice, the report is laser-printed, transferred to a transparency, and projected onto a screen so all participating can view the patient data while the discussion is progressing. The report may be produced at any other time as well. The report is temporary and does not become part of the permanent patient record. The header information contains patient identifying information (name, patient number, and room number), the name of the attending physician of record, the patient's sex, height (in cm), weight (in kg), body surface area (BSA, in m^2), basal energy expenditure (BEE, in kcal), the admitting diagnosis (as entered by the admitting clerk as free text), and the admitting date. Three physiology-based severity of illness calculations also appear in the header. The multiorgan failure (MOF) score is a locally developed severity of illness scoring algorithm.

MELVIN A   24699   W715

LAB DATA - CBC

DATE	TIME	VALUES								
		WBC	RBC	HGB	HCT	MCV	MCH	MCHC	RDW	PLAT
22APR	06:05BL	7.6	3.18	9.1	27.3	85.9	28.6	33.3	12.4	155
		WBC	RBC	HGB	HCT	MCV	MCH	MCHC	RDW	PLAT
21APR	07:17BL	10.1	3.37	9.5	28.8	85.5	28.2	33.0	12.4	126
		WBC	RBC	HGB	HCT	MCV	MCH	MCHC	RDW	PLAT
20APR	04:11BL	10.8	3.51	10.0	30.4	86.6	28.5	32.9	12.6	115
		WBC	RBC	HGB	HCT	MCV	MCH	MCHC	RDW	PLAT
19APR	04:04BL	10.6	3.80	10.9	32.4	85.2	28.7	33.7	12.4	127
		WBC	RBC	HGB	HCT	MCV	MCH	MCHC	RDW	PLAT
18APR	22:56BL	13.6	3.83	11.5	32.6	85.0	30.0	35.3	12.9	115
		WBC	RBC	HGB	HCT	MCV	MCH	MCHC	RDW	PLAT
18APR	18:10BL	15.7	3.98	11.5	33.6	84.5	28.9	34.2	12.3	135
		WBC	RBC	HGB	HCT	MCV	MCH	MCHC	RDW	PLAT
17APR	10:10BL	7.8	5.20	14.3	44.3	85.2	27.5	32.3	12.5	258

MELVIN A   24699   W715

LAB DATA - SMA-7

DATA	TIME		NA+	K+	CL-	CO2	BUN	GLUC	CREAT
22APR	06:05	SE	136	4.1	104	27	21	104	.9
21APR	07:17	SE	138	3.9	109	27	25	100	1.1
20APR	04:11	SE	140	4.0	112	24	22	161	.9
19APR	04:04	SE	138	4.0	114	20	23	194	1.1
18APR	22:56	SE	136	3.9	113	17	24	200	1.1
18APR	18:10	SE	133	3.7	107	21	24	148	1.2
17APR	10:10	SE	136	4.5	104	26	17	247	1.0

MELVIN A   24699   W715

LAB DATA-SMAC REPORT

DATE	TIME		VALUE									
			NA+	K+	CL-	CO2	CA++	PO4	GGT	GLUC	CREAT	BUN
17 APR	10:10	SE	136	4.7	101	28	8.7	3.4	36	233	1.2	19
			URIC	CHOL	PROT	ALB	T.BILI	D.BILI	ALK	LDH	SGOT	SGPT
			5.6	145	6.5	3.3	.7	.2	72	114	19	17

Figure 4.3. Routine HELP system laboratory data report.

The rest of the report is organized by organ system. After the title of each system is an integer value (e.g., the 0 following cardiovascular). The integer value is an automatically calculated organ failure score for that system. The MOF score appearing in the header is the sum of the individual values. Results present in the computerized database are printed in the report automatically. Occasionally, blank spaces are left for results that are not contained in the system to be handwritten in. All laboratory values and many physiologic values have an associated time printed with them.

Cardiovascular is the first system and contains cardiac output, blood pressure, cardiac enzyme, and ECG results, if present. An interpretation of the cardiac output data, "LV parameters within normal limits," can be seen (see Chapter 21 on Hemodynamic Monitoring for further details). In the upper right, two lines are set aside for the results of the cardiac physical examination to be written in.

Respiratory data is the next section. Blood gas results and interpretations and the patient ventilator settings are reported. The results of computerized respiratory therapy charting follow (computerized blood gas reporting is discussed in Chapter 19 and computerized respiratory therapy charting is discussed in Chapter 14).

The neurological section displays a Glasgow Score (automatically calculated from computerized nurse charting results). Space is left for other neurological findings to be handwritten in. The coagulation, renal, metabolic, GI, and infection sections that follow display laboratory and physiologic results relevant to that organ system rather than the results from any single laboratory panel. This is feasible since all patient data is stored in a central file. For example, under the renal section, fluid intake and output data are presented along with weight data and serum and urine chemistry results. In most hospitals multiple locations would have to be searched manually to find all of these data relevant to the patient's kidney function.

After the organ system data is a listing of current medications with units, route, and dose received for the past 24 hours. Microbiology data completes the report.

Data from multiple sources (e.g., patient monitors, clinical laboratory, pharmacy, respiratory therapy charting, nurse charting, etc.) are used in the creation of this report. Such a report would not be possible without an integrated database.

7-Day Report. Whereas the ICU Rounds Report gives a detailed description of the patient for the previous 24 hours, the 7-day Summary (Figure 4.5) gives an overview of key patient care parameters for a period of up to 7 days. Unlike the Rounds Report, the 7-day Report becomes a permanent part of the patient's medical record.

Physiological data are displayed graphically first. Systolic and diastolic blood pressures (represented by the letters "S" and "D," respectively), heart rate ("*"), and temperature ("E" or "C") are displayed on a single set of axes (the "E" designates that the temperature reading was taken from an ear probe, "C" indicates a core temperature reading). These are data that have been captured automatically by the system (from patient monitors) every 15 minutes. Next, total doses of medications administered are presented, followed by intake/output data, net balance (intake minus output), and the patient's weight. Nutritional information follows and includes nonprotein energy, total energy, grams of protein, fat, and carbohydrate intake, nonprotein energy per gram of nitrogen, and grams of nitrogen intake. The data are displayed for each of the 7 days.

Other Reports. The preceding two reports have been described to give a sense of the benefits of data integration and the amount of data that can be displayed in a limited space. Other important clinical reports include the nursing shift report and nursing comments report. Both of these computerized reports are the result of computerized nurse charting, which is described in detail in Chapter 13. Both become part of the permanent patient

---

Figure 4.4. LDS Hospital ICU Rounds Report. *Cardiovascular section:* CO = cardiac output, CI = cardiac index, HR = heart rate, SV = stroke volume, SI = stroke index, VP = central venous pressure (M indicates a measured reading), MSP = mean systolic pressure, MP = mean pressure, SVR = systemic vascular resistance, LWI = left ventricular work index, PW = pulmonary wedge pressure, PA = pulmonary artery pressure, PVR = pulmonary vascular resistance, RWI = right ventricular work index, SP = systolic pressure, DP = diastolic pressure, LACT = lactate dehydrogenase, CPK = creatine phosphokinase. *Respiratory section:* BE = base excess, HB = hemoglobin, CO/MT = carboxyhemoglobin/methemoglobin, PO2 = oxygen partial pressure, SO2 = oxygen saturation, O2CT = oxygen content, %O2 = fraction inspired oxygen, AVO2 = arterial–venous oxygen difference, VO2 = minute oxygen consumption, C.O. = cardiac output, A-a = arterial–alveolar gradient, QS/QT = shunt fraction, PK/ = peak airway pressure, PL/PP = plateau pressure/PEEP, MR/SR = machine rate/spontaneous rate, VENT = ventilator model, MODE = ventilatory mode, VR = ventilator respiratory rate, Vt = tidal volume, O2% = fraction inspired oxygen, PF = peak flow, IP = inspired pressure, MAP = mean airway pressure, PK = peak pressure, PL = plateau pressure, PP = positive end expiratory pressure (PEEP), m-Vt = machine tidal volume, c-Vt = corrected tidal volume, s-Vt = spontaneous tidal volume, MR = machine rate, SR = spontaneous rate, TR = total rate, m-VE = machine minute ventilation, s-VE = spontaneous minute ventilation, t-VE = total minute ventilation, Cth = compliance, Pc = cuff pressure, MIP = mean inspiratory pressure, MEP = mean expiratory pressure, MVV = minute ventilation, PK flow = peak flow. *Neuro and psych section:* DTR = deep tendon reflexes, BABIN. = babinski, ICP = intracranial pressure. *Coagulation section:* PT = protime, PTT = partial thromboplastin time, FSP = fibrin split products. *Renal section:* IN = total intake, OUT = total output, NGOUT = nasogastric output, NET = net intake, WTCHG = weight change, S.G. = specific gravity, AGAP = anion gap, UOSM = urine osmolality, UNA = urine sodium, CRCL = creatinine clearance. *Metabolic section:* KCAL/N2 = kilocalories per gram of nitrogen, UUN = urine urea nitrogen, N-BAL = nitrogen balance. *GI, liver, and pancreas section:* HCT = hematocrit, TOTBILI = total bilirubin, ALKPO4 = alkaline phosphatase, GGT = gamma glutamyl transferase, GUAIAC = stool guaiac, DIRBILI = direct bilirubin.

L D S   H O S P I T A L   I C U   R O U N D S   R E P O R T
DATA WITHIN LAST 24 HOURS

NAME: DEBRA A.   NO. 110745.   ROOM: E60   DATE: JUN 12 15:23
DR. RICHARDS, C. DAVID   SEX: F   AGE: 37   HEIGHT: 165   WEIGHT: 81.30   BSA: 1.89   BEE: 1542   MOF: 0
ADMT DIAGNOSIS: GUNWOUND TO STOMACH   ADMIT DATE: 10 JUN 90

CARDIOVASCULAR: 0   EXAM:

TIME	CO	CI	HR	SV	SI	VP	MSP	MP	SVR	LWI	PW	PA	PVR	RWI
JUN 12 04:00	8.60	4.54	124	69	37	6.0M	98	82	9	45	8	14	.7	4.0

LV PARAMETERS ARE WITHIN NORMAL LIMITS

	SP	DP	MP	HR	LACT	CPK	CPK-MB	LDH-1	LDH-2
LAST VALUES	106	61	79	133	( )	( )	( )	( )	( )
MAXIMUM	154	93	116	141					
MINIMUM	79	49	62	106					

RESPIRATORY: 0

	pH	PCO2	HCO3	BE	HB	CO/WT	PO2	SO2	O2CT	%O2	AVO2	VO2	C.O.	A-a	QS/QT	PK/	PL/PP	MR/SR
12 04:01 V	7.41	40.0	25.1	1.1	9.5	1/ 1	42	75	10.0	40						38/	24/ 5	16/ 0
12 04:00 A	7.43	36.7	24.1	.8	9.6	1/ 1	71	93	12.6	40				124		38/	24/ 5	16/ 0

SAMPLE # 8, TEMP 38.0, BREATHING STATUS : ASSIST/CONTROL
NORMAL ARTERIAL ACID-BASE CHEMISTRY
MODERATELY REDUCED O2 CONTENT

11 17:00 A	7.45	34.2	23.6	.8	10.6	2/ 1	71	92	13.7	50	2.78	187	34	0/	0/ 5	20/ 0

------ machine settings ------ | ------------ patient values ------------

VENT	MODE	VR	Vt	O2%	PF	IP	MAP	PK	PL	PP	m-Vt	c-Vt	s-Vt	MR	SR	TR	m-VE	s-VE	t-VE	Cth	Pc
B-I	A/C	12	700	40	55	-60	34	22	5	800	684	16					10.9			40.2	

	VC	VT	VE	MIP
	55	245	8.8	-60

12 04:10   INTERFACE: OROTRACH; ALARMS CHECKED; POSITION: SEMI-FOWLER; PATIENT CONDITION: CALM; SUCTIONED;
12 04:10   THERAPIST: WARNOCK, ROSALEE, RRT

DATE	TIME	HR	VR	VT	VC	VE	MIP	MEP	MVV	PK FLOW	THERAPIST	EXAM:
06/12/90 07:29	114	36	245								MCEWEN, PAT	

NEURO AND PSYCH: 0
GLASCOW 13 (16:00)   VERBAL _____   EYELIDS _____   MOTOR _____   PUPILS _____   SENSORY _____
DTR _____   BABIN. _____   ICP _____   PSYCH _____

(Continued.)

```
COAGULATION: 0
   PT: 11.6  (05:00   ) PTT:      42 (05:00   ) PLATELETS:  77 (12:17   ) FIBRINOGEN: 0(00:00) EXAM: _____
FSP-CON:  0  (00:00   ) FSP-PT:    0 (00:00   ) 3P:            (00:00   )

RENAL, FLUIDS, LYTES: 0
   IN  3127 CRYST 2587  COLLOID        BLOOD  500 NG/PO   NA  147 (12:17) K   4.1 (12:17) CL  117 (12:17)
  OUT  5172 URINE 2765  NGOUT   650  DRAINS 40  310 OTHER  CO2 19.0 (12:17) BUN  17      CRE 1.2 (12:17)
  NET -2045 WT   81.30  WT-CHG         S.G. 1347   1.020   AGAP 15.1       UOSM     UNA       CRCL _____

METABOLIC --- NUTRITION: 0
  KCAL    377 GLU 91 (12 12:17)  ALB   .0 (00 00:00)  CA  .0 (00 00:00)  FE  .0 (00 00:00) GGT   TIBC 0 (00 00:00)
  KCAL/N2   0 UUN .0 (00 00:00)  N-BAL .0           PO4 .0 (00 00:00)  MG  .0 (00 00:00)       CHOL 0 (00 00:00)

GI, LIVER, AND PANCREAS: 0                                                                    EXAM:
  HCT 30.8 (12 12:17)  TOT BILI .0 (00 00:00)  SGOT 0 (00 00:00)  ALKPO4 0 (00 00:00) GGT  0 (00 00:00)
  GUAIAC (   )         DIR BILI .0 (00 00:00)  SGPT 0 (00 00:00)  LDH    0 (00 00:00) AMYL 0 (00 00:00)

INFECTION: 0
  WBC 3.6(12:17 ) TEMP 39.0 (11/21:15) DIFF 18 B, 59P, 13L, 6M, 4E (12:17) GRAM STAIN: SPUTUM ____ OTHER ____

SKIN AND EXTREMITIES:
  PULSES _____ RASH _____ DECUBITI _____

TUBES:
  VEN _____ ART _____ SG _____ NG _____ FOLEY _____ ET _____ TRACH _____ DRAIN _____
  CHEST _____ RECTAL _____ JEJUNAL _____ DIALYSIS _____ OTHER _____

MEDICATIONS:
  MORPHINE, INJ               MGM IV      20   IMIPENEM CILASTATIN (PRIMAXIN), INJ  MGM IV  2000
  ACETAMINOPHEN, ELIXIR        ML NG      40   DOPAMINE, INJ                        MGM IV   117
  MIDAZOLAM (VERSED), INJ     MGM IV  17.500   FAMOTIDINE (PEPCID), INJ             MGM IV    40
  CEFOTAXIME (CLAFORAN), INJ  MGM IV    1000   POTASSIUM PHOSPHATE, INJ             MEQ IV    40
```

```
-ROUTINE CULT-  **-PRELIMINARY REPORT.-**        10JUN 22:
   SOURCE:   SPUTUM
   STAIN:    NUMEROUS GRAM NEGATIVE BACILLI, NUMEROUS WBCS, FEW YEAST,
   RESULT:   STREPTOCOCCUS ALPHA HEMOLYTIC    LIGHT GROWTH
   RESULT:   GRAM NEG. BACILLI     HEAVY GROWTH
   RESULT:   ENTERIC BACILLI    LIGHT GROWTH

-ROUTINE CULT-  **-PRELIMINARY REPORT.-**        10JUN 22:00
   SOURCE:   DRAINAGE,
   STAIN:    FEW GRAM POSITIVE COCCI, FEW GRAM POSITIVE COCCI IN CHAINS,
             NUMEROUS WBCS,
   RESULT:   GRAM NEG. BACILLI     MODERATE GROWTH
   RESULT:   GRAM NEG. BACILLI     MODERATE GROWTH
   RESULT:   STREPTOCOCCUS NON HEMOLYTIC   MODERATE GROWTH

-ROUTINE CULT-  **-PRELIMINARY REPORT.-**        10JUN 22:00

   SOURCE:   URINE,  CATHETERIZED

-ARD CULTURE-  **-PRELIMINARY REPORT.-**         10JUN 20:00

   SOURCE:   BLOOD,
(END)
```

Figure 4.4. *Continued.* See page 42 for caption.

DOROTHY F.

210150   E603

P HR TEMP
SD * C E

	SUNDAY JUN 03 90	MONDAY JUN 04 90	TUESDAY JUN 05 90	WEDNESDAY JUN 06 90	THURSDAY JUN 07 90	FRIDAY JUN 08 90	SATURDAY JUN 09 90

235 41
220 40
205 39
190 38
175 37
160 36
145 35
130 34
115
100
85
70
55
40
25
10

	Unit	Route	JUN 03	JUN 04	JUN 05	JUN 06	JUN 07	JUN 08	JUN 09
MORPHINE, INJ	MGM	IV	200.0	208.0	113.0	164.0	172.0	299.0	321.0
ACETAMINOPHEN, ELIXIR	ML	NG	20	20	20	20	20	20	20
MIDAZOLAM (VERSED), INJ	MGM	IV	260.0	250.0	331.0	324.0	326.0	326.0	310.0
GENTAMICIN, INJ	MGM	IV				175.0	225.0	150.0	150.0
IMIPENEM CILASTATIN (PRIMAXIN), INJ	MGM	IV	1000	1000	1000	1000	1000	1000	1000
AMPHOTERICIN B, INJ	MGM	IV	60.0	60.0	60.0	60.0	60.0	60.0	60.0
FLUCONAZOLE (DIFLUCAN), INJ	MGM	IV	100	200	100				
NYSTATIN, POWDER	APPLICTOPIC		4	4		4	4	4	4
ACYCLOVIR 5% (ZOVIRAX), OINT	APPLICTOPIC		4	4	3	4	4	4	1
NYSTATIN TABS	VAG	VAG	1	1	1	1	1	1	1
NEOSPORIN, OINT	APPLICTOPIC		4	4	3	4	4	2	
ALBUTEROL (PROVENTIL), INHALATION SOLUTION	MGM	INHAL	15.00	15.00	15.00	17.50	15.00	10.00	15.00
VERAPAMIL, INJ	MGM	IV			5.0				
FUROSEMIDE, INJ	MGM	IV	80	40	60	80	80	80	80
SUCRALFATE (CARAFATE), TAB	MGM	NG	6000	5000	6000	6000	6000	6000	6000
AMPHOJEL, LIQUID	ML	NG	120	120					
DIPHENHYDRAMINE (BENADRYL), INJ	MGM	IV	25.0	25.0	25.0	25.0	25.0	25.0	25.0
HYDROCORTISONE NA SUCCINATE (SOLU-CORTEF), INJ	MGM	IV	150	150	150	150	150	150	150
LEVOTHYROXINE (SYNTHROID), INJ	MGM	IV	0.100	0.100	0.100	0.100	0.100	0.100	0.100
ARTIFICAL TEARS (LACRIL), SOLUTION	GTTS	OTIC		1					
ARTIFICAL TEARS (LACRIL), SOLUTION	GTTS	OPTH	4	3	2	2	2	3	2
PLATELETS (RANDOM DONOR)	ML	IV							250
PACKED RBC	ML	IV		400	400				
FRESH FROZEN PLASMA	ML	IV		540	480				
NORMAL SALINE, INJ	ML	IV	1398	1188	1413	891	1195	1425	1331
NORMAL SALINE, INJ	ML	IRRIG	9900	5200	4700	700	4400	4700	3850
NACL 0.45, INJ	ML	IRRIG	397	381	394	3900	383	386	409
AMINOSYN RF, INJ	ML	IV	662	635	657	380	639	643	681
DEXTROSE 70%, INJ	ML	IV				633			
POTASSIUM	MEQ	IV	32.7	38.1	39.4	33.3	10.6	21.1	44.9
CALCIUM	MEQ	IV	6.1	5.8	6.3	7.6	7.7	7.7	8.2
MAGNESIUM	MEQ	IV	10.6	10.2	10.5	10.1	11.0	11.6	12.7

(Continued.)

Figure 4.5. Seven-Day Summary Report (see text for description).

Parameter	Units							
ZINC	MGM IV	5.3	5.1	5.3	5.1	5.1	5.1	5.5
COPPER	MGM IV	1.1	1.0	1.1	1.0	1.0	1.0	1.1
MANGANESE	MGM IV	0.5	0.5	0.5	0.5	0.5	0.5	0.5
CHROMIUM	MCG IV	10.6	10.2	10.5	10.1	10.2	10.3	10.9
ACETATE	MEQ IV	69.8	73.7	76.2	55.5	25.6	26.1	44.9
SULFATE	MEQ IV	10.6	10.2	10.5	10.1	11.0	11.6	12.7
GLUCONATE	MEQ IV	6.1	5.8	6.3	7.6	7.7	7.7	8.2
ELECTROLYTE VOLUME	ML IV	50.8	52.1	54.5	55.7	42.0	32.6	43.4
POTASSIUM CHLORIDE, INJ	MEQ IV	35.6	40.0	40.0	40.0	80.0	0.0	60.0
D5W, INJ	ML IV	300	300	300	400	500	300	300
FAT EMULSION 20% (LIPOSYN), INJ	ML IV	200	200	200	200	200	200	200
SODIUM	MEQ IV	37.1	35.6	36.8	22.2	0.0		
MAGNESIUM SULFATE 50%, INJ	GM IV		80.0	40.0		1.00		
POTASSIUM ACETATE, INJ	MEQ IV			20.0		40.0		
POTASSIUM PHOSPHATE, INJ	MEQ IV							
INSULIN REGULAR, INJ	UNITS IV	26	25	26	25	26	26	27
MVI-12, INJ	ML IV	9.3	8.9	9.2	8.9	8.9	9.0	9.5
INTAKE (ML): BLOOD			400	880				250
COLLOID			540					
NON-BLOOD IV		3018	2767	3028	2568	2969	2995	2974
NG DRUG		140	140	20	20	20	20	20
IRRIGATION		9900	5200	4700	4600	4400	4700	3850
TOTAL		17878	15147	15828	14388	8589	7805	7094
OUTPUT (ML): INSENSIBLE LOSS		873	663	667	852	767	654	541
FOLEY CATH URINE		4155	3220	2095	3175	3940	5192	4420
WOUND DRG. 4		8765	4885	3554	4131	3675	3750	2650
INTEST. DECOMP. TUBE		670	1280	1275	1025	245	150	200
JEJUNOSTOMY			335	925	825	83		
WOUND DRG. 2		50	570	125	25	10		
WOUND DRG. 3		400	130	525	420	55		
ILEOSTOMY DRG.		150		150	115	20		
STOOL		1200	1730	1250	1475	410		
NG TUBE DRG.		850	1130	400	475	125	200	100

PHLEBOTOMY OUTPUT

TOTAL	3	3	7			3	
	17116	13943	10973	12518	9330	9949	7911
NET BALANCE (ML):	762	1204	4855	1870	-741	-2144	-817
WEIGHT (KG)	56.2	55.0	57.1	58.6	57.8	55.7	55.6
NUTRITIONAL: NP ENERGY KCAL (IV)	2027	1962	2015	1975	2005	1981	2073
TOTAL ENERGY KCAL (IV)	2110	2042	2097	2055	2085	2062	2157
PROTEIN GM	20	20	21	20	20	20	22
FAT GM	40	40	40	40	40	40	40
CHO GM	479	460	475	463	473	465	492
NP ENERGY/N2 KCAL/GM	506	654	503	493	668	495	518
N2 IN GM	4	3	4	4	3	4	4

DOROTHY F.   # 210150   E603

TIME OUT: JUN 12 90 15:28   PROCESS TIME: 00:17

(END)

Figure 4.5. *Continued.*

record. Several other reports are produced by the HELP system and are described in the relevant chapters.

4.2.2 Surveillance

Organized display of patient data is but one benefit that can be accrued from a computerized hospital information system. Another benefit is the ability to examine patient data automatically and, with the use of "medical knowledge" coded into computer logic, to ascertain when adverse conditions may exist. This is an example of how computerized medical decision making may be used on the HELP system. Results of HELP system decision logic are conveyed to hospital personnel for review and necessary action. At this time all HELP system decisions are transmitted to a human intermediary; that is, no action is automatically initiated as a result of a HELP system decision. In systems parlance, these are "open-loop," rather than "closed-loop," systems. Successfully transmitting the results of computerized medical decisions to the appropriate personnel in a manner that permits the appropriate action to be taken (i.e., integrating the computer into the normal work routine) is an issue of man–machine interface and is, perhaps surprisingly, one of the most difficult issues faced by designers of computerized hospital information systems. This section describes some of the mechanisms that have been used at LDS Hospital in solving this problem. All of the following are described in more detail in their respective chapters. They are described here briefly to give an overview of the various approaches.

Computerized Laboratory Alerting System (Chapter 17). A HELP system application has been developed to scan automatically all laboratory values stored to the system for potentially life-threatening results (e.g., potassium < 3.0 meq/dl, glucose > 500 mg/dl). In the program development, an important question was who should receive the message that such a result existed, and how should the communication be made. After experimenting with a number of alternatives, the decision was made to display the result on the computer terminal to anyone reviewing the laboratory data of the patient with the result in question. If the person reviewing the alert is anyone other than the patient's physician, it is their responsibility to inform the physician. Laboratory data review is the most frequently used HELP system application and was thus chosen as the program into which the alert display was integrated. Displaying the word "Alert" on a terminal was considered but discounted because of potential anxieties caused to patients and family.

Computerized Medication Monitoring (Chapter 15). All patient medication orders at LDS Hospital are entered into the HELP system by nurses or pharmacists. At the time of order entry a HELP system program automatically compares the order with other data in the patient database to see if the order might generate a potential drug–drug, drug–laboratory, or drug–allergy interaction. If so, a message is immediately displayed to the person entering the order. It is the responsibility of the pharmacist to notify the attending physician of the alert.

Computerized Infectious Disease Monitor (Chapter 18). The computerized infectious disease monitor (CIDM) uses microbiology and other patient data present on the HELP system to identify a variety of potentially adverse situations including (1) nosocomial infections, (2) unusual sensitivity patterns, and (3) infections not being treated with appropriate antibiotics. The program prints a report daily just before Infectious Disease Rounds. The report is then available for infectious disease personnel to review and take appropriate action.

Another module of the infectious disease monitor identifies (1) preoperative patients who require an antibiotic order but for whom no antibiotic is ordered, and (2) patients who are receiving prophylactic antibiotics beyond 48 hours after surgery without evidence of infection. In the first case a suggestion of an antibiotic order is made and in the second case discontinuation is suggested. Both of these types of suggestions are printed for review by a clinical pharmacist who is responsible for contacting the attending physician, if necessary.

Respiratory Therapy Alerts (Chapter 14). Respiratory therapy charting at LDS Hospital is done via the computer. As a result, a wide variety of patient data related to respiratory status and ventilator management exists within the HELP system. Programs have been written to examine automatically the data for adverse conditions related to patients' respiratory therapy. The results of this program are printed each morning for review by the director of Respiratory Therapy. Additional reports are printed for management purposes for the Department of Respiratory Therapy. These reports are discussed further in Chapter 14.

4.2.3 Electronic Data Transfer

Two examples exist of electronic output by the HELP system. A copy of the transaction file (see Chapter 3) is sent nightly to the hospital's IBM AS 400 financial computer. The AS 400 manages patient billing procedures. The HELP system also sends a copy of patient demographic data gathered at admission to the laboratory computer for management purposes.

4.2.4 Outputs for Research Purposes

The clinical data stored in the HELP database serve a variety of purposes while patients are receiving care at LDS Hospital. The sum of the patient data accumulated over a long period of time has also been useful for a variety of research purposes. The Strato [5] program was written to facilitate the searching of archived data by user-defined parameters. Sets of patients may be retrieved based on name, number, dates of admission, physician, laboratory values in a certain range, or complex combinations of the above. Patient data files retrieved with Strato are downloaded to PCs for further statistical analysis.

Also, certain computerized logic (notably Bayesean-based inferencing) depends on accurate estimations of a priori and conditional probabilities of the presence of patient findings. Frequently experts' estimates or published values (when available) are used. Researchers at LDS Hospital have used probabilities calculated from the HELP database to modify statistics in diagnostic programs and have measured the benefits of using the revised probabilities [6].

4.3 Conclusions

The critical role of the man–machine interface in the success of medical information systems cannot be underestimated. Developing methods to acquire relevant patient data and deliver the analyzed data to the appropriate personnel at the appropriate time has consumed most of the effort in the development of the HELP system. Many of the issues relate to human engineering rather than computer science. Machine–machine interfaces have also had to be developed for the HELP system to be successful.

References

[1] Gardner RM. Patient-monitoring systems. In: Shortliffe EM, Perrault LE, eds. *Medical Informatics*. New York; Addison–Wesley;1990.
[2] Gardner RM. Computerized management of intensive care patients. *MD Comput* 1986;1:36–51.
[3] Bradshaw KE, Gardner RM, Clemmer TP, Orme JF Jr, Thomas F, West BH. Physician decision-making – evaluation of data used in a computerized ICU. *Intl J Clin Monit Comput* 1984;1:81–91.
[4] Gardner RM, Tariq H, Hawley WL, et al. Medical information bus: The key to future integrated monitoring. *Intl J Clin Monit Comput* 1989;6:205–209. Editorial.
[5] Bouhaddou O, Haug PJ, Warner HR. Use of the HELP clinical database to build and test medical knowledge. *Symposium for computer applications in medical care (SCAMC)* 1987;11:64–67.
[6] Haug PJ, Clayton PD, Shelton P, et al. Revision of diagnostic logic using a clinical database. Proceedings of the American Association of Medical System Informatics. 1987;6:238–242.

5
Decision Support on the HELP System

One of the prime motivations for the development and continued expansion of the HELP system has been to assist decision makers at LDS Hospital through improved data management. Decision making is ubiquitous in all medical settings and is performed by both clinical and administrative personnel. Since the HELP system manages clinical as well as administrative data, it is able to assist clinicians as well as administrators in their tasks. This chapter begins with an introduction to the concepts of data, information, and knowledge in the medical setting and shows how the HELP system manages each of these to assist decision makers in their tasks.

5.1 Data, Information, and Knowledge

In his excellent book, *Clinical Information Systems*, Blum [1] notes that there are three types of computerized health-care applications: (1) data-oriented, (2) information-oriented, and (3) knowledge-oriented. Data are raw, uninterpreted elements used by decision makers. Examples of data in the hospital setting are laboratory results, physiologic signals, and patient charges. Computers are able efficiently to capture, store, organize, retrieve, and display otherwise unmanageable amounts of patient and institutional data. Blum identifies data-oriented computerized medical applications as those in which the computer is used for its ability to perform mathematical computations and perform simple repetitive tasks. He lists clerical tasks, automated capture of patient physiologic data from monitors, automation of the work process in the clinical laboratory, and maintenance of financial accounts as examples of computerized data-oriented applications.

In contrast, information is a set of data elements that have been organized in a manner that will convey meaning. A temperature measurement is a single data element but if displayed with other measurements from the same patient over the course of a hospitalization, one measure of the patient's health status can be inferred. Blum identifies information-oriented computerized applications as those in which data are retrieved (usually from a database) and organized and presented to human users for analysis and subsequent action.

Knowledge-oriented computerized medical applications are those in which medical knowledge is represented in the computer and is used to infer new data from patient data that are already present. The question of how best to represent medical knowledge within the computer has been a research question at LDS Hospital and at many other places for many years and undoubtedly will be for many more years to come. Different approaches to medical knowledge representation on HELP are discussed later in the chapter. The results of knowledge-based applications on HELP are translated into messages directed toward health-care providers and used to try to improve the patient care process. Examples of medical knowledge that have been encoded into HELP include (1) algorithms for determining when infections are hospital acquired, (2) guidelines for determining when laboratory results may be life threatening, and (3) guidelines for determining when adverse drug reactions may occur.

The ability of the HELP system to produce informative reports and perform automated decision making depends to a large extent on the fact that data from a wide variety of sources are stored in the integrated patient database described in Chapter 3. Figure 5.1 shows the various types of data stored by the HELP system and the various applications that make use of the data either for decision making or for reports. As can be seen, the widespread use of data from multiple application permits functions to be carried out that would be impossible without an integrated system.

5.2 Data Management on the HELP System

The HELP system is responsible for many administrative, financial, and clerical duties at LDS Hospital. Admission-discharge-transfer (ADT, Chapter 7) functions such as patient registration, procedure scheduling, and recording of demographics are performed using HELP. Admission-discharge-transfer and census reports are produced daily to assist admitting clerks and managers. Medical supplies are ordered for patients with the HELP system's order entry program (Chapter 11) rather than through a paper-based system, and patient charges are automatically captured and added to the patient bill. Surgery times are entered into the computer and operating room charges are automatically calculated and added to the bill. Medical records discharge data are entered into the system and case-mix, length of stay, and accounting data are automatically generated. An automated medical records chart tracking system is in place.

The HELP system also assists with the large volume of clinical data present in a tertiary care setting. Laboratory results are entered either auto-

Applications

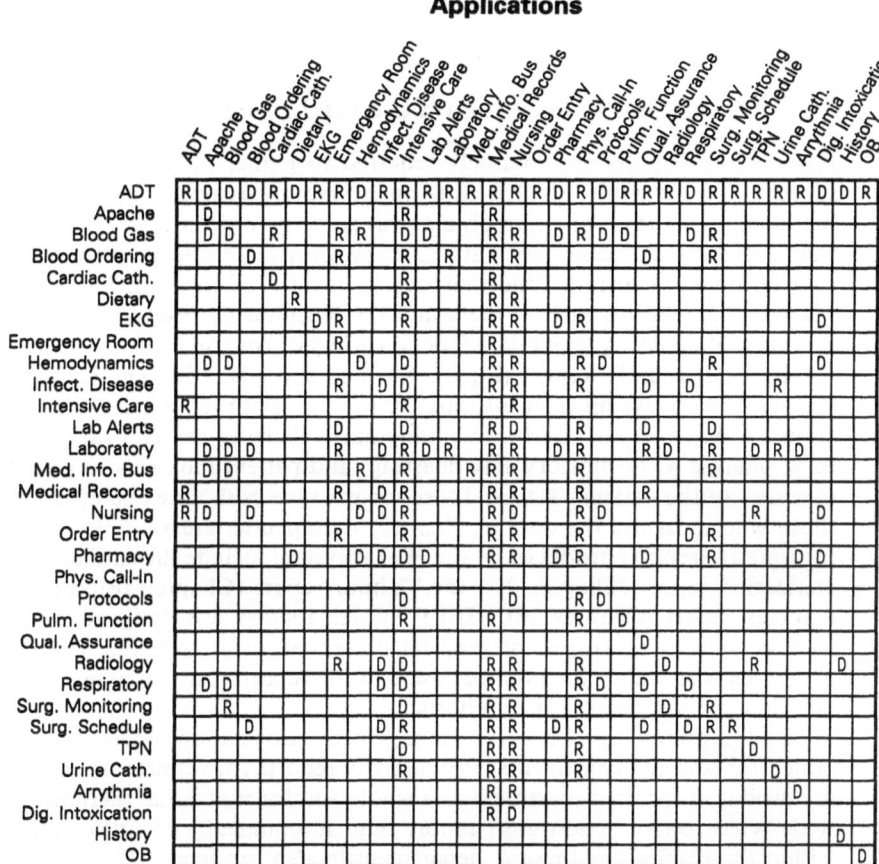

Data Types

R indicates that the application uses the data type in the creation of Reports.
D indicates that the application uses the data type in Decision Logic.

Figure 5.1. HELP system data integration chart.

matically or manually into the laboratory system (Chapter 16). Routine laboratory reports are produced by the laboratory system, but a copy of the patient data is electronically transferred to HELP for the inclusion in comprehensive patient reports. Data from patient monitors (Chapter 21) are automatically captured by the HELP system every 15 minutes and are displayed either alone or with other data in the form of informative reports. Computerized nurse charting programs (Chapter 13) permit nursing activities to be recorded and patient acuity and nursing charges to be automatically calculated. Nurse charting programs also allow the patient's clinical status to be determined. HELP manages a trauma log in the emergency room and captures data from word processors in the Radiology Department.

5.3 Information Processing on the HELP System

Health-care workers are routinely required to gather data from multiple sources. For example, a physician might require temperature readings, blood counts, x rays, and microbiological cultures to determine if an infection is present. A nurse may want to know previous hemodynamic values to determine if a current abnormal measurement is of concern. The HELP system assists health-care workers at LDS Hospital by producing clinical reports that contain patient data from a variety of sources.

The ICU Rounds Report (Figure 4.4) and the 7-Day Patient Summary (Figure 4.5) are examples of informative reports. Both reports contain data from a wide variety of sources including nurse charting, the clinical laboratory, pharmacy, respiratory care charting, and patient monitors to give an overview of the patient's status. These reports save both time and effort for those who must use the data in planning therapeutic interventions. Other informative reports produced by the HELP system include the cardiac catheterization report (Chapter 27), the 72-hour report (Chapter 16), and the diabetes report (Chapter 16). HELP system reports are designed by the LDS Hospital Department of Medical Informatics staff working with the personnel who will be the eventual users of the reports. At this time, HELP does not support user-defined reports.

In addition to clinical reports, management information reports are generated for several departments. Productivity reports for respiratory therapy are produced for the department as a whole and for individual therapists as well. A corrective action alert report that automatically identifies errors within the respiratory therapy department is also produced (Chapter 14). Nursing acuity is automatically calculated and printed in reports that assist with staffing decisions (Chapter 13).

The distinction between data-oriented and information-oriented applications can at times be arbitrary. A function designed simply to present one type of data (e.g., a single patient's medications) would probably be categorized as data-oriented, but is informative as well since it allows the data items to be viewed collectively.

5.4 Use of Computerized Medical Knowledge on the HELP System

5.4.1 Introduction

Early in his experiences with the computer in the cardiovascular physiology laboratory, Dr. Homer Warner realized that the computer could be used to interpret and make decisions about the data as well as acquire it (see Chapter 1). The first uses of computerized interpretations of clinical data at LDS Hospital included diagnosis based on clinical findings, diagnosis based on history, ECG analysis, and blood gas interpretation. As the HELP system has expanded over the years, so has the role of computerized medical decision making within the system. The approach of the designers of the HELP system continues to be that the best use of the computer for decision making is the analysis, interpretation, and manipulation of clinical data in a way that assists health-care workers in their routine tasks. Currently active HELP system medical decision-making programs include (1) alerting to potentially adverse conditions in patients receiving respiratory care therapy, (2) identification of life-threatening laboratory results, (3) alerting to potentially adverse drug orders, and (4) a program that identifies patients with nosocomial (hospital-acquired) infections. Other applications (such as automated diagnosis) continue in an experimental mode only.

It should be noted that although this section refers to "computerized decision-making programs," the output of HELP system computer programs in no case affects patient care directly. Rather, the results of the "decision-making" programs are suggestions or comments that are directed toward health-care workers to assist in the care process. Computerized medical decision making programs are also sometimes referred to as "medical expert systems" since they attempt to emulate the outputs and/or cognitive efforts of medical experts. An excellent review of computerized medical decision-making programs other than the HELP system is contained in reference [2].

5.4.2 Acquisition of Medical Knowledge on the HELP System

Although the obstacles may not be immediately apparent to the lay person or to the novice in medical informatics, knowledge acquisition may be the most difficult and time-consuming aspect of the development of a computerized decision-making system. Defining the exact logic to be used in generating automated decisions taking into account all the possible expected situations and all possible exceptions can be painstaking. Deciding on the specific numerical values to be used (such as the values that will differentiate high or low measurements of a particular test, or probability estimates for statistically based programs) increases the effort required. When the knowledge is acquired from a panel of experts rather than a single source the likelihood of accurate results may be higher; however, any disagreements between the expert participants must be resolved. Finally, after it has been acquired, the knowledge must be translated into a computer language for testing and validation. Any aspect of the logic found to be invalid must be reanalyzed and corrected. The time-intensive nature of knowledge acquisition has led some to refer to a "knowledge acquisition bottleneck" [3] and, whereas automated tools are being developed, most knowledge acquisition currently still occurs manually. Knowledge for HELP decision-making programs is gained from local consensus development, established published guidelines, database analysis, and automated knowledge gathering tools.

Local Consensus Development. Several formal methods exist for obtaining a consensus opinion from a panel of experts. Examples include the Delphi method and the nominal group method [4]. Consensus development for HELP knowledge has frequently used similar but less formal techniques. Typically, a committee is formed consisting of experts in the medical domain of interest and faculty from the Department of Medical Informatics. The Medical Informatics faculty function as knowledge engineers.[1] A literature search is performed to define the state of the art. Discussions ensue (which may last for weeks or even years) in which (1) the relevant knowledge is abstracted from the literature, and (2) the knowledge as stated in the literature is criticized by the domain experts in terms of (a) their own extensive clinical experience, (b) any details relevant specifically to the Salt Lake City community, and (c) the exact needs of the decision-making program under consideration. The knowledge engineers ensure that the logic as stated by the experts is consistent and can be converted into a computer-processable form in whatever knowledge representation scheme will be used. As construction of the knowledge base[2] approaches completion, a prototype computer program is constructed to demonstrate the functioning of the logic. The program is reviewed by the committee and any deficiencies in the logic are identified and corrected. Examples of HELP programs that have used this approach include the blood ordering program (Chapter 26) and the ventilator management program (Chapter 23).

Use of Published Guidelines. Another source of knowledge for HELP system programs is accepted published guidelines from other sources. An example is Centers for Disease Control guidelines [5], which have been implemented in HELP in the CIDM (Chapter 18) to help diagnose nosocomial infections. The same Centers for Disease Control guidelines had previously been used manually by infectious control practitioners (ICPs). The CIDM automatically gathers from the HELP database the same data that the ICPs had collected manually previously and affects the same logic. Health-care guidelines meant to be used manually may be computerized as long as the necessary data elements are present on the HELP system.

Database Analysis. The HELP database itself may function as a source of medical knowledge. Many computerized decision-making programs (both

---

[1]The process of obtaining knowledge from experts and converting it into computer processable form is known as knowledge engineering.
[2]The set of medical knowledge contained in a computerized decision making program is referred to as the knowledge base (similar to data in a database).

in and out of the HELP system) use probabilistic estimates of the occurrence of diseases and associated findings to perform inference. Although these estimates are frequently obtained from experts and/or the medical literature, analyses of the HELP database have also been used as a source of probabilistic data. An example is an experimental HELP program that infers pulmonary diagnoses based on a priori likelihoods of diseases and the sensitivities and specificities of certain findings [6]. The initial estimates of the statistics were based on a literature review and on expert opinion. When the probabilities were revised based on an analysis of the database, the program was found to be more accurate in its predictions.

Automated Knowledge-Gathering Tools. As mentioned above, automated tools can play a role in the knowledge gathering process. QSTN, PMOKFE, and PROTOKOL are HELP tools that are either currently being used or are in the process of development.

QSTN, described in detail in Chapter 4, is clinically active and allows procedural knowledge (in terms of screen formats and a hierarchical arrangement of data-gathering screens) to be gathered. QSTN can be used to design data entry applications with minimal training. It has been used to design data entry screens for the computerized nurse charting program (Chapter 13) and the order entry program (Chapter 11). The pharmacy medication ordering knowledge frame editor (PMOKFE) [7] is described in detail in Chapter 15. PMOFKE is experimental and is intended to gather intelligent medication ordering rules from pharmacists for use in medication ordering programs. Finally, PROTOKOL (described in Chapter 35) is an experimental program used to generate computerized protocol logic from graphical input.

5.4.3 Knowledge Representation in the HELP System:
The HELP Frame Language

Computerized decision-making programs require medical knowledge to be represented in some fashion within the computer. Research efforts worldwide have resulted in many different knowledge representation schemes and, indeed, the frontiers of computer science have been advanced through those efforts. An overview of the different approaches to medical knowledge representation is contained in reference [2].

Medical knowledge has been represented in the HELP system procedurally, through the use of rules, and statistically. Over the years these schemes have been implemented using HCOM (the HELP compiler language described in Chapter 1) and the TAL and PAL procedural programming languages (described in Chapter 3). In the late 1980s researchers from several institutions, including LDS Hospital, began collaborating to develop a modular representation of medical knowledge. At LDS Hospital the modular representation is known as the HELP Frame Language.

Introduction.[3] The frame is used as the basic unit of knowledge representation. The structure of the frame has been designed to be sufficiently universal to contain knowledge required to implement not only expert systems, but almost all traditional hospital information systems (HIS) functions including ADT, order entry, and results review. The design incorporates a two level format for the knowledge. The first level as ASCII records is used to maintain the knowledge base while the second level, converted by special knowledge compilers to standard computer languages, is used for efficient implementation of the knowledge applications.

Implementation of HISs at large hospitals is now becoming routine. Initially, HISs were conceived of as administrative and financial systems. They allowed the hospital to admit patients and track the orders and charges on those patients. As clinical systems became available, either as stand-alone systems to be networked to the central HIS or integrated within the HIS, new challenges were placed on the design of the HIS. The software architecture of an HIS generally follows one of two models. The first is a traditional time-sharing model where multiple independent processes perform the applications and share the hardware resources of the system. The second model, which has become increasingly more popular, is referred to as a transaction model. In the transaction model the system supports a general screen driver and a series of application servers. By defining low-level generic servers, a transaction-oriented system greatly reduces the frequent process start-up overhead often observed in the time-sharing model. With the advent, however, of medical expert systems requiring efficient interaction of the system with the user and a medical knowledge base, the advantages of the transaction model are somewhat reduced by potentially unacceptable processing times. This section discusses a new software architecture model referred to as the knowledge-driven model, which permits efficient execution of traditional HIS functions as well as newer expert system technology.

Functional Requirements. The design of any new software architecture should ensure that the functional requirements of the applications incorporated in the HIS are adequately met. In our analysis of HIS applications, we have found that the following functions comprise the necessary and sufficient functions used to create todays HIS applications:

1. data acquisition, including screen formatting
2. data review and editing, including hard copy generation
3. database interaction, including storage and retrieval
4. data analysis, including support of multiple expert systems

---

[3]The remainder of section 5.4.3 is adapted from: Pryor TA, Design of a knowledge-driven hospital information system (HIS). *Symposium for computer applications in medical care (SCAMC)* 1987;11:60–63. Used with permission, IEEE Publishing.

With the knowledge driven model, the content of the knowledge base has been extended to incorporate the logical structures necessary to perform all four of the HIS functions enumerated above. The frame, or medical logic module (MLM) [8], has been chosen as the basic unit of knowledge representation. Through proper definition of the frame structure we are attempting to develop a highly modular knowledge base in which applications are defined in terms of structured groups of frames. Thus, a single frame may be used for multiple applications. For example, a frame for acquisition of vital signs may be used by a nurse charting application or a medical expert system controlling hypertension management.

Frame Structure. We have chosen a generic frame structure that, depending on the frame type (data acquisition, review, diagnoses, management, etc.), allows the application developer (frame writer) to enter knowledge into frame-specific slots. The generic structure of the frame has slots for entry of variable definitions, declarative knowledge, and knowledge documentation. Figure 5.2 is an abbreviated example of a diagnostic frame.

The variable definition section of the frame allows the user to define local knowledge variables used in the declarative section of the frame. The user defines a variable by linking it to one or more data descriptors present on the system data dictionary. The HELP system data dictionary (see PTXT, Chapter 3) defines only basic elements that are captured either through an automated machine interface or from a computer terminal. The variables defined in the frame, however, may be complex entities that are the result of searches and analyses of the patient's database. For example, a variable may be limited to a simple entity within a fixed-time window. Within that time window the defined variable may be the first instance or the last instance or the average of the recorded element. Constraints to other events within the time window may also be stated in the variable definition. Additional information about the variable may also be declared in this section of the frame. This additional information may, for example, declare relationships between the variable and a disease. The relationships could include statistics such as sensitivities and specificities, fuzzy logic associations, or associations similar to the concepts of import and evoking strength in the Internist-I program. Because of the volume and complex nature of the variables, we have chosen to include them in the frame rather than have a separate complex data dictionary. In so doing, the developer and reviewer of the frame are able to see immediately the variables being used in the declarative knowledge and understand their meaning. In Figure 5.2, the variables chest-pain, cough, and so forth, have been declared. Statistics relating those variables to the disease pneumonia are also included. The text in parentheses refer to the text associated with the simple entities defined in the data dictionary.

Also within the variable definition section of the frame is a slot used to enter evoking logic. This logic defines the criteria for the automatic evoking

---

Title: Pneumonia diagnosis (7.141.1)

Type: Diagnosis

Author: Peter Haug.

Date: 12/12/86

Message: "<disease_prob (val; #.##)> Pneumonia (history)".

Variables: chest_pain as (DO YOU HAVE CHEST PAIN?),
           cough as (HAVE YOU HAD A COUGH WITH THIS ILLNESS?),
           fever_or_chills as MAX(fever, chills)
           where fever is (HAVE YOU RECENTLY HAD A FEVER?)
           and chills is (HAVE YOU HAD CHILLS RECENTLY?)
           if Exist (fever) or Exist (chills),
                                    •
                                    •
                                    •
Statistics: for fever_or_chills with **TPR**(YES, 0.85;   NO, 0.15),
           and **FPR**(YES, 0.3;   NO, 0.7),
           for cough with **TPR**(YES, 0.9;   NO, 0.1),
           and **FPR**(YES, 0.2;   NO, 0.8),
                                    •
                                    •
                                    •

Evoking Criteria: If chest_pain EQ YES or fever EQ YES or chills EQ YES or cough EQ YES.
Logic: disease_prob = 0.014.
           If Exist(fever_or_chills) then disease_prob = Bayes(disease_prob, fever_or_chills).
           If Exist(cough) then disease_prob = Bayes(disease_prob, cough).
                                    •
                                    •
                                    •
           If disease_prob LT 0.014 then finish.
Ask: Patients(fever, chills, cough) Heirarchical.
Urgency: 5/9
Gold Standard: If ICD_pneumonia and pneumonic_infiltrate
References: Harrison's Principals of Internal Medicine. Braunwald E, et al (editors)
Validation: Tested experimentally (DD method)

Figure 5.2. Example of abbreviated diagnostic frame for pneumonia written in HELP Frame Language. The frame is processed if the criteria in the evoking criteria slot is met. The ask slot indicates which information the frame may interactively collect. The urgency slot indicates the relative urgency of this disease. The urgency slot indicates the relative urgency of this disease. The gold standard slot specifies criteria available by the time of discharge, which would prove the existence of the disease modeled. References refer to literature support for the frame and validation indicates the degree of evaluation to which the frame has been subjected.

of the frame when the data contained in the evoking logic is stored in the patient's data file. The example in Figure 5.2 has the evoking criteria that would be used to data drive the knowledge of this frame.

Because of the need to have formatted screens, special variables called windows are also defined in the variable definition section of the frame. These windows define the screen format of the terminal display and link the entered data to the previously defined knowledge variables.

The second section of the frame, the knowledge section, contains the frame's declarative knowledge. The contents of the knowledge section are frame dependent. That is, depending on the type of frame being created, the knowledge slots available for entry are changed. For example, if the frame were to be used as part of an expert system for disease diagnosis, the knowledge slots would require entry of the associational model knowledge. This could include if–then rules for the diagnosis, Bayesean statistics, or any other appropriate scoring logic. If, on the other hand, the frame were an order entry frame, knowledge concerning the critiquing of the order would be requested. As new applications arise we anticipate that newer frame types will be required that define new slots within the knowledge section.

The knowledge documentation section of the frame is used to enter information regarding the author of the frame, revision dates, references to pertinent literature articles, level of validation of the knowledge, and gold standards. The level of validation of the knowledge is intended to provide the developer with a slot where he can declare the clinical status and possible utility of the frame. For example, if the frame was merely the developer's first educated guess without any serious validation studies, this would be indicated in this section. Thus, anyone using the frame would understand the level of confidence that one might give to the performance of the frame. Likewise, if the frame had been tested on a large population with satisfactory results, the frame user would be aware of those tests and results. The gold standard slot is intended to contain information regarding any gold standard data that may be used to compare the results of the frame. For example, if the frame suggests a new treatment for some disease, the gold standard slot could contain a definition for the successful outcome of that therapy.

Many of the slots we have added to the structure of the frame are intended for use by knowledge management routines. These routines scan the knowledge base to report to the systems manager critical information on the use and state of the knowledge base. For example, linkages between the data dictionary and the knowledge frames will be maintained, allowing rapid alteration of the frame variables when associated entities in the data dictionary are changed.

Procedural Control and Meta-Frames. The frames, as described here, contain the declarative knowledge of the application. To generate an actual HIS application an additional frame type is required. This frame, referred to as the meta-frame, contains procedural control knowledge necessary to implement an efficient application. For the meta-frame, the knowledge section consists of slots controlling the execution of those frames required in a single application. A slot is used to enter the procedural control knowledge of the application. This knowledge may be as simple as the declaration of the initial frame to be executed in a goal-driven application or as structured as an ordered list containing the frames and their execution sequence for the application. A second slot defines the inference engine for the application. Again this can be as simple as nothing (i.e., execute the frames according to the control logic and exit the application), or a hypothesis driven frame scoring algorithm similar to Internist-I, which determines the frames to be executed based on an iterative process of data collection and frame evaluation. Logic pertaining to the storage of results in the database are contained in another slot of this section.

The example in Figure 5.3 is a meta-frame for control of an arrythmia protocol. This frame contains the logic for control of a branching protocol. Execution of the frame will cause execution of the declarative knowledge frames of the protocol under the proper conditions.

Frame-Based Applications. Having developed a design for a frame structure, a conceptual model for knowledge acquisition has been developed. Using this model we are now able to create a knowledge base sufficiently universal to support the applications of an HIS. Examples of the applications we are currently developing using this frame structure include:

1. *Physician ordering.* This application will use the frames to capture knowledge concerning the test/procedure to be ordered. Critiquing knowledge will also be present in the frames. In using the frame-based applica-

---

Title: Arrythmia^protocol

Message: Patient has criteria for arrythmia management protocol"

Frame Type: Meta (protocol)

Evoking Criteria: If Acute^MI and not NSR

Procedural Logic: Evoke IV^Dextrose,
        If Hypokalemic then Evoke KCL,
        If rhythm = PVC then Evoke PVC^management and exit,
        If rhythm = Bigeminy then Evoke PVC^management and exit,
        If rhythm = Vtach then Evoke Vtach^management and exit,
        If rhythm = Afib or rhythm = Aflutter then Evoke
        Afib^flutter^management and exit,
        If rhythm = Sinus^tach then Evoke Stach^management and exit,
        If Heart^rate < 60 then Evoke Brady^management.

Figure 5.3. An example of a meta-frame used to implement a protocol for the management of arrhythmias. The logic section of the frame controls the execution of declarative frames of this application.

tions, the entry of an order will cause the appropriate frame to be executed, assisting the physician in entry of the order.

2. *Nursing care plans*. In this application the frames will contain the knowledge defining the problem and the creation of an appropriate working care plan. For example, many of these frames will be data driven from the physician order frames described above.

3. *Monitoring/Alerting*. These applications will define frames that are data driven by the receipt of data into a patient's file. They will contain knowledge about the data being stored to determine if alerts need be transmitted to the appropriate personnel.

4. *Diagnosis*. Application programs here will be driven by frames containing models of the diseases being diagnosed.

5. *Treatment Planning*. These applications are similar to the nurse care plan application in that knowledge about different treatments will be entered into the frames and executed as the appropriate conditions arise.

This list of applications is only a small fraction of the HIS applications that lend themselves easily to the frame model, but hopefully the list gives a flavor of the direction of our system. Our knowledge base will be stored as a set of ASCII records. In this form the frames are easy to access, modify, display and transmit to other centers involved in this research [9]. Because of the design of the frames and their stored form they are also machine independent.

Transformation into Executable Formats. The final step of our HIS design is the implementation of the operational form of the applications. In our design the frames will be transformed into a traditional computer language for execution. We are currently exploring two modes of implementation, a compiled form and an interpreted mode. In both of these modes an intermediate computer language is used as a target language by an application generator. The application generator is driven by a meta-frame, which contains the list of frames and the procedural knowledge of the application. On identifying to the application generator the appropriate meta-frame, the application generator assembles the required set of frames and inserts the necessary procedural logic, resulting in a single application program in the target language. The derived application program can then either be interpreted or compiled. Because of the overhead associated with either process start-up of compiled applications or interpretation of code, the decision to compile or interpret appears to be application dependent. For example, in an application such as hypothesis-driven disease diagnosis where the application would be active for many minutes, the limiting factor is execution speed and not start-up time. In such cases compiled code is more efficient, but in a data-driven application where only a single frame is to be executed the processing time may be minimal compared to the start-up time. Here the interpreted mode is optimal. In either case the important concept is the creation of application programs. The ability to create application modules from the general knowledge base is key to our design and assumes that regardless of the size of the knowledge base, use of the knowledge will be limited to specific applications synthesized from a subset of the knowledge present.

We are currently mid-way through the development of the necessary knowledge editors and compilers. Design of the frame structure has been completed. Simulation studies involving hand-compiled frames into an application program written in a target language (Pascal and PAL) have been tested. Our testing has involved actual implementation of some applications on HELP. These implementations to date have shown the design described in this chapter to be feasible.

5.5 Summary

Computerized HISs hold the potential to assist health-care workers in multiple ways. HELP assists workers at LDS Hospital through data management, information processing, and automated decision support. HELP can perform these tasks in large part because of the integrated patient database because all patient data are available to all system programs. Many HELP decision support programs are part of the normal work routine at LDS Hospital and many have been shown to positively influence patient care. The HELP applications introduced in this chapter are all discussed in detail in later chapters in the book.

References

[1] Blum BI. *Clinical Information Systems*. New York; Springer–Verlag. 1984.
[2] Reggia JA, Tuhrim S. (ed.) *Computer-Assisted Medical Decision-Making*. New York; Springer–Verlag. 1985.
[3] National Library of Medicine Long Range Plan. Report of Panel 4 – Medical Informatics. U.S. Department of Health and Human Services; December, 1986.
[4] Fink A, Kosecoff J, Chassin M, et al. Consensus methods: characteristics and guidelines and use. *Am J Pub Health* 1984;74:979–983.
[5] SENIC Project: Algorithms for diagnosing infections – Appendix E. *Am J Epidemiol* 1980;111:635–643.
[6] Haug PJ, Clayton PD, Shelton P, et al. *Revision of diagnostic logic using a clinical database*. Proceedings of the AAMSI Conference 1987;6:238–242.
[7] Prokosch HU, Hulse RK, Wall M, Wong TW, Pryor TA. New decision support concepts for the pharmacy application in the HELP system. EFMI Special Topic Congress; September, 1988; Hannover, West Germany.
[8] Clayton PD, Pryor TA, Wigertz OB, Hripcsak G. Issues and structures for

sharing medical knowledge among decision-making systems: The 1989 Arden Homestead retreat. *Symposium for computer applications in medical care SCAMC* 1989;13:116–121.

[9] Clayton PD, Pryor TA, Wigertz OB, et al. Sharing medical knowledge for automated decision making. *Symposium for computer applications in medical care (SCAMC)* 1988:12;591–594.

6
Current Usage of the HELP System

The HELP system plays an integral role in the clinical, administrative, and financial functioning of LDS Hospital. As is described in detail in other chapters of this book, the system is used for patient registration, surgery scheduling, charge capture, order communications, management of clinical patient data (including laboratory, pharmacy, bedside monitoring, nursing, and respiratory care charting data), results review, and other functions, and new functions are constantly being added. The HELP system has required a significant effort to develop it and it has impacted the manner in which care is delivered at LDS Hospital. This chapter examines some of the ways in which routine medical practice at LDS Hospital has been affected by HELP.

6.1 Evaluation of the HELP System

The designers of HELP have always had an interest in the impact of computerization on the practice of medicine. The prevalent attitude in the Department of Medical Informatics is that computerization is not an end in itself but rather a means to improved care. "So what?" is a comment frequently heard from proponents of evaluation studies in the department. Many of the HELP modules have been evaluated and descriptions of evaluation studies appear in the individual chapters. Some of the evaluation studies are mentioned here to give an overview of HELP's impact on medical care. The reader is referred to the individual chapter for references and further details.

HELP system programs that have automated manual procedures have been studied to measure how well they perform the intended functions. For example, the CIDM (Chapter 18) was designed to identify automatically nosocomial infections based on data contained in the computerized patient file. Previously, identification of nosocomial infections had been done by infection control practitioners (ICPs). An evaluation study revealed that the CIDM correctly identified more patients with nosocomial infections than did the ICPs. The CIDM has now replaced the ICPs for this purpose at LDS Hospital. Similarly, the accuracy of HELP's blood gas interpretation program (Chapter 19) was studied. Seven expert physicians were asked to interpret a series of blood gases. Since the physicians' interpretations varied greatly, the statistic measured was the computer's agreement with the majority of the physicians. The computer agreed with most of the physicians as often as did any of the individual expert physicians.

Some HELP applications have been shown to have a positive impact on the quality of medical care. The computerized laboratory alerting system (CLAS, Chapter 17) was shown to decrease the time spent in a life-threatening situation, decrease the total hospital length of stay, and increase the frequency of appropriate treatment for patients with certain life-threatening alerts. A program that identifies patients with inappropriate perioperative antibiotic orders (Chapter 18) increased the frequency of appropriateness of the orders from 40% to 58%. More importantly, the post-operative wound infection rate in these patients dropped significantly from 1.9% to 0.9%. Other programs that have been evaluated with respect to their quality assurance functions are the medication monitoring program (Chapter 15) and the blood ordering program (Chapter 26). Quality assurance in general on the HELP system is described in Chapter 9.

The HELP system has also been shown to save money at LDS Hospital. The corrective action alert (Chapter 14) report informs the Respiratory Therapy Department of inaccuracies in billing for therapeutic oxygen. In a single year this report helped save the department $80,000. A postoperative antibiotic monitoring program (Chapter 18) reduces the average number of postoperative antibiotic doses from 19 to 13 with potential savings of almost $90,000 per year. The computerized respiratory therapy charting program (Chapter 14) was found to increase respiratory therapists' productivity by 10%. An early evaluation of bedside terminals, however, has found no measurable change increase in nurses' productivity using bedside terminals and the computerized nurse charting program (Chapter 13).

Special purpose programs, such as a natural language processor for extracting coded findings from chest x rays (Chapter 29) and an experimental diagnostic program based on patient history data (Chapter 33) have been evaluated and found to perform quite well.

6.2 User Attitudes

The impact of the HELP system on the end users, that is, the personnel of LDS Hospital (e.g., clerks, nurses, physicians, ancillary staff, etc.), has been a careful consideration of the HELP system designers. Several informal questionnaires have been administered to assess the attitudes of users and many of these are mentioned in the chapters that follow. Some of the more important measures of user attitudes are mentioned here.

After the computerized respiratory therapy charting module was installed (Chapter 14), the therapists' attitudes toward the new system was measured. More therapists (64%) preferred the computerized charting than preferred the manual method (20%). Three quarters of the therapists felt the new charting method would make their job easier.

Two questionnaires were administered housewide to attending physicians and nurses to determine their attitudes toward general system functions. Both physicians and nurses felt that the computer's ability to display laboratory results and blood gas results was very important. Both groups felt that the computer's ability to generate and display alerts based on the patient data was also important. When asked what was important for the future, both physicians and nurses indicated they wanted the knowledge base to be expanded to allow for a greater number of alerts. Physicians also wanted access to the pharmacy knowledge base for advice during prescribing. Both groups felt a greater level of user support (i.e., training) should be provided by the hospital so that users could obtain maximum benefit from the system.

6.3 Security

Maintaining confidentiality of medical records is a difficult issue even for paper-based systems. Standards for confidentiality of computerized records have not developed yet. In general, the ease of access to records decreases as the security level increases and concern for patient privacy must be balanced with the providers' needs to obtain the data to provide care.

Currently, accessing the HELP system at LDS Hospital does not require a security log-on code. Security is maintained by physical placement of the terminals in areas staffed by hospital personnel. A scheme that will require users to identify themselves through a computer-assigned five-character log-on code is being devised. The scheme will require users to identify themselves to the system to access data. Which users are looking at what data will also be able to be determined and inappropriate examination of data can be followed up on. (This is something that cannot occur with paper-based records, i.e., a benefit of computerized records.)

Once in place, the log-on scheme will yield benefits other than simply increased security. Since the HELP system will know who the user is, individualized screen formats and access privileges to certain applications will be devised. Utilization patterns by individual or group (e.g., physician, nurse, etc.) will be able to be determined.

6.4 Future Directions of the HELP System

The lessons learned from the existing HELP implementation serve as stepping stones to the continued development of this unique system. Future developments for HELP are planned in four distinct areas: (1) applications development, (2) database development, (3) hardware considerations, and (4) standards.

Future applications developments will consist of continued use of the HELP Frame Language described in Chapter 5. The knowledge-based approach to building applications has proven very successful and will result in newer applications using this approach. Examples of these newer applications under development include the new order entry system where all of the ordering knowledge is contained in the knowledge base, a knowledge driven problem-oriented patient management system, an anesthesiology charting system, and others. In these new applications the use of user-specific knowledge coupled to specific application drivers will be the rule rather than the exception.

A new direction is planned for the HELP database management system. Currently HELP uses a locally developed database system (see Chapter 3). Investigation is now underway to determine if the HELP database can be structured to function under commercially available structured query language (SQL) database software. The movement of the HELP system to the SQL standard database structure will not restrict the data models that can be supported by HELP but will use the benefits of commercial software to enhance the maintenance and performance of PTXT. Additional features are also planned for the HELP data models, of which aliasing is the most notable. Aliasing permits data elements to assume a different context in different applications.

The third area of future research is with the hardware platforms on which HELP resides. Currently HELP functions only on Tandem Computer systems. A new model being explored will allow processing of the medical knowledge frames on PCs. Using this model, the knowledge base could be targeted to run either on the Tandem or on a PC. This distributed processing model would greatly enhance the performance characteristics of HELP. The distributed processing model would result in more of a client/server model where the patient data and the applications would reside on a database server and would be downloaded to a PC for processing.

The final major area of future direction is in the development of standards for the creation of a medical knowledge base. This effort (being conducted in conjunction with other research centers) is attempting to develop a standard medical knowledge language (see Chapter 5) through which medical informatics research centers could share medical knowledge. Using this standard language, validated knowledge from one center could be transmitted to another center for implementation on their system. Since this will help to circumvent the knowledge acquisition bottleneck (see Chapter 5), HELP's knowledge base should be able to grow at a much greater rate than at present. An expanded knowledge base will decrease the time required to realize further the potential benefits of HELP.

6.5 Summary

This chapter has reviewed some system-wide issues related to the implementation of the HELP system at LDS Hospital. The remaining chapters of the book examine each of the HELP system's applications in detail.

II
Administrative Functions on the HELP System

Although developed initially for purely clinical purposes, the ability of the HELP system to manage administrative functions was quickly realized. Today, administrative aspects of LDS Hospital managed by the HELP system include ADT functions, certain medical records functions, order communications, and quality assurance. This part of the book examines these and other administrative functions of the HELP system.

7
Admit-Discharge-Transfer Functions

The HELP system is used at LDS Hospital to schedule patients for elective admissions, to record demographic, insurance, and clinical information in preadmitting and admitting procedures, to keep track of changes in the patient's location during the hospitalization, and to document the patient's discharge from the hospital. Similar data are gathered when patients come to LDS Hospital for emergency services, outpatient radiology and outpatient clinic services, and for short-stay (1 day) procedures. The ADT data are used to generate a computerized census nightly. The patient demographic data captured during the admission process are stored in a variety of long- and short-term computerized patient files and are used by many other HELP system programs.

7.1 Patient Types and the Master Patient Index

Within the HELP system, patients receiving care at LDS Hospital are designated as one of the following:

1. inpatients
2. short-stay patients (patients admitted for a surgical procedure who are scheduled to be discharged without spending a night in the hospital)
3. AM admissions (patients who will be admitted on the morning of a scheduled surgery but who will spend at least 1 night in the hospital)
4. outpatient clinic patients
5. outpatient radiology patients
6. emergency room patients

A list of all patients (either inpatient or outpatient) who have been admitted to LDS Hospital is kept in a HELP computerized file known as the Master Patient Index, or MPI. The LDS Hospital MPI is comprehensive since 1965 and includes the records of many patients admitted in the 1950s and 1940s as well. Each record in the MPI includes data such as the patient's name, sex, birthdate, medical record number, whether they are deceased or not, their social security number, next of kin, the microfilm volume number (if their chart has been microfilmed), the radiology number (described below), the dates of their first admission to LDS Hospital, and the dates of their five most recent admissions.

Patients other than clinic and radiology outpatients have their records maintained in the Hospital's Medical Records Department and thus receive a medical records number. Because some outpatients do not receive a medical records number, and in order to keep outpatient and inpatient radiographs together, a radiology filing number was introduced in 1979. The radiology number is generated by the computer using the patient's name and date of birth. The number for any two patients will be different unless they have the same name and birthdate. Since the radiology number is calculated by the computer but the medical records number must be generated manually, the radiology number is available earlier (sometimes many hours earlier) than the medical records number. The radiology number is stored in the computer as part of the MPI.

With the advent of IHC Net (a plan to link electronically all hospitals managed by IHC, see Chapter 1 for more detail), a new unique identifier, assigned sequentially, will be created for Net-wide use. Even when use of this identifier is implemented, the radiology number will continue to be used at LDS Hospital and both will appear in the MPI.

7.2 Scheduling and Admitting Programs

Approximately one third of all inpatient admissions to LDS Hospital are elective. When a physician's office calls to arrange for an admission, an admitting clerk enters into a scheduling program the diagnosis, demographic data, laboratory tests that need to be performed, and any intended procedures. The clerk then contacts the patient directly to complete the data gathering and arrange for preadmission laboratory tests to be performed. When direct contact is made with the patient (the so-called preadmission procedure) the computer accesses the long-term patient file to see if the patient has been admitted previously (see above) and either creates a new record or links to a pre-existing one.

When a patient is admitted the MPI is searched (based on the patient's name and date of birth) to determine if he/she has ever been admitted previously. The search uses a Soundex program that retrieves names that sound similar to the one that is entered, even if the spelling is different. Soundex programs retrieve names that "sound" like the one entered into the computer. The programs function by eliminating vowels and equating consonants that may sound alike (such as "c" and "k"). If the name "Smythe" is entered, the names "Smith" and "Smooth" will be retrieved. Similarly, "Smth" may be used to retrieve any of these. The search also takes into account name changes, such as those that occur with marriage. If no matching record is found, a new one will be created. If more than one possible match is found, the admitting clerk determines which (if any) is the correct record and links the present patient record with the previous record. The MPI (also known as the long-term patient file) is used by HELP computer programs (e.g., scheduling, chart location) that must determine if patients have been admitted at any time in the past. The MPI is updated at the time of discharge to include the discharge date.

When a prescheduled patient arrives, data from the scheduling program are transferred to the admitting program. At this time all the data must be completed, including insurance and guarantor information. When a patient is admitted on an emergency basis, no scheduling data exist and all relevant data must be gathered at the time of admission. A Quick Admit mode can be used that requires only a minimum of information for the patient to be entered into the system and a patient number (see below) to be generated. The Quick Admit is for severely traumatized or "John Doe"-type patients who may not have any identifying data available at the time of admission. Outpatients at LDS Hospital (emergency room, outpatient clinic, outpatient radiology, and short-stay) are all entered into the admitting program and are thus also all linked to the MPI.

7.3 Patient Numbers

Every time a patient is admitted to LDS Hospital (either inpatient or outpatient), he/she is assigned a computer-generated 8-digit patient number (occasionally also called account number or encounter number). The patient number is used to identify the patient during that particular stay or encounter. For example, a patient coming to the emergency room twice on a given day would receive different patient numbers for the two visits. The patient number is the number used by the HELP computer index data in the computerized patient data file (see Chapter 3 for further discussion).

The first digit in the patient number indicates the department where care was rendered (e.g., numbers beginning in 6 indicate outpatient clinic care, numbers beginning in 2 indicate inpatient care, etc.). The digits in a valid patient number are chosen so that they satisfy a specific mathematical relationship. Any time a patient number is entered into the HELP system the digits are checked to make sure the number is a valid patient number. This is done to minimize mistakes resulting from mistyping errors (e.g., retrieving incorrect patient data as a result of transposition of digits).

Admit—Discharge—Transfer Report
LDS Hospital
Inpatient Patient List
06/14/1990

NEW ADMISSIONS

ID			Room
210821	Shirley Marie M	Richards, David	E732
210831	Robert Benjamin	Jackson, Christo	W827
110819	Sherri Ann	Stevens, Mark H	E805
210643	Ann	Stevens, Lawrenc	N3AM
110382	Lori Lea	Rosenberg, Thoma	N3AM
210635	Julie	Havlik, Kevin L.	E420
110689	Boy Christine	Allred, Gerald L	E420
210683	Christine Lynn	Rasmussen, E. Ke	E412
210807	Warde Mitchell	Reilly, William	E841
110507	Boy Lisa	Fox, Jesse N.	E420
210501	Lisa Rae	Rasmussen, E. Ke	E402
110834	Lee	Walsh, Kevin J	W714
210606	Becky NMI	Rasmussen, E. Ke	E409
110612	Boy Becky	Sanders, Gill O.	E420
210817	Janice	Allred, Gerald L	E420
110823	Francom	Rasmussen, E. Ke	E405
210823	Delbert Leroy	Nielsen, Robert	W838
210552	Ann Marcella	Rasmussen, E. Ke	E403
110558	Boy Ann	Schmidt, Jeffrey	E420
210622	Annette Cecila	Ennis, Harry H.	N3AM
210804	Mary Ann	Jones, Kent W.	W716
210811	Charles Stewart	Rosenberg, Thoma	N3AM
210798	Gneiting	Sorensen, Sherma	W735
210814	Dora Elizabeth	Moders, Frederic	E741
110837	Weston Eugene	Evans, Burtis	W804
210831	Irene Lavin	Allred, S. Willi	W335
210523	Boy Marla	Allred, Gerald L	E420
110529	Marla	Rasmussen, E. Ke	E410
210427	Maryann	Lambert, Joseph	E420
210427	Lisonbe	Heywood, E. Reed	E411
110811	Nelson	Millar, Roger Cl	N3AM
210819	Valerie	Lambert, Joseph	E420
110817	Gail NMI	Belnap, Legrand	W651
110827	Sherry	Bentley, Louis F	E432
110837	E	Sorensen, Sherma	E743
210826	K.	Muir, Mark W.	E818
110747	Pamela Piercey	Stevens, Mark H	W636
110822	Janene	Allred, Gerald L	E420
110824	Angela Jean	Rasmussen, E. Ke	E405
110339	Lloyd Howard	Wilkinson, Craig	E704
210772	Tammy Gay	Allred, S. Willi	W309
110819	Donald Leroy	Mangelson, Ned L	E835
110798	Kellie Jan	Vincent, G. Mich	E738
110811	Jennifer	Sanders, Gill O.	E420
110829	Wallace Newell	Jones, Kent W.	W720
210819	Leanne	Johnson, Gary H.	E407
210820	Mildred Johnson	Richards, Kent F	W608
110830	Kristine Peterse	Hebertson, Richa	E406
110807	Jeremy NMI	Mausberg, Lionel	WSIC
210541	Lisa Fleitmann	Bond, Robert E.	W845
210796	Carolyn Oswald	Sorensen, Bruce	N3AM
210814	Mohulamu NMI	Motoki, David S.	N3AM
210813	Daphne Williams	Pearl, James E.	W809
210813	Frederick Henry	Harris, Alvin E.	W650
210831	San Irene	Sellers, Daniel	W318
110529	Koloti Vihi	Swensen, Swen	E636

059 TRANSFERS

110819	Sherri Ann	Robert E.	210484	E805	W653	E612	E639
210762	Emmanuel Heber	Lohr S.	110736	E707	W740	E633	W754
210643	Ann	Maryann	210427	N3AM	W641	E411	W423
110382	Lori Lea	Nelson	110811	N3AM	W312	W707	E710
110758	Dora B.	Nelson	110811	E743	W706	N3AM	W707
110641	Linnell	Jack NMI	110747	E410	W437	E643	W842
210683	Christine Lynn	Valerie Lynn	210813	E412	W410	E412	W409
210501	Lisa Rae	Sherry L.	210811	E402	W443	E401	W427
210606	Becky NMI	Janene Raby	210741	E409	W441	E405	W435
110765	John Joseph	Tanya Powell	210208	E734	W727	E402	W425
110823	Francom	Jennifer Lynn	210792	E405	W431	E406	W445
210823	Delbert Leroy	William H.	210722	W338	E633	E632	W637
210552	Ann Marcella	Leanne	110829	E403	W415	E407	W408
210622	Annette Cecila	Brenda Ann	110723	N3AM	W451	E645	W616
110763	Rinaldo	Carolyn Oswald	210541	E641	W723	N3AM	W614
110782	C	Mohulamu NMI	210796	E740	E702	N3AM	W649
210811	Charles Stewart	Lila Bean	110785	N3AM	W354	W714	E715
210700	Dean	M.	210591	E715	W712	E711	W725
210765	Estel NMI	Albert NMI	210774	E744	W719	E704	W738
110529	Marla Anderson			E410	W416		

039 DISCHARGES

210726	Wilma E.	Brinton, Gregory	V.	210733	W839	Richards, Kent F	W639
110768	Kathleen Louise	Stinson, James B	NMI	210456	W637	Miller, David K.	W614
110771	Mildred Rogers	Petersen, S. Kei	Anitra Wati	210705	W804	Whitesides, Alan	W704
110382	Lori Lea	Rosenberg, Thoma	Ralph Mason	110680	W312	Ford, Clynn R.	W736
210799	James Taylor	Lee, Richard F.	Jennifer NMI	110631	E803	Richards, Kent	W320
210732	Daryn James *NI*	Moench, Louis A.	Smith	110587	N599I	Bertin, Kim	W350
210533	John Kaspar	Mangelson, Ned L	Jewel	210600	W735	Ford, Clyde D.	E845
110784	Alton	Lee, Richard E.	EH	110769	E816	Stinson, James B	W641
110663	Deborah Ann	Nelson, John C.	Rebecca Rosetta	210643	E830	Tobler, Gilbert	W312

Figure 7.1. Admit—discharge—transfer report.

LDS Hospital Census Report
for the Day of 06/14/90

NURSING DIV	OCC PRE	ADMITS	TXFR IN	TXFR OUT	DSCH TDAY	COMP	OCC TDAY	BED COMP	VAC	OCC %
AM ADMIT	0	7	0	7	0	0	0	0	0	0
WEST 3	43	4	2	0	9	0	40	46	6	86
WEST 6 NORTH	19	3	4	0	4	0	22	23	1	95
WEST 6 SOUTH	22	1	2	0	3	0	22	23	1	95
WEST 7	36	4	10	2	8	0	40	46	6	86
WEST 8	33	5	1	1	8	0	30	45	15	66
EAST 8 NORTH	9	1	0	0	1	0	9	17	8	52
EAST 8 SOUTH	26	4	0	1	5	0	24	29	5	82
NORTH 6	0	0	0	0	0	0	0	21	21	0
TOTAL	188	29	19	11	38	0	187	250	63	75
STICU	8	0	0	1	0	0	7	12	5	58
MSICU	11	1	2	5	0	0	9	12	3	75
CCU	12	5	0	4	0	0	13	16	3	81
TICU	10	1	3	4	0	0	10	16	6	62
LAFICU	0	0	0	0	0	0	0	2	2	0
E6SL	0	0	0	0	0	0	0	0	0	0
TOTAL	41	7	5	14	0	0	39	58	19	67
EAST 4	6	10	0	14	0	0	2	0	0	0
WEST 4	23	0	14	0	9	0	28	37	9	75
WEST 4 GYN	10	0	1	0	1	0	10	11	1	90
TOTAL	39	10	15	14	10	0	40	48	10	83
ADULT PSYCHIATRY	6	0	0	0	1	0	5	16	11	31
DAYSPRING	14	0	0	0	0	0	14	20	6	70
WSCH	10	1	0	0	0	0	11	0	0	0
TOTAL	30	1	0	0	1	0	30	36	17	83
GRAND TOTAL	298	47	39	39	49	0	296	475	179	62
NURSERY-E420	1	11	0	12	0	0	0	0	0	0
NURSERY-W434	17	0	13	0	7	0	23	0	0	0
NURSERY-E430	0	0	0	0	0	0	0	0	0	0
NURSERY-E431	4	0	2	0	1	0	5	0	0	0
NURSERY-E432	7	1	0	3	0	0	5	0	0	0
NURSERY-E433	4	0	0	0	0	0	4	0	0	0
NURSERY-E434	0	0	0	0	0	0	0	0	0	0
TOTAL	33	12	15	15	8	0	37	91	0	41
TRANS CARE CTR	14	1	0	0	1	0	14	14	0	100
REHABILITATION	15	1	0	0	0	0	16	19	3	84
TOTAL	15	1	0	0	0	0	16	19	3	84
EMERGENCY	0	20	0	0	0	0	20	0	0	0

TOTAL SHORT STAY PATIENTS = 22

Figure 7.2. LDS Hospital Census Report. OCC PRE = previous occupancy, TXFR IN = transfers in, TXFR OUT = transfers out, DSCH = discharges, OCC TDAY = occupancy today, BED COMP = bed complement, VAC = vacancies, OCC % = percent occupancy.

7.4 Transfers and Discharges

In addition to the patient's admission, all transfers and the discharge are recorded in the electronic ADT file. The ADT program automatically informs the hospital's billing computer when the patient is discharged. An ADT Report (Figure 7.1), listing all admissions, discharges, and transfers, is generated daily for use by the Admitting Department, Medical Records Department, and other departments throughout the hospital.

7.5 Room Trace and Census

As part of the admission procedure, the room to which the patient is admitted is entered into the system. This location, and all subsequent locations to which the patient may be transferred, are recorded in a room trace file. The room trace file is the means by which HELP computer programs determine a patient's location in the hospital at a given time. Also, retrospective searches use the room trace file to determine which patient data were collected while the patient was on a given unit (e.g., ICU). The room trace records of patients undergoing surgical procedures is updated to indicate when they entered the operating room, when they entered recovery, and when they were transferred to the postoperative recovery unit.

The HELP computer uses admission, discharge, and transfer data to generate a daily census (Figure 7.2). An admitting clerk uses the ADT Report (Figure 7.1) to verify the census each night.

8
Medical Records Functions

8.1 Overview

The patient medical record at LDS Hospital is the traditional paper-based medical record. Patient data maintained on the HELP system complement rather than replace the traditional record. Certain routine medical reports (e.g., blood gas data, respiratory care charting, radiology reports, etc.) consist of data contained entirely within the HELP database and are printed by HELP for inclusion in the permanent medical record.

The Medical Records Department at LDS Hospital uses the HELP system to assist with a variety of administrative functions. A list of incomplete charts is maintained on HELP with notices automatically sent to delinquent physicians at regular intervals, records of patients' ICD-9-CM codes are kept in an electronic file for quality assurance and utilization review-related purposes, and a chart locator program helps keep track of charts when they leave the department. The Medical Records Department uses reports of admissions and discharges (based on data entered by the admitting department) to assist them in their work. Several special purpose reports are produced to assist with day-to-day work.

8.2 Medical Records Data Maintained on the HELP System

A study [1] was performed to analyze what proportion of the LDS Hospital patient medical record is maintained on the HELP system in electronic form (and is thus available for computer-assisted decision making). Patient charts from various medical services were examined and each page of each chart was determined to consist of (1) handwritten data (e.g., a physician's or nurse's progress note), (2) typed data (e.g., dictated reports), (3) computerized data (i.e., a report produced by the

HELP system), (4) a form (e.g., a consent form), or (5) other (e.g., photographic data, ECG recordings, etc.). It was determined that for the hospital as a whole approximately 25% of the chart consists of data contained entirely within the HELP system (Figure 8.1). For nursing divisions that make use of computerized nurse charting functions, the computerized fraction of the chart was as high as 40%. The major computerized categories of the chart were computerized nurse comments, shift reports, laboratory results, and respiratory care charting notes. The major handwritten categories were handwritten nurses' notes, physicians' notes, physicians' orders, and graphic flow sheets.

The fraction of the chart that is computerized is expected to rise as the typed portions of the chart (i.e., radiology reports, history and physical, operative reports, discharge summaries, histology reports, pathology reports, etc.) are added to the HELP system's free-text file and as all of the nursing units begin using computerized nurse charting.

8.3 Manipulation of the Medical Record

The medical record that has been accumulated for the patient during the hospital stay is sent from the nursing unit to the Medical Records Department on the morning after discharge (Figure 8.2). The charts received are compared with a computer-generated list of the previous day's discharges

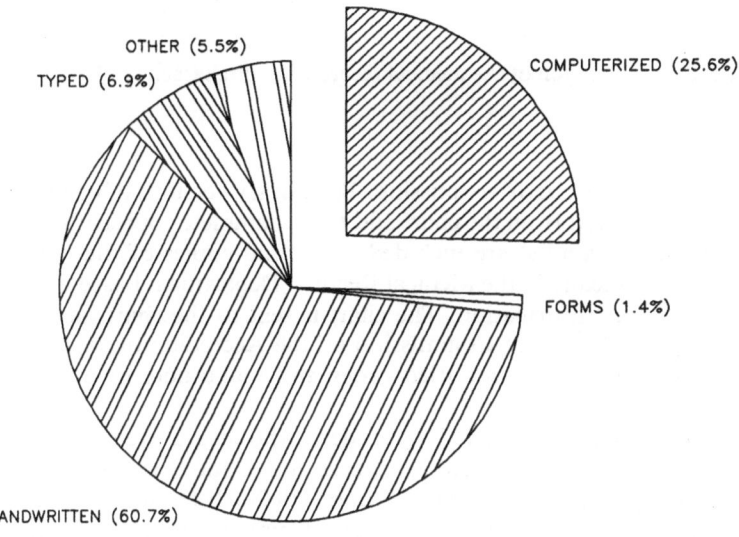

Figure 8.1. Proportion of LDS Hospital medical record maintained in electronic (computerized) form.

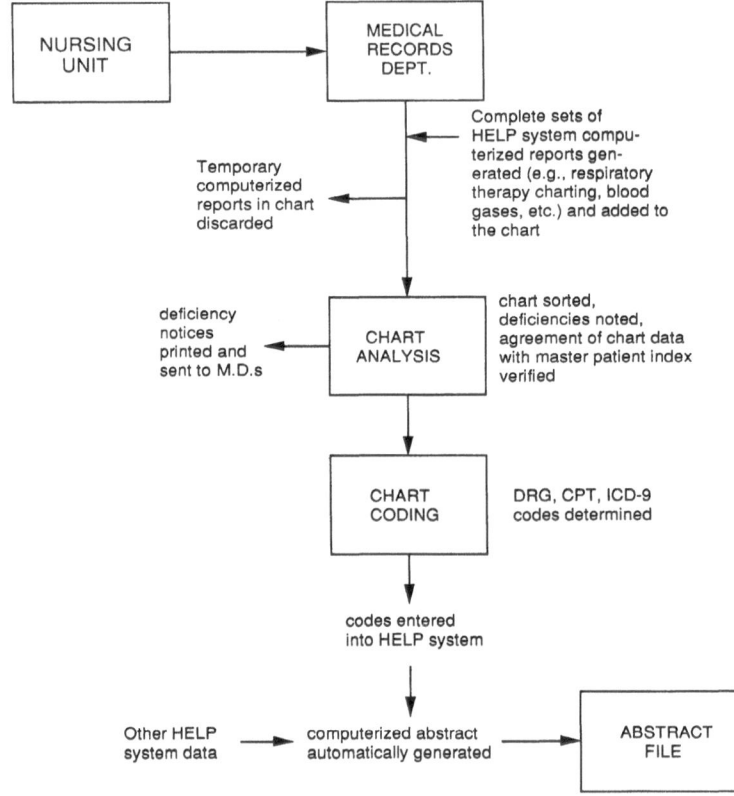

Figure 8.2. Manipulation of the medical record after discharge.

to determine if any charts are missing. Temporary versions of computerized reports are removed from the chart and final, complete copies of the reports (e.g., blood gas results) are included. The final copies of the reports are printed automatically in the Medical Records Department based on computerized discharge information. The chart is sent to the chart analysts who put the chart into a Joint Commission on the Accreditation of Healthcare Organizations (JCAHO)-specified order and make note of any charting deficiencies.

The chart is then reviewed by coders who use a freestanding Data General computer running 3M Corporation's Code 3 software to determine the proper diagnosis-related group (DRG), International Classification of Diseases, Version 9 (ICD-9), and Common Procedural Terminology, 4th Edition (CPT-4) codes. The Code-3 system provides information on reimbursements and utilization outliers and generates alerts when diagnoses are considered nonspecific (and thus insufficient for reimbursement purposes).

The record is then transferred to data entry personnel who enter the DRG, ICD-9, and CPT-4 codes into the HELP computer system. Using these data, an abstract of the hospital stay (Figure 8.3) is generated and a copy is placed in the chart. An electronic copy of the abstract is added to the computerized Abstract File which is used in retrospective analyses of hospital use. For Medicare and Medicaid patients the printed abstract must be signed by the physician as accurately representing the patient's diagnoses and the procedures performed.

The specific charting deficiencies and the name of the corresponding physician are subsequently entered into the computer. A printout of the outstanding deficiencies is entered into the front of the chart and is used as a worksheet until charting deficiencies have been eliminated. Computer-generated notices are sent at regular intervals to physicians who are delinquent in completing their charting deficiencies (Figure 8.4).

8.4 Chart Location

When a chart is requested from the Medical Records Department the patient's name is entered into the computer and the MPI is searched. The search program uses the same Soundex program used on admission (see Chapter 7). When the chart leaves the department the medical record number, date, and requesting department are entered into the computer. This information is useful in tracking the location of charts. If a chart is deficient, this fact will be displayed when it is being signed out. The information is erased when the chart is returned to the department.

8.5 Reports and Other Utilities

The following HELP system reports assist in work management in the Medical Records Department:

1. correspondence programs: to keep track of requests for copies of the medical record
2. deficiency reports: reports of current deficient charts can be sorted by physician, dates, time deficient, and other parameters
3. pathology report: used in determining diagnoses
4. admission and discharge lists (Figures 8.5 and 8.6)
5. surgery schedule

LDS Hospital Patient Abstract

110244   MITCH L   Med Rec #: 5507
Admit Date :   5/24/90 11:00   Sex   : MALE
Disch Date :   5/26/90 12:11   LOS: 2   Pri Pay : CASH
Birth Date :   12/ 2/59   Age: 30   Sec Pay : CASH
Admit Type :   Service : 58 ORTHOPEDIC   ORS
Disposition : HOME OR SELF CARE (01)   Sec Serv :
Physician :   1137 ROSENBERG, THOMAS D.   Admitting Physician: 1137 ROSENBERG, THOMAS D.
Consultants: 5407.41
Total cost :
DRG   :   219 LOWER EXTREM & HUMER PROC. EXC. HIP, FOOT, FEMUR AGE > 17 W/   061590

              DIAGNOSIS   SG
Admit dx   718.86 OTHER JOINT DERANGEMENT, NOT ELSEWHERE CLASSIFIED, INVOLVING LOWER LEG

Prin Disch dx   717.83 OLD DISRUPTION OF ANTERIOR CRUCIATE LIGAMENT
                717.7 CHONDROMALACIA OF PATELLA
                717.9 UNSPECIFIED INTERNAL DERANGEMENT OF KNEE
                736.42 GENU VARUM (ACQUIRED)

                                                   SURGERIES
1   5/24/90   1137   81.45   OTHER REPAIR OF THE CRUCIATE LIGAMENTS
2   77.27   WEDGE OSTEOTOMY OF TIBIA AND FIBULA
3   80.6   EXCISION OF SEMILUNAR CARTILAGE OF KNEE
                        78.07   BONE GRAFT OF TIBIA AND FIBULA
                        77.79   EXCISION OF OTHER BONE FOR GRAFT, EXCEPT FACIAL BONES
                        83.88   OTHER PLASTIC OPERATIONS ON TENDON
                        80.26   ARTHROSCOPY OF KNEE

Figure 8.3. Abstract of hospital stay.

LDS Hospital Medical Records Department

673 Med.Rec.#	Doctor XXX Patient name	Disch.Date	Delinquent (Age of chart)	Chart location
1899	Boy Deon	5/28/90	18	WEST 6
1369	Maxine Lloyd	4/28/90	48	
3499	Shawnia Rae	4/22/90	34	
2293	Larue Miller	5/26/90	20	
2293	Larue Miller	6/ 4/90	11	
5349	Clark H.	6/ 5/90	10	
4263	Patrick Maurice	5/16/90	30	
5388	Ellis P.	4/ 7/90	69	
4575	Charles F.	4/30/90	46	
5365	Steve NMI	3/ 1/90	106	
1718	Mario J.	4/21/90	55	
3975	Julie King SS	4/23/90	53	
2120	Gail Dyer	6/11/90	4	
5053	Marlayne Thomas	2/17/90	118	
5504	Sarah Jean	5/24/90	22	
5500	Renae Wall	5/12/90	34	
5508	Edna Julia	5/29/90	17	
5343	Dale Alma	2/10/90	125	
4899	Mary Margaret	5/ 4/90	42	
5389	Gertrude Helen	4/28/90	48	
5399	Lucille Wagoner	5/ 7/90	39	MED REC/CORSP
786	Melva Butler	3/23/90	84	
5086	Gay Gilbert	3/20/90	87	MED REC/CORSP
3021	Smith NMI	4/16/90	60	
3021	Smith NMI	4/28/90	48	
5021	Richard James	5/27/90	19	
4223	Janice L.	4/15/90	61	
3596	Darrell Keith	9/ 4/89	284	
5380	Maureen Kathleen	3/26/90	81	
4696	Marie	5/11/90	35	
5505	Samuel James	5/25/90	21	
1850	Lillian Gladys	4/22/90	54	
2404	Iolanda Tasso	1/25/90	141	
2404	Iolanda Tasso	5/23/90	23	
5379	Powell	3/22/90	85	
5394	Nancy C.	4/28/90	48	
4220	Deane	4/20/90	56	
3977	Rockey L ER	6/ 5/90	10	
5510	Kaylynne Keller	6/ 9/90	6	
5355	Larry Alan	2/22/90	113	
2480	Laurie Lynne	2/22/90	113	
5221	Cloyd Grant	6/ 8/90	7	
5505	Shari Dianne	5/26/90	20	
5507	Janet Ruth	5/30/90	16	
4867	Joanne Gerber	4/ 7/90	69	
5349	Van E.	4/ 4/90	72	
4164	Clara Baca	4/13/90	63	
1758	Allan Dean	6/ 4/90	11	

TOTAL: 48
END OF REPORT

Figure 8.4. Physician deficiency list.

ADMISSIONS FOR 7/ 2/90

110582	0	Haddox	Maternity
110742	0	Gayle	Newborn
110746	0	Marie	Maternity
110748	0	Leisha	Newborn
110806	0		Gastric Tumor
110832	0	Ann	USI; Ventral Hernia
110900	0	Angella	Newborn
110900	0	Klass	Maternity
110914	0	NMI	Uterine Fibroids
111004	0	Margaret	Newborn
111013	0	Steven George	Anterior L L45 Herniated Lumbar Disc
111065	0	Willard	Retinal Detachment; Diabetic Retinopathy Right Eye
111067	0	Mark (Ted)	Cleft Lip
111133	0	Jeannie	Newborn
111150	0	S	
111155	0	W.	Unstable Angina
111158	0	NMI	R/O P E
111160	0	Kimberly Sue	Newborn
111162	0	D.	Necrosis of Penis
111162	0	Girl Kristi	Newborn
111163	0	Roy A.	
111165	0	Anna Lee	Abdominal Pain
111165	0	NMI	Depression
210487	0	Kristine	Newborn
210576	0	Jennifer	Newborn
210740	0	Danielle	Newborn
210773	0	Joyce	Cystitis and SUI
210894	0	Dawn	Maternity
210894	0	Christine	Newborn
210894	0	Patricia	Newborn
210987	0	Gayle NMI	Cholelithiasis
211101	0	NMI	Anemia; Low Platelet Count; ITP
211110	0	Ferron Watkins	Cholecystitis; Multiple Incisional Hernias
211128	0	Afton	New Onset Diabetes
211135	0	Deon	Newborn
211138	0	Carolyn	Newborn
211144	0	Shawna	Newborn
211154	5501	Sue Newcomb	Maternity
* 211144	8906	Carmen Beatrice	Obstructive Jaundice
* 111146	5007	Arthur	DKA
D210344	0509	Christine	Maternity
D210800	9809	Louise	Diverticulitis
*D211083	9513	Joyce	Ovarian Cancer
210742	9216	Lindstrom	Maternity
211150	8217	William R.	Fracture Left Elbow
111107	1220	Martinez	Menometrorrhagia; Uterine Enlargement and Prolapse

Figure 8.5. Portion of admission list for a single day. Asterisks indicate nonscheduled admissions. First number listed is encounter number, second number is medical records number. Note some patients have not yet had medical records number assigned.

*D111164	0223	Deron Anderson	Acute Leukemia w/Sepsis
211100	6927	Frewin	Pelvic Mass
211136	1030	Fred Joseph	CHF
*D111151	8230	William	CHF, SOB
111159	8031	Wagstaff	S/P R Hip Revision
* 211141	0033	May Holfeltz	CVA
211150	2434	Van	CHF; Aortic Stenosis
210736	4838	Gayle Greenwood	Maternity
* 111133	5040	B	Cholelithiasis
111144	1441	Jones	Maternity
111123	6441	Victor Pardoe	BPH
111121	3842	Smith	Cholelithiasis
111150	5953	Lee	Maternity
211107	6456	Gae	Breast Cancer Right Breast
* 111160	9257	Davis	Coronary Artery Disease
* 111147	0160	Melvin Glen	FX L Hip
111141	4461	Johnstun	Maternity
210977	3664	Michael Fred	L-5, S-1 Radiculopathy
* 111157	8164	Bronson	GI Bleed
* 111140	1166	Schafer	L Olecranon FX
110493	0368	Rae	Maternity
* 211134	7869	Charles Joseph	R/O Myocardial Infarction
* 111163	0071	Lenard NMI	R/O Renal Transplant Rejection
111117	8272	Lorenzo	Coronary Artery Disease
211142	0779	S.	Uncontrolled Hypertension
111114	2080	NMI	Spinal Stenosis
211157	9681	Walker	Maternity

Figure 8.5. *Continued.*

This list is not exhaustive but rather is meant to give an idea of the types of reports generated to assist with the departmental work. A PC LAN is in place in the Medical Records Department to assist with case mix and utilization analyses.

Reference

[1] Kuperman GJ, Gardner RM. The impact of the HELP system on the LDS Hospital Medical Record. *Symposium for computer applications in medical care (SCAMC)* 1990; 14:673–677.

Discharge List 6/14/90

M.R.#	PATIENT NAME	LOS	CAR.		PHYSICIAN	CHARGES	ACCOUNT #
796	Virginia	1	0757	497	Reilly, William F. Jr.	325.94	210779
5513	Wilma E.	5	0707	279	Brinton, Gregory S.	6748.37	210726
5345	John	1	3131	1520	Laser, Jeffrey A.	3004.57	210763
2595	Kathleen Louise	2	0757	519	Stinson, James B.	6230.16	110768
3711	Mildred	3	0707	438	Petersen, S. Keith	3282.91	110771
5094	Lori Lea	1	0103	1137	Rosenberg, Thomas D.	370.62	110382
4543	Dale Webb	1	0757	325	Rigby, Odell F.	169.13	210538
3641	Carma	1	2424	7186	Christensen, Brent J.	961.93	210747
	Farelyn	1	0000	5098	Rasmussen, Gary Lynn	290.73	21064
4319	James Taylor	1	0401	735	Lee, Richard E.	549.16	210799
5222	Daryn James *NI*	5	8301	898	Moench, Louis A.	3031.68	210732
	Pares L.	1	0757	588	Preece, Michael J.	2619.21	210754
	Robert Andrew	1	0104	690	Bennion, Jerald H.	169.13	210821
5513	John	3	0707	738	Mangelson, Ned L.	3558.15	210533
454	Alton	1	0707	735	Lee, Richard E.	1643.78	110784
3455	Deborah Ann	6	0101	647	Nelson, John C.	5830.51	110663
4944	Frank, Garcia	3	0707	739	Dahl, Douglas S.	2968.33	110738
5514	Boy Julie	1	2828	1629	Nielsen, Douglas B.	257.61	210498
4811	Julie Davis	1	2828	1391	Reiser, A. Hamer III	1117.65	110504
3318	Helyn	7	0401	722	Knibbe, W. Patrick	4405.68	210696
	James M.	1	5050	1137	Rosenberg, Thomas D.	169.13	110817
	Anne	1	5023	2564	Gemmell, Jeanne	200.53	110618
5514	Boy Elizabeth	2	2727	8002	Unknown, Pediatrician	445.81	210552
4997	Elizabeth	2	2727	102	Farnsworth, Kent W.	1554.15	210552
3195	Margaret	1	0707	340	Millet, F. Jackson	222.46	210762
2675	Norden	7	0757	442	Hart, Paul D.	4906.97	110699
1217	Richard Kent	1	0101	632	Wilson, Merrill L.	692.57	110806
437	Kenneth F.	1	5050	919	Swensen, Laird S.	1788.26	210362
4874	Krpan	2	0707	1520	Laser, Jeffrey A.	10367.30	210755
	E.	1	0101	5382	Motoki, David S.	1288.19	210796
2170	J Harold	6	0707	1878	Herrod, John N.	3597.97	110728
2215	Beverly	1	5050	632	Wilson, Merrill L.	701.17	210716

	Patient	LOS	Insurance	Physician No.	Physician	Charges	Account No.
3006	Hyer	4	1950	2232	Pead, William G.	2007.97	110746
98	Rita	1	0757	2542	Stevens, Mark H	594.49	210780
3175	Swomitra	2	4848	56	Paulos, Lonnie	2656.00	210528
5514	Wendy	2	0402	277	Swensen, Paul H.	443.61	110457
5514	Sue	2	0000	2074	Atwood, Lynn	1373.92	210451
5398	Guy Richard	44	0707	111	Box, Terry D.	40402.07	110014
4298	Roy P.	1	0707	5501	King, Kurt V.	891.21	110813
5380	Steffani	1	0505	648	Sanders, Gill O.	181.50	110829
3656	James	9	0707	913	Samuelson, Kent M.	10448.95	210475
5513	Ray NMI	3	2828	799	Sorensen, Sherman G.	4359.50	210744
4656	V.	5	0707	66	Richards, Kent F.	6004.74	210733
3550	Debra NMI	21	2030	630	Miller, David K.	21006.04	210456
3837	Jana	1	5050	304	Price, Richard R.	1506.16	210760
5384	E. Floyd	1	0757	799	Sorensen, Sherman G.	2105.48	110784
1572	Davies	6	0707	913	Samuelson, Kent M.	2725.19	110720
4819	Anitra	6	1912	2052	Whitesides, Alan N.	10886.17	210705
5507	Ralph	1	0707	635	Ford, Clynn R.	1368.75	110680
914	Ruth	1	0757	1155	Thomson, John W.	345.45	210800
3891	Jennifer NMI	10	0401	3573	Richards, Kent	18639.45	110631
	Janet	1	5109	2564	Gemmell, Jeanne	708.37	110792
2956	Smith	4	0789	1551	Bertin, Kim	6830.43	110587
3012	Jewel	12	0707	877	Ford, Clyde D.	7822.77	210600
3690	Hushang	1	0707	519	Stinson, James B.	592.36	110769
5514	Rebecca	2	2121	681	Tobler, Gilbert G.	2381.59	210643
5514	Chris K.	1	5050	580	Lazarus, Harrison	1748.42	110797
5514	Linda	2	0000	56	Paulos, Lonnie	2478.94	210528
	E. Richard	1	0757	1011	Horton, Steven C	3043.43	110723
	Evan John	1	2121	690	Bennion, Jerald H.	635.72	210785
5514	Boy Jane	2	0000	2584	McElligott, Kathleen	430.45	110715
3672	Jane	2	0000	876	Skidmore, Thomas C.	1253.85	210709
2284	Karen Elaine	1	0104	919	Swensen, Laird S.	3364.36	110372
2613	James Russell	1	2121	730	Bourne, Hal H.	240.47	110686
5232	Christine Ann	1	2727	102	Farnsworth, Kent W.	1327.47	210574
5514	Girl Christine	1	2729	648	Sanders, Gill O.	257.61	110580

Figure 8.6. Discharge list showing patient, length of stay, insurance carrier, physician, charges, and account number.

9
Computerized Data in Quality Assurance

9.1 Overview

The ability of the HELP computer system to integrate data from a variety of sources is valuable in the execution of quality assurance activities at LDS Hospital. Although the Quality Assurance (QA) Department has primary responsibility for monitoring and evaluating the quality of hospital care, several clinical departments use the computerized data to perform internal quality assessments. Quality assurance capabilities have been designed into many HELP system modules or, indeed, have been the motivation for their creation. Computerized clinical data are automatically stored during the hospital stay and are available for retrospective or concurrent QA review.

Specifically, computerized data are used in QA review of the blood bank, pharmacy, infectious diseases, pathology, nutritional support, respiratory therapy, and certain laboratory test results. In many instances the computer performs some QA screening functions itself, thus reducing the work that would otherwise have to be done manually. A Tandem gateway product allows data from the HELP system to be downloaded to a PC network in the QA department for further analysis.

9.2 Quality Assurance

Quality assurance in the hospital setting is a process of monitoring and evaluating clinical practice that is generally achieved through completion of the following nine steps [1]:

1. Responsibility for the process is assigned to an individual or group
2. The scope of the activity to be monitored and evaluated is defined
3. Important aspects of the activity are determined
4. Indicators (i.e., measurable standards of quality used to determine appropriateness and quality of the activity) are identified

5. Criteria (specific standards of high quality care used to measure indicators) are established
6. Data are collected and analyzed
7. Actions are taken to resolve identified problems
8. Actions are assessed and improvement is documented
9. Relevant information is communicated to supervisory QA programs

The HELP system is of most assistance in step 6 listed above. Patient data, stored in electronic form, are easily gathered, analyzed, and reported by the computer according to the criteria determined by medical committees. These data (which would otherwise have to be gathered manually) are reviewed by the QA department and appropriate cases are referred to medical committees for action. Longitudinal data are maintained for trend analyses and reporting to local and national agencies. Research is being performed to see if the computer can be used to correlate clinical variables with outcome and contribute to the establishment of indicators and criteria. Specific applications of the HELP system to the QA process are given below.

9.3 Blood Ordering Program

Physicians at LDS Hospital enter orders for blood products directly into the HELP computer. At the time the order is entered, the computer asks for a reason for the order (e.g., if the request is for packed cells, the computer will inquire if this is for bleeding, anemia, preoperative, etc.). When possible, the computer verifies that the reason entered is consistent with other data in the electronic patient file (e.g., if the reason given is anemia, the computer will check the last hematocrit value, or, if preoperative is given as the reason, the computer will check the surgery schedule). If a discrepancy between data in the file and the stated reason exists, the computer informs the physician. If an override is necessary, the order may still be entered but the physician must enter a (free-text) reason. The computer automatically keeps a record of all blood orders. Instances in which a physician overrides the computer's programmed criteria are documented and reviewed by QA personnel who use the medical record to investigate the override. The blood ordering program and related QA functions are described in detail in Chapter 26.

In a 3-month period in 1988 19% (871/4357) of all blood orders failed to meet criteria as determined by the computer. After review by QA personnel, only 17 (0.375%) were found to be true exceptions to accepted criteria and were referred to medical committees. This effort represents review of 100% of all blood orders at LDS Hospital. The computer serves to screen out more than 80% of the orders and confine the work of the QA department to cases with a higher likelihood of nonconformance.

9.4 Pharmacy

Medication orders at LDS Hospital are written by physicians, transcribed by unit clerks, and entered into the HELP system by clinical pharmacists and nurses. The computerized medication monitoring system uses logic criteria to review automatically every drug order and, with data from other parts of the computerized patient file, determines if the new order may result in an adverse reaction [2]. Data from the patient file that are used in the logic includes (in addition to the drug data) laboratory, blood gas, ECG, spirometry results, vital signs, and records of surgical procedures. Interactions that generate alerts are (1) drug-drug, (2) drug-allergy, (3) drug-lab, (4) drug-disease, (5) drug-diet, (6) drug-dose, and (7) drug-interval. The alerts are reviewed by the clinical pharmacist who determines (from the clinical situation) if the alert is an "informational" alert or one that requires action on the part of the physician. In the case of action alerts, the pharmacist contacts the physician directly. The pharmacy module of the HELP system and the computerized medication monitoring system are described in detail in Chapter 15.

The system has been operational since 1975. In 1987 to 1988 approximately 100 action alerts were generated per month (Figure 9.1). At that time, physician compliance (i.e., changing therapy) when presented with an action alert was greater than 97% (Figure 9.2).

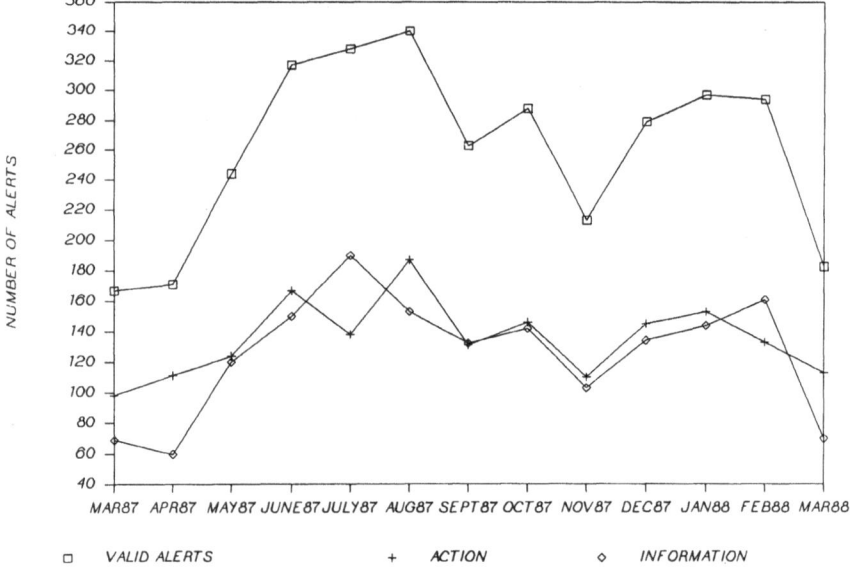

Figure 9.1. Graph showing number of alerts (informational and action-oriented) generated per month by medication monitoring system (see text for further details).

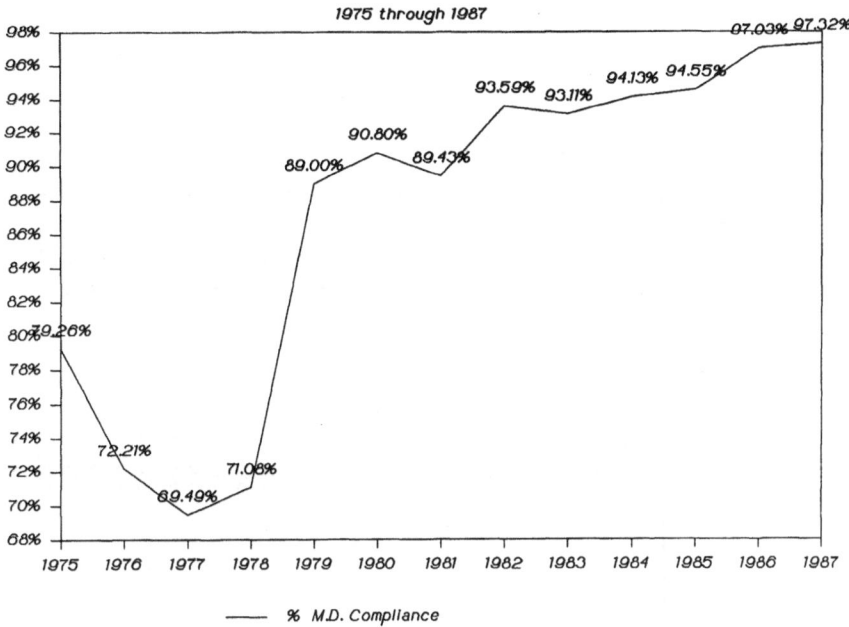

Figure 9.2. Graph showing compliance of physicians with action-oriented alerts generated by medication monitoring system.

The computerized medication monitor is an example of concurrent quality assurance review. The QA department also receives a list from the pharmacy each month detailing how often each of the drug alerts is being generated by the medical staff as a whole (Figure 9.3). If any type of alert is generated frequently, the QA department conveys this information to the medical committees. The information may then be used to direct policy or continuing education decisions.

9.5 Infectious Disease Monitoring

The computerized infectious disease monitor (CIDM) is a module of the HELP system that uses microbiology data (already in the system), other data in the patient file, and decision logic to identify automatically:

1. nosocomial infections
2. infections at a normally sterile body site
3. unusual sensitivity patterns
4. patients receiving inappropriate antibiotic therapy
5. patients who could be receiving less expensive antibiotics

LDS HOSPITAL PHARMACY DRUG APPROPRIATENESS MONITOR SUMMARY FOR MAY-JUNE 1988

TYPE	DESCRIPTION OF MONITOR	NUMBER	TOT MAY-JUNE88	ACT	COMP	INFO
ALLERGY	BENZODIAZEPINE ALLERGY	304105	8	2	2	6
ALLERGY	CALCIUM CHANNEL BLOCKER ALLERGY	303016	3	2	2	1
ALLERGY	CODEINE ALLERGY	300501	2	1	1	1
ALLERGY	DIGITALIS ALLERGY	303005	1	1	1	0
ALLERGY	ETHACRYNIC ACID ALLERGY	303704	1	0	0	1
ALLERGY	INDOMETHACIN CROSS REACTION TO ASA ALLERGY	300813	1	1	1	0
ALLERGY	MORPHINE/OPIUM ALLERGY	300400	2	1	1	1
ALLERGY	OXYCODONE ALLERGY	300105	2	2	2	0
ALLERGY	PHENOTHIAZINE ALLERGY	304104	3	0	0	3
ALLERGY	PROPOXYPHENE ALLERGY	300103	1	1	1	0
ALLERGY	PSYCHOTHERAPEUTIC AGENT ALLERGY	304101	3	1	1	2
ALLERGY	RESERPINE ALLERGY	303010	1	0	0	1
ALLERGY	SULINDAC CROSS REACTION TO ASA ALLERGY	300806	1	1	1	0
ALLERGY	THEOPHYLLINE ALLERGY	304401	2	0	0	2
DOSE	AMIKACIN DOSE MAY BE EXCESSIVE	311201	3	2	2	1
DOSE	GENTAMICIN DOSE MAY BE EXCESSIVE	311301	19	8	8	11
DOSE	TOBRAMYCIN DOSE MAY BE EXCESSIVE	311401	2	0	0	2
DOSE	TOBRAMYCIN DOSE MAY BE SUBTHERAPEUTIC	311402	1	0	0	1
DOSE	VINCRISTINE	305200	2	0	0	2
DRUG INTERACTION	ANTACIDS AND DIGOXIN	303901	88	11	11	77
DRUG INTERACTION	CIMETIDINE AND PHENYTOIN	305501	2	0	0	2
DRUG INTERACTION	COUMADIN AND CIMETIDINE	302510	2	2	2	0
DRUG INTERACTION	COUMADIN AND QUINIDINE	302502	4	1	1	3
DRUG INTERACTION	COUMADIN AND THYROID	302504	2	0	0	2
DRUG INTERACTION	DIGOXIN AND QUINIDINE	305000	16	12	12	4

CATEGORY		NUMBER	TOT	ACT	COMP	INFO
DRUG INTERACTION	ERYTHROMYCIN AND THEOPHYLLINE	305300	9	6	6	3
DRUG INTERACTION	PROPRANOLOL AND DIABETIC AGENTS	303400	5	0	0	5
MONITORING	AMIKACIN AND NO RENAL STUDIES LAST 3 DAYS	310804	1	0	0	1
MONITORING	COUMADIN, NO DAILY PT DURING 1ST WEEK THERAPY	311600	6	2	2	4
MONITORING	DIGOXIN LEVEL OUT OF THERAPEUTIC RANGE	310400	12	9	9	3
MONITORING	GENTAMICIN AND NO RENAL STUDIES LAST 3 DAYS	310801	38	17	17	21
MONITORING	K SPAR DIURETIC,K SUPP/RENAL IMPAIR,NO K LEVELS	311700	20	11	11	9
MONITORING	K SUPP,RENAL IMPAIR, NO K LEVELS LAST 24 HRS	311800	12	4	4	8
MONITORING	LITHIUM,DIURETIC/NA DIET,NO LEVELS IN WEEK	311500	1	0	0	1
MONITORING	TOBRAMYCIN AND NO RENAL STUDIES LAST 3 DAYS	310803	12	5	5	7
PROPHYLAXIS	ANTIBIOTIC ORDERED LONGER THAN 24 HR POST OP	315000	29	29	29	0
THERAPEUTIC USE	CONCURRENT IV AND ORAL THEOPHYLLINE	304500	3	2	2	1
THERAPEUTIC USE	CYCLOSPORINE LEVEL OUT OF THERAPEUTIC RANGE	312300	2	1	1	1
THERAPEUTIC USE	DIGOXIN AND HYPOKALEMIA	310300	15	13	13	2
THERAPEUTIC USE	K SPARING DIURETIC AND HYPERKALEMIA	310500	1	0	0	1
THERAPEUTIC USE	K SUPP IN HYPERKALEMIA	311900	26	22	22	4
THERAPEUTIC USE	KCL SUPP WITH ELEVATED CHLORIDE	312200	16	10	10	6
TOTAL			380	180	180	200

FOLLOWUP SUMMARY MAY-JUNE 1988

WHOM CONTACTED		COMPLIANCE FORM		NON-COMPLIANCE	
HOUSESTAFF	20	LAB TEST ORDERED	18	MD DISBELIEF	0
ATTENDING	22	DRUG D/CD	62	OTHER	0
RN	37	DOSE CHANGED	21	RISK VS BENEFIT	0
NOTE LEFT	100	SCHEDULE CHANGED	9		
OTHER	157	NEW DRUG	19		
		MONITORED PT	52		
		OTHER	2		

Figure 9.3. Report detailing number of alerts generated by medication monitoring system for a 2-month period. NUMBER = HELP decision frame number, TOT = total number of alerts, ACT = number of action-oriented alerts, COMP = number of action-oriented alerts complied with, INFO = number of information-oriented alerts.

6. reportable diseases
7. patients receiving postoperative prophylactic antibiotics longer than indicated
8. patients whose preoperative antibiotic orders deviate from accepted practice

The alerts are printed in a report that is reviewed daily by infectious disease specialists and pharmacists. Figure 9.4 shows an example of the daily Infectious Disease Monitor Report. The figure shows two alerts. The first alert is an indication of a patient with a possible nosocomial wound infection and the second is an indication that a less expensive suitable antibiotic is available. For each alert the message itself is stated followed by patient identifying data and the name of the physician. The admission date, admission diagnosis, and previous admission dates (from the long-

```
        INFECTIOUS DISEASE MONITOR REPORT FOR 12 MAY 1988
                          FOR PAST 24 HOURS
                  PRINT TIME:   5/12/88.13:20

***** ANTIBIOTIC ALERT *****
***** PATIENT WITH POSSIBLE NOSOCOMIAL WOUND INFECTION *****
@PAT:  000700   DOE, JANE                      72  F    W807    MR#: 444444
DOC:  0422 SMITH, JOHN                  SERVICE:
ADMITTED: 05/07/88.13:32     ADMIT DIAG: END STAGE RENAL DISEASE
PREV. ADMIT  04/22/1985  PREV. DSCH  04/24/1985
SURGERY:   CLEAN CONTAMINATED       SURGEON 9890
05/05/88.08:15   KIDNEY TRANSPLANT - RECIPIENT (LIVING-RELATED)
CURRENT ANTIBIOTICS
  05/11/88.20:42 AMPICILLIN 1000 MG, INJ    Q6H
CULTURE RESULTS            -INTERIM REPORT-             ROUTINE CULTURE
  SOURCE: WOUND, LEFT QUADRANT                   COLLECTED: 05/07/1988.21:14
  STAIN:   MOD WBCS
           GRAM-POSITIVE COCCI
  RESULT: STAPHYLOCCUS AUREUS
  SENSITIVE TO: Nafcillin, Cefazolin, Cefotaxime, Vancomycin, Imipenem,
                Gentamicin
  RESISTANT TO: Ampicillin, Penicillin-G

***** SEPTRA WOULD BE A LESS EXPENSIVE ANTIBIOTIC *****
***** GALL BLADDER ASSOCIATED CULTURE *****
@PAT:  000566     DOE, FREDA                   90 F    E645    MR#: 147081
DOC:  0670 SMITH, JOHN                  SERVICE:
ADMITTED: 05/03/88.14:19     ADMIT DIAG: FX HIP
PREV. ADMIT  10/15/1982  PREV. DSCH  10/17/1982
SURGERY:     Clean                      Surgeon 670
  05/05/88.08:34   TOTAL HIP
ADDITIONAL PROCEDURES:
CURRENT ANTIBIOTICS
  05/10/88.17:16 CEFOXITIN (MEFOXIN)  1000 MGM, INJ          Q  8 HRS
  05/12/88.09:00 TETRACYCLINE   500 MGM, INJ                 Q  6 HRS
CULTURE RESULTS            -PRELIMINARY REPORT-        ROUTINE CULT
  SOURCE: OF SPECIMEN, GALL BLADDER             COLLECTED: 05/10/1988.09:35
  STAIN: NUMEROUS GRAM POSITIVE COCCI IN CHAINS
  RESULT: ENTEROCOCCUS    HEAVY GROWTH
  RESULT: HAFNIA ALVEI    LIGHT GROWTH
  SENSITIVE TO: Amikacin,Gentamicin,Imipenem,Piperacillin,
                Ticarcillin,Tobramycin,Trimethoprim-sulfamethoxazole,
  INTERMED. TO: Tetracycline,
  RESISTANT TO: Ampicillin,Cefazolin,Cefoxitin,Cefuroxime,
                Cephalothin,Ceftazidime,
CULTURE RESULTS            -PRELIMINARY REPORT-       ANAER CULT
  STAIN: NUMEROUS GRAM POSITIVE COCCI IN CHAINS
  RESULT: NO ANAEROBES ISOLATED
```

Figure 9.4. Sample output from Infectious Disease Monitor Report. Names have been falsified (see text for details).

term patient file) are listed. Descriptions of surgeries performed on the current admission, allergies, current therapeutic regimens, and culture results are all listed next as supporting evidence for the alert. All data come from the HELP computerized patient database. Blank space is left for additional notes.

Alerts related to the inappropriate use of antibiotics are reviewed by pharmacists and conveyed to the physicians. In the case of inappropriate postoperative antibiotic use, the pharmacist may write a "stop" order, which will become active in 24 hours unless the physician writes an order to continue the medication. The CIDM is described in detail in Chapter 18.

The CIDM's impact on medical practice at LDS Hospital has been studied. The CIDM has been shown to be more efficacious than infection control practitioners in identifying nosocomial infections [3]. Using HELP to identify inappropriate timing of perioperative antibiotic administration has led to a decrease in wound infection rates from 1.8% to 0.9% ($p < 0.02$) [4] and a reduction in the number of average postoperative prophylactic doses from 19 to 13 ($p < 0.001$) [5]. Also, over a 1-year period physicians changed therapy 125 times when presented with antibiotic suggestions generated by the computer regarding appropriateness of therapy [6].

The QA department also uses data from the HELP system to monitor clean wound rates and infection rates of particular sites by nursing unit and medical service. In the future physician-specific data may be retrieved directly from the HELP system.

9.6 Pathology

Hospital regulations require that all tissue removed surgically undergo microscopic examination by a pathologist. For each specimen the pathologist dictates a descriptive report and determines a diagnosis and related SNOP code. The diagnosis and the code are later entered into the HELP system by clerical personnel.

QA personnel review the medical records of patients whose surgical specimens are diagnosed as "No abnormality" for the purpose of determining the appropriateness of the surgical procedure. To facilitate the review, the HELP computer produces a report listing all specimens with SNOP codes corresponding to normal diagnoses. The HELP computer thus eliminates the need for manual identification of the cases to be reviewed.

9.7 Nutritional Support

The ordering of total parenteral nutrition is done via the HELP system and monitoring of nutritional therapy is assisted by HELP system programs. These functions are described in detail in Chapter 22.

The Nutritional Support Services (NSS) team consists of two dietitians and three nurses and is responsible for monitoring the quality of care related to the administration of parenteral and enteral nutrition. The goal of NSS is to identify emerging nutritionally related problems and inform involved health-care personnel (physicians, nurses) so corrective action may be taken.

A variety of specialized reports is produced by the HELP system for use by the NSS. Specialized laboratory reports printed in the NSS office allow review of nutritionally related test results. An example of a typical laboratory report used by the NSS is shown in Figure 16.4 in Chapter 16 describing the HELP laboratory system. The 7-day summary reports are used to follow intakes and outputs, weights, and complete nutritional data. An example of the 7-day report is shown in Figure 4.5 with the nutritional information appearing toward the end of the report.

9.8 Respiratory Therapy

Respiratory therapists at LDS Hospital enter patient treatment data directly into the HELP computer. The data entered include the date and time of treatment, patient position, type of interface to the respiratory equipment, medication received, patient condition during treatment, vital signs, ventilator settings, equipment changes, and cough and sputum quality. From the computerized respiratory care charting data and other data in the patient file, the HELP system produces two types of quality assurance reports [7]. The Medical Directors Alert Report (Figure 9.5) is produced daily and contains alerts related to physiologic values that are out of range (or not recorded) and may result in a poor outcome if not corrected. Examples include prolonged elevated FiO_2, excessively high PEEP/CPAP pressures, and low PaO_2 values without a change in therapy. This report is reviewed each morning by the medical director of respiratory therapy.

The QA Report (Figure 9.6) is produced weekly and is reviewed by respiratory therapy's QA coordinator. Alerts are produced for this report if there is deviation from accepted policies and procedures on aspects of care such as ventilator management, aerosolized bronchodilator treatments, and extubations. A monthly summary of this report is prepared for review by the medical director of respiratory therapy and by the hospital's respiratory care QA committee. Computerized respiratory therapy charting at LDS Hospital is described in detail in Chapter 14.

9.9 Computerized Laboratory Alerting System

The computerized laboratory alerting system (CLAS) is described in detail in Chapter 17. The goal of the CLAS is to notify as rapidly as possible health-care personnel of laboratory values that indicate potentially life-

```
                    RESPIRATORY CARE ALERT
             REVIEWS 24 HOURS FROM 07:00 MAY 19
                     10:18 MAY 20, 1988


       N690              DOE, JANE
    05/19         USN EQUIPMENT NOT CHANGED   - 89986

       E433              DOE, JOHN
    05/19         VENT EQUIPMENT NOT CHANGED  -  9330

       E608              DOE, EGBERT
    05/20 06:05   PEEP: 14 CM
    05/20 04:09   FIO2 > .60 FOR AT LEAST 16 HRS
    05/19 19:49   CUFF PRESSURE: 31 CM
    05/19 17:23   PEAK PRESSURE: 65 CM
    05/20         VENT EQUIPMENT NOT CHANGED  - 83771

       E716              DOE, FREDA
    05/19 20:45   pH: 7.56   PCO2: 25.5   PO2: 55   SaO2:   91   FIO2: 50
    05/19 22:20   pH: 7.59   PCO2: 22.2   PO2: 45   SaO2:   87   FIO2: 60
    05/20 00:15   pH: 7.59   PCO2: 23.0   PO2: 48   SaO2:   88   FIO2: 70
    05/20 01:25   pH: 7.55   PCO2: 26.3   PO2: 60   SaO2:   93   FIO2: 70
    05/20 04:38   FIO2 > .60 FOR AT LEAST 16 HRS

       E433              DOE, FRANKLIN
    05/19         VENT EQUIPMENT NOT CHANGED  - 75442

       E707              DOE, ELEANOR
    05/19 19:30   FIO2 > .60 FOR AT LEAST 16 HRS

       E433              DOE, WASHINGTON
    05/19         VENT EQUIPMENT NOT CHANGED  - 75442

       E607              DOE, ROOSEVELT
    05/19 19:35   pH: 7.61   PCO2: 27.7   PO2: 47   SaO2:   83   FIO2:100
    05/20 01:00   PEAK PRESSURE: 73 CM
    05/20 01:00   FIO2 > .60 FOR AT LEAST 16 HRS

       W438              DOE, LINCOLN
    05/19 13:52   APNEA ALARM MONITORING

       E433              DOE, JACKSON
    05/19         VENT EQUIPMENT NOT CHANGED  -  9330

       E635              DOE, EISENHOWER
    05/20 06:05   pH: 7.28   PCO2: 51.0   PO2:148   SaO2:   96   FIO2:  3

       E610              DOE, FREDERICKSON
    05/19 20:00   CUFF PRESSURE: 30 CM
```

Figure 9.5. Example of Respiratory Therapy Medical Director's Daily Alert Report showing patient room number, patient name, date, time, data for alert, and alerts. Names have been falsified.

threatening situations. A panel of medical experts determined criteria for 11 alerting conditions related to high, low, and falling sodium values, high, low, and falling potassium values, high and low glucose values, low potassium with the patient on digoxin, metabolic acidosis, and falling hematocrit. Alerts are generated when test results are transferred from the laboratory computer to the HELP system and are displayed when a health-care worker reviews the alerting patient's laboratory data on a computer terminal.

A study [8] performed before and after the CLAS was implemented on an acute care unit showed that for certain alerts (those having to do with sodium, potassium, and glucose, called Group 1 in the study) the frequency of appropriate treatment for the alerting condition increased significantly

BRONCHODILATOR GIVEN WITH PRE HR > 130

1. 2311868   05/12 09:10   132/124/124   DOE,JOHN
        BRONCHODILATOR: ALUPENT
        COMMENT: PT STATES SHE DOESN'T NEED TO COUGH AT THIS TIME AND WOULD
                 PREFER TO WAIT TILL SHE FEELS A NEED TO.
2. 40202798 05/13 16:30   138/142/147   DOE,JANE
        TACHYCARDIA, R.N. NOTIFIED; BRONCHODILATOR: ALUPENT

BRONCHODILATOR GIVEN WITHOUT PRE, DUR OR POST HR

1. 40210015 05/12 06:45   84/106/ 0   DOE, ALLEN
        BRONCHODILATOR: ALBUTEROL
2. 40210023 05/12 08:15   0/ 73/ 83   DOE, GEORGE
        BRONCHODILATOR: ALBUTEROL
3. 2563880   05/14 03:35   0/120/120   DOE, BETTIE
        BRONCHODILATOR: ALBUTEROL

ADVERSE REACTION DURING BRONCHODILATOR THERAPY

1. 40202798 05/13 16:30   138/142/147   DOE, MARY
        TACHYCARDIA, R.N. NOTIFIED; BRONCHODILATOR: ALUPENT
2. 2312734   05/15 10:20   116/116/116   DOE, FRED
        HYPOTENSION; BRONCHODILATOR: ALBUTEROL
        COMMENT: RN AWARE OF BLOOD PRESSURE

FIO2 NOT ANALYZED WITHIN FIRST 4 HOURS OF SHIFT

1. 2312734   05/14 AM SHIFT - CPAP   DOE, KEOKI
2. 2309615   05/14 AM SHIFT - HPN   DOE, ELISE
3. 40207169 05/15 AM SHIFT - VENT   DOE, KEOKI

PARAMETERS NOT DONE 3 HOURS PRIOR TO EXTUBATION

1. 2566180   05/13 04:00   DOE, JERRY
        LAST PARAMETERS PERFORMED BEFORE EXTUBATION: 05/13 00:50
        COMMENT: PER ABGS ET PARAMETERS
2. 40209694 05/14 08:30   DOE, KIM
        PARAMETERS NOT DOCUMENTED PRIOR TO EXTUBATION

ABNORMAL HUMIDIFICATION TEMPERATURE

1. 2566750   05/12 22:15   TEMP: 33.0   DOE, ABRAHAM
              05/12 19:50   TEMP: 33.0
        COMMENT: < CUFF PRESSURE TO 19 C ADEQUATE SEAL
2. 2564326   05/13 12:06   TEMP: 30.0   DOE, CARMEL
              05/13 09:57   TEMP: 32.0

EQUIPMENT NOT CHANGED AFTER 24 HOURS OF USE

1. 3940053   05/13 PM SHIFT - VENT   DOE, HARRY
2. 2313732   05/13 AM SHIFT - VENT   DOE, FRANK
3. 2321487   05/13 AM SHIFT - VENT   DOE, SAM

CUFF PRESSURE NOT CHECKED WITHIN FIRST 4 HOURS OF SHIFT

1. 2320281   05/12 PM SHIFT   DOE, SALLY
2. 2321180   05/16 PM SHIFT   DOE, DERRICK
3. 40207169 05/16 PM SHIFT   DOE, GEORGE

CUFF PRESSURE > 27 CM H2O

1. 2566180   05/12 19:30   CUFF PRESSURE: 28   DOE, RALPH
        COMMENT: < CUFF PRESSURE TO 18 C ADEQUATE SEAL
2. 2564326   05/13 00:36   CUFF PRESSURE: 42   DOE, RALPH
        COMMENT: PER DR ORDER ON RETURN FROM O.R. CUFF PRESSURE < TO 22 C
                 ADEQUATE SEAL

TRACH CARE CATHETERS NOT CHANGED

1. 2312734   05/13
2. 2320281   05/13
3. 3940053   05/13

Figure 9.6. Weekly Respiratory Therapy Quality Assurance Report showing alerting conditions, patients with the alerting conditions, and relevant data from computerized respiratory care charting. Names have been falsified.

($p = 0.017$) from 68.1% to 83.8% and the time spent in the life threatening condition decreased significantly ($p = 0.0125$) from 30.4 hours to 15.7 hours. The average length of stay for Group 1 alerting patients decreased significantly from 350.6 hours to 211.9 hours after implementation of CLAS. Time in the life-threatening condition was also decreased significantly (44.3 hours to 25.6 hours) for patients generating a metabolic acidosis alert. Alerts were reviewed an average of 3.6 hours after they were generated.

The CLAS is an example of health-care workers performing QA concurrently. The CLAS helps health-care personnel respond more rapidly to dangerous clinical situations.

9.10 Other Functions

The QA department maintains a PC-based LAN. A link has been established between this network and the HELP computer system to facilitate data review. At present QA personnel use routine HELP programs and files such as laboratory review, long-term files, insurance information, administrative reports, and ICD-9 codes to assist them in their day-to-day routine. Research is underway to examine the ability of the system to perform functions currently performed manually such as review of length of stay information, management of incidence reports, and identification of high-risk, high-cost, and high-volume practices.

Physician-specific data may be gathered to help produce physician profiles necessary for reappointment and hospital accreditation purposes. Due to its sensitivity, this material will reside on the department's local network rather than the HELP computer.

9.11 Continuous Quality Improvement

Continuous quality improvement (CQI) is an approach to quality management that has been used for many years in industrial settings. It has been claimed that the successful implementation of CQI techniques in Japan has been largely responsible for Japan's recent industrial rise. Proponents of continuous quality improvement in the industrial arena include Deming [9], Juran [10], and Crosby [11]. CQI attempts to lead to increased efficiency by decreasing inappropriate variation in the execution of a process. The goals of CQI are (1) to find the most efficient way to carry out a process and (2) eliminate the waste caused by mistakes in carrying out the process. In industry, CQI techniques have been shown to increase the desired results with lower resource expenditures.

Continuous quality improvement is carried out by the following steps: First, the process to be improved is defined and a team that thoroughly understands the process is assembled. Any person or group that has an expectation of the outcome of the process is determined and their expectations are identified. Such people or groups are known as customers of the process. High quality is defined to be present when groups' or customers' expectations are met or exceeded. It is important to note that quality is thus a value judgment and cannot be imposed or dictated. The second step in CQI is thorough documentation of the process to be improved. Desired outcomes (based on expectations) and the factors (called key process factors) in the process that causally determine whether the desired outcomes will be achieved are listed. Measurable specifications that can be used to determine whether or not the desired outcomes and the key process factors have been achieved are identified.

The third step of CQI involves the collection and analysis of data. Data are collected for the specifications and are displayed in a statistical process control (SPC) chart (Figure 9.7). An SPC chart displays the results of data collection for one of the specifications either between individuals or for a single individual over time. Statistically determined control limits (the dashed lines on the control chart) set the range within which an individual measurement can be expected to fall based on random variation from the

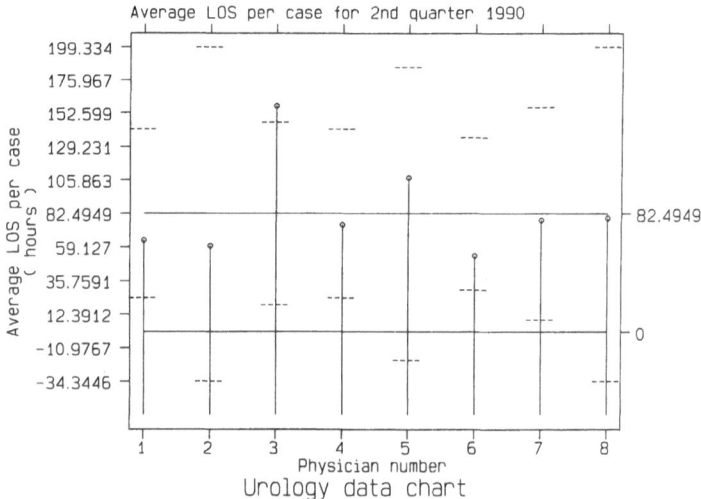

Figure 9.7. Example of an SPC chart.

overall mean. If a measurement falls outside of the control limits, this is known as assignable (rather than random) variation. The presence of assignable variation indicates that the process is not being carried out consistently and warns that this aspect of the process needs examining. For (an industrial) example, if one department in a company has assignably fewer absences, perhaps the manager has superior personnel management skills. If the output of a single machine is significantly different from the output of other similar machines, perhaps the first machine is out of adjustment. Data are collected and analyzed continuously over time. As adjustments

Table 9.1. Data elements gathered by HELP for TURP CQI project

Data element	Module collecting data
1. Name of surgeon	Surgery scheduling
2. Length of stay	ADT
3. Age	ADT
4. Admitted thru ER?	ADT
5. 2nd surgical procedure?	Surgery scheduling
6. Same-day surgery?	ADT/surgery scheduling
7. Surgical time	Surgery scheduling
8. Mortality	ADT
9. Septicemia?	Infectious disease
10. Urinary tract infection?	Infectious disease
11. Cancer found?	Pathology
12. Atelectasis present?	Medical records abstract
13. Thrombophlebitis present?	Medical records abstract
14. # urine cultures	Laboratory
15. # ABGs	Laboratory
16. # CBCs	Laboratory
17. # CXRs	Laboratory
18. # ECGs	ECG module
19. # Protimes	Laboratory
20. # PTTs	Laboratory
21. # SMACs	Laboratory
22. # SMAC-20s	Laboratory
23. # Urinalyses	Laboratory
24. IVP/Pyelogram performed?	Radiology module
25. Amount oxygen administered	Respiratory care
26. Type and screen performed?	Blood bank
27. More than one type & screen	Blood bank performed?
28. Units of blood ordered	Blood bank
29. Units of blood dispensed	Blood bank
30. Grams of tissue removed	Pathology at surgery
31. Foley catheter management	Nurse charting data (removal times, etc.)
32. Temperatures	Nurse charting

are made and the process better understood, inappropriate variation disappears. If assignable variation reappears, it will be revealed in the SPC charts and may indicate a disturbance in the process. A key aspect of CQI is that arbitrary standards are not set since the process can always be improved.

Recently, researchers have claimed [12,13] that the techniques of CQI can be applied to medical care to yield an increase in quality (in terms of improved outcomes) with a simultaneous decrease in costs. As described above, carrying out CQI techniques requires identifying and documenting a process and subsequently gathering large amounts of data concerning the process on a continual basis. At LDS Hospital, research has begun to see if the HELP system can be used to gather the data required by the CQI techniques. The pilot process being examined is transurethral resection of the prostate (TURP), a surgical treatment for prostatism. The LDS Hospital urologists are participating and have documented the factors involved in carrying out a TURP. These factors are listed in Table 9.1. Almost all of the necessary data elements were contained in the HELP system prior to the start of the project. Pre- and post-hospitalization functional status and symptom status paper-based questionnaires will be administered to correlate long-term outcomes with process variables. Data on all the variables will be reported to practicing urologists in SPC charts such as the one in Figure 9.7 showing the average length of stay for TURP patients for each of the urologists in a given 3-month period. The data will be presented blindly (i.e., each of the urologists knows the number that refers to them but does not know the numbers referring to any of the other urologists) and nonpunitively. The point of the charts is not to chastise outliers but rather to stimulate discussion among the urologists of the best way to carry out care. CQI theory states that the physicians will use the information and their own best interests for the patients to provide the highest quality and most efficient care. Experiments are going on currently at LDS Hospital to determine if the CQI techniques can impact the cost and quality of medical care.

References

[1] Van Schoonhoven P, Berkmann EM, Lehmann R; Fromberg R, ed. Medical staff monitoring functions – Blood usage review. Joint Commission on Accreditation of Hospitals; 1987.

[2] Hulse RK, Clark SJ, Jackson JC, Warner HR, Gardner RM. Computerized medication monitoring system. *Am J Hosp Pharm* 1976;33:1061–1064.

[3] Evans RS, Larsen RA, Burke JP, et al. Computer surveillance of hospital-acquired infections and antibiotic use. *JAMA* 1986;256:1007–1011.

[4] Larsen RA, Evans RS, Burke JP, et al. Improved perioperative antibiotic use and reduced surgical wound infections through the use of computer decision analysis. *Infect Control Hosp Epidemiol* 1989;10(7):316–320.

[5] Evans RS, Pestotnik SL, Burke JP, et al. Reducing the duration of prophylac-

tic antibiotic use through computer monitoring of surgical patients. *DICP Ann Pharmacother* 1990;24:351–354.

[6] Pestotnick SL, Evans RS, Burke JP, et al. Therapeutic antibiotic monitoring: surveillance using a computerized expert system. *Am J Med* 1990;88:43–48.

[7] Elliot CG, Simmons D, Schmidt CD, et al. Computer-assisted medical direction of respiratory care. *Resp Manage* 1989:19;31–35.

[8] Bradshaw KEH. *A Computerized Laboratory Alerting System to Warn of Life-Threatening Events.* Salt Lake City, Utah: University of Utah; 1988. Dissertation.

[9] Deming WE. *Out of the Crisis.* Cambridge; Massachusetts Institute of Technology Center for Advanced Engineering Study; 1986.

[10] Juran JM, ed. *Quality Control Handbook, 3rd edition.* New York; McGraw-Hill;1984.

[11] Crosby PB. *Quality is free.* New York; McGraw-Hill;1979.

[12] Berwick DM. Continuous improvement as an ideal in health care. *New Engl J Med* 1989;320:54–59.

[13] Laffel G, Blumenthal D. The case for using industrial quality management science in health care organizations. *JAMA* 1989;262:2869–2873.

10
Surgery Scheduling

10.1 Overview

Operating room times for patients undergoing surgery at LDS Hospital are scheduled using the HELP system [1]. The scheduling program allows data such as the patient's name, age, and room number, the names of the surgeon and the anaesthetist, the procedure to be performed, and the choice of operating room to be used to be entered for any date and time in the future. Using the name of the surgeon and the procedure to be performed, the computer determines which one of 3700 predefined case carts needs to be assembled for each scheduled surgery. The list of carts to be assembled for the following day's surgeries (including the contents of the cart) is automatically printed each day at 2 PM in the central supply department. The surgery scheduling data are also used in computerized decision making involving determination of appropriate use of antibiotics.

Operating room (OR) times (e.g., time of entering OR, starting anaesthesia, ending anaesthesia, etc.) are recorded by OR nurses on all patients while the surgery is in progress. After the surgery, data on the procedure, supplies used, and OR time used are automatically transferred to the financial computer for billing purposes.

10.2 Scheduling

The scheduling program manages the 18 ORs, two extracorporeal shock wave lithotripsy suites, and two cardiovascular catheterization laboratories at LDS Hospital. The system has the capacity to store data on a surgery planned for any OR, time, and date in the future. The system also has capacities for editing data, moving previously entered cases to other time/room slots, and searching for a particular record.

Any given day's schedule is considered "final" at 2 PM on the previous day. At that time copies of the schedule (Figure 10.1) are automatically printed in the admitting department, central supply, and on certain nursing divisions. Cases scheduled after 2 PM are entered into the scheduling program but are considered "add-on" cases. The list of add-on cases for any day may be reviewed on any HELP system terminal (Figure 10.2).

10.3 Case Carts

The HELP system maintains a computerized list of the supplies and instruments used by each surgeon for each procedure he/she performs. Currently, there are more than 3700 such case cart definitions. When the operating room schedule for any given day is finalized (2 PM of the previous day), a list of the carts to be assembled is automatically printed in the central supply department. The list contains the individual items required for each cart, as well as the item's location (aisle and bin number) in the department. Requests for carts for add-on and emergency surgeries are printed on a separate printer to indicate urgency.

10.4 Decision Support

Studies [2–4] were done using computerized surgery scheduling, microbiology, and pharmacy data to examine patterns of preoperative antibiotic use at LDS Hospital. Also studied was the effects of computer-generated reminders to physicians when their practice differed from accepted standards. The computerized alerts generated in these studies relied on surgery scheduling data and knowledge of the type of surgery for which the patient was scheduled.

First, a computerized list was compiled by medical experts and information scientists of the surgical procedures for which patients should receive preoperative antibiotics. Then, by examining surgery scheduling and pharmacy data present in the HELP system, it was easy to determine that only 40% of surgical patients requiring preoperative antibiotics were receiving them "on time" (i.e., within 2 hours before the start of surgery).

A program (approved by the Pharmacy and Therapeutics Committee) was then implemented, whereby the computer would automatically examine the next day's surgery schedule at 3:30 PM each day. If the computer identified a patient that required preoperative antibiotics for whom none was ordered, an alert message would be sent to a clinical pharmacist who would then leave a note reminding the physician of that fact on the patient's chart.

After implementation of this program, the number of cases with appropriate timing increased to 58% from 40% and cases with late administration dropped from 27% to 14%. More importantly, postoperative wound infection rates dropped significantly from 1.9% to 0.9% ($p < 0.02$) [3].

L D S   H O S P I T A L   OPERATING ROOM SCHEDULE FOR THURSDAY 8/18/88   REVISED AS OF   8/18/88. 17:32

OR	PTRM	TIME		AGE	PROCEDURE	SURGEON	ANESTH
1.1	W312	AM(0530)	0730	49F	REMOVAL OF HARDWARE/R ANKLE	HILLYARD,R	SMITH,KD
1.2	W309	AM(0830)	CALL	30F	EXPL R WRIST,BONE GRAFT & STAPLES	GREENE,MH	OTTO,WK
1.3	OR01		CALL	25M	BIL FEMORAL RODDING	GREENE,MH	OTTO,WK
2.1	W303	SS(0530)	0730	40M	DEBRIDE,MANIP,POSS ADHESIOLYSIS R KN	ROSENBERG,TD	BRADWAY,JA
2.2	W327		CALL	14F	ARTHROSCOPY, PATELLA REALIGNMENT	ROSENBERG,TD	BRADWAY,JA
2.3	W315		CALL	20M	R SCOPE,ACL,AUTOGRAFT,POSS MENISECTO	ROSENBERG,TD	BRADWAY,JA
2.4A	OR02		CALL	21F	GROSSE KEMPF NAILING/ LT. FEMUR	TOBLER,GG/HUNTER,G	TOWNSEND,DW
3.1	DSCH	SS(0530)	0730	40M	L OLECRANON BURSECTOMY	TOBLER,GG	DILLON,BJ
3.2	W354		CALL	39M	LUMBAR FUSION,THORA-COLUMBAR	TOBLER,GG	DILLON,BJ
3.3	N3SS	SS(0900)	CALL	65M	REMOVE HARDWARE L ANKLE	RASMUSSEN,GL	WALTON,CW
3.4	OR03	AM(1000)	CALL	22F	STONE HOFFMAN PROCEDURE/L	RASMUSSEN,GL	WALTON,CW
4.1	E632		0730	47F	CRANIOTOMY/L FRONTAL	SORENSEN,BF	OSTERSTOC,JJ
4.2	W616		CALL	56F	POSTERIOR CERVICAL LAMINECTOMY/DECOM	SORENSEN,BF	OSTERSTOC,JJ
4.3	W605		CALL	41M	LUMBAR LAMINECTOMY	RICH,JC	OSTERSTOC,JJ
4.4	OR04		CALL	44M	LUMBAR LAMINECTOMY	RICH,JC	OSTERSTOC,JJ
5.1	W306	AM(0530)	0730	56F	L 1ST MT JOINT REPLACEMENT HAMMER TOES L 1ST & R 2ND	RASMUSSEN,GL	WALTON,CW
5.2	W307	AM(0700)	CALL	27M	R SUPRA MALLEOLAR OSTEOTOMY/POSS FUS	RASMUSSEN,GL	WALTON,CW
6.1	N3SS	SS(0530)	0730	72F	L BR BX POSS MASTECTOMY	NOYES,RD	RAO,SA
6.2	W611	AM(0630)	CALL	32F	CHOLE & GRAMS	NOYES,RD	RAO,SA
6.3	DSCH	SS(0830)	CALL	80F	WIDE EXC CA BACK & FLAP	NOYES,RD	LOCAL +,
6.4A	W629		CALL	34F	EXPLORATORY LAPAROTOMY	PRICE,RR	RAO,SA
7.1	DSCH	SS(0550)	0730	73F	EXC BASAL CELL FOREHEAD FS	HUNTER,G	LOCAL +,
7.2	DSCH	SS(0630)	CALL	66M	EXC SQUAMOUS CELL LOWER LIP *LIP SHAVE	HUNTER,G	WELLING,E
7.3	N3SS	SS(0750)	CALL	34F	TISSUE EXPANDER R ARM	HUNTER,G	TOWNSEND,DW
7.4	N3SS	SS(0830)	CALL	60M	SPLIT THICKNESS SKIN GRAFT/L LEG	HUNTER,G	TOWNSEND,DW

					Procedure		
7.5	N3SS	SS(0930)	CALL	71M	EXC BASAL CELL L CHEEK AND EAR	HUNTER,G	TOWNSEND,DW
8.1	W639	AM(0530)	0730	43M	BIL SUBCUTANEOUS MASTECTOMY	PATTON,L	OTTO,WK
8.2	N3AM	AM(0700)	CALL	12M	VENTRAL HERNIA & SCAR REVISION	PATTON,L	OTTO,WK
10.1	N669		0730	45F	CYSTO-CYSTOGRAM	BOURNE,HH	LOCAL,
10.2	E816		CALL	41F	EXPLORE L URETER	BOURNE,HH	SMITH,KD
10.3	PACU		CALL	72M	PYELOPLASTY/CYSTO RETRO	BOURNE,HH/SORENSON	SMITH,KD
10.4A	OR10		CALL	77M	TURP R RETRO	BOURNE,HH	MCALLISTE,DD
11.1	E630		0730	31M	KIDNEY TRANSPLANT - RECIPIENT (LIVIN STEVENS,MH/RICHARD		BLOLAND,E
11.2A	W837		CALL	69F	PLACEMENT OF TENCKHOFF CATHETER/REMO PRICE,RR		BLOLAND,,LOC+
12.1	W640		0700NB	27F	KIDNEY TRANSPLANT - DONOR (LIVIN	BELNAP,LP/REES,WV	TOWNSEND,DW
12.2	W617		CALL	76F	CHOLE & GRAMS/COMMON DUCT EXPL	BELNAP,LP	TOWNSEND,DW
13.1	DSCH	SS(0530)	0730	28F	L SCALENE NODE EXC	WILSON,ML	PACE,LJ
13.2	E704		0900NB	40M	DRAIN EMPYEMA/DECORDACATION L LUNG	COLLINS,MP	PACE,LJ
14.1	W653		0730	76M	VENTRAL INCISIONAL HERNIA REPAIR	JONES,KW	HURST,SN
14.2A	OR14		0900NB	39F	LIVER TRANSPLANT	BELNAP,LP	HURST,SN
15.1	E707		0730	68F	MITRAL VALVE REPLACEMENT/REDO	DOTY,DB	SUTHERLAN,HH
16.1	E709		0730	69M	DCAG CS	RICHARDS,LS	SEAGER,P
17.1	E841	AM(0700)	0730	26F	R SALPINGO OOPHERECTOMY	NELSON,JC	MCALLISTE,DD
17.2	E834	AM(0930)	CALL	65F	ANTERIOR REPAIR	NELSON,JC	MCALLISTE,DD
					*POSS VAG VAULT SUSP		
17.3	DSCH	SS(0530)	CALL	30F	REMOVE VAGINAL SEPTUM	NELSON,JC	MCALLISTE,DD
17.4	DSCH	SS(0600)	CALL	24F	LAPAROSCOPY, POSSIBLE LAPAROTOMY	NELSON,JC	MCALLISTE,DD
18.1	E835	AM(0700)	0730	36F	TOTAL ABDOMINAL HYSTERECTOMY	GEMMELL,J	RICHARDS,WL
18.2	DSCH	SS(0530)	CALL	22F	D & C	GEMMELL,J	RICHARDS,WL
18.3	DSCH	AM(0600)	CALL	15F	LAPAROSCOPY/DIAG, POSS LASER	GEMMELL,J	RICHARDS,WL
18.4	DSCH	SS(0700)	CALL	48F	L BREAST BX,NL @ 0800,POSS AXILLARY	MUIR,MW	RICHARDS,WL

Figure 10.1. Example of a final operating room schedule for a single day.

```
SATURDAY     8/20/88
             PTRM           PATIENT          PROCEDURE                           SURGEON

4.2   ADD    E602           JONES, JAMES     REDUCE LAMINA C4-5/ANTERIOR FUSION  SANDERS,JM
17.1  ADD    E820  0700NB   JONES, MARY      EXPL LAP  ECTOPIC                   STEELE,MM
ADD                         JONES, SAMUEL    CHOLE & GRAMS                       SWENSEN,S

END OF ADD/ONS - RETURN->
```

Figure 10.2. Add-on list for surgical schedule. Names have been falsified.

10.5 Operating Room Times Program

All surgical patients have the times of their "surgical milestones" (e.g., entering OR, start of anesthesia, end of anesthesia, etc.) recorded in the computer by a nurse using the OR times program.

All the ORs have HELP system terminals present. The entry of the OR times is facilitated by the use of function keys. The nurse needs to press only a single key to indicate a milestone is occurring. This may be followed by a press of the enter key to indicate the event is occurring at the current time. If the event occurred previously, the proper time of occurrence may be entered instead. Analysis of data shows that more than 90% of milestones are entered within 15 minutes of the occurrence of the event. The OR times data are immediately available for review on HELP system clinical terminals (see Figure 10.3).

10.6 Data Review

Data from the OR times program are immediately available for review on hospital terminals. Nurses on the postoperative nursing units can follow the patient's progress without having to call to the ORs. This is helpful in anticipating the patient's return to the division as well as keeping the patient's family apprised of the progress of the surgery.

Physicians may follow the progress of preceding cases to help determine

```
SATURDAY     8/20/88 A.M.
RM    PATIENT          DOC 7   8   9   10  11  12  1   2   3   4   5   6   7
17.1  SMITH, JOHN      /MMS/OAS-SR--R-------------------------------------------

H = ARRIVE HOLD   O = ENTER OR    A = ANESTH ST    S = SURG START   R = ARRIVE RR
M = MED/SEND      O = LEAVE OR    A = ANESTH END   S = SURG END     R = LEAVE RR

RETURN IF FINISHED ->
```

Figure 10.3. Display from operating room times program. Name has been falsified.

when a later case will begin. In addition to terminal review, physicians may use a PC and modem to review this data from their home or office. (Remote physician access is discussed in detail in Chapter 31.)

10.7 Cost Capture

At the end of surgery, a nurse indicates whether the intended procedure was actually performed, and if not, what procedure was performed. Information on additional equipment used (e.g., prostheses, additional disposables, etc.) is also entered into the computer. Information on the procedure, the equipment used, and the length of time of use of the OR is automatically transferred to the financial computer for use in determining the patient's bill.

References

[1] Clayton PD, Delaplaine KH, Jensen RD, Bird B, Evans RS. Integration of surgery management and clinical information systems. *Symposium for computer applications in medical care (SCAMC)* 1987;11:393–395.
[2] Evans RS, Gardner RM, Burke JP, et al. A computerized approach to monitor prophylactic antibiotics. *Symposium for computer applications in medical care (SCAMC)* 1987;11:241–245.
[3] Larsen RA, Evans RS, Burke JP, et al. Improved perioperative antibiotic use and reduced surgical wound infections through the use of computer decision analysis. *Infect Control Hosp Epidemiol* 1989:10(7);316–320.
[4] Pestotnick SL, Evans RS, Burke JP, et al. Therapeutic antibiotic monitoring: surveillance using a computerized expert system. *Am J Med* 1990;88:43–48.

11
Miscellaneous
Administrative Functions

11.1 Order Communications

11.1.1 Overview

A generalized order communications program has been developed for the transmission of requests within LDS Hospital (Figure 11.1). The program allows nurses and unit clerks to enter requests for ECGs, blood gases, and medical supplies. Use of the computer to order laboratory tests is planned for the near future. The requests are sent electronically to the appropriate hospital department. The same order entry program is used for the entry of information on patients requiring isolation precautions. The computer uses the entered data to determine automatically the appropriate isolation requirements and then automatically transmits a request for the necessary supplies to the central supply department. The order entry program is menu driven and has been created using the general questionnaire asking program (GQAP, a HELP system software development tool discussed in Chapter 4). The HELP system is also used by physicians, nurses, and clerks to enter orders for blood products, total parenteral nutrition, and radiology studies. These are discussed in detail in separate reports and will not be covered here.

11.1.2 Electrocardiograms

After choosing "ECG" from the general order menu, the nurse or unit clerk enters which type of cardiac study is desired. The choices are various routine ECGs, Holter studies, echocardiograms, and exercise thallium studies. Next, a (free-text) reason for the test is requested followed by a priority level (e.g., stat, ASAP, routine AM or PM, etc.). After verification, the order is automatically printed out in the ECG department for inclusion in the schedule. When a study is ordered "stat," the technician must be paged as well.

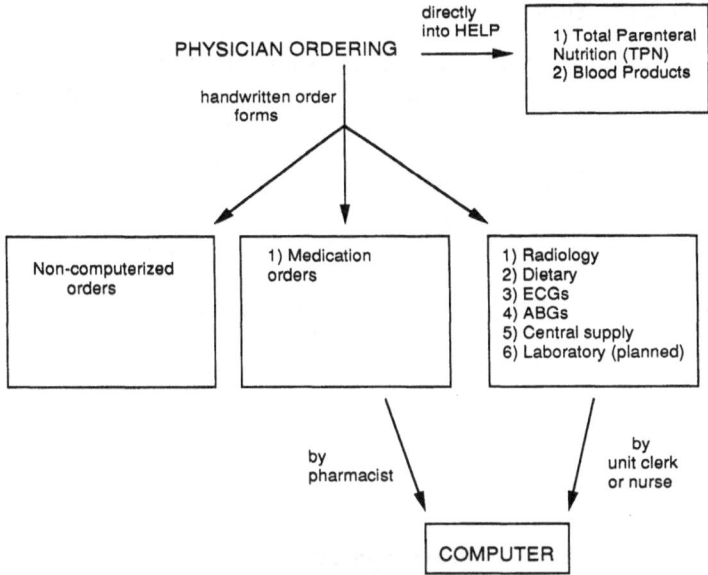

Figure 11.1. Physician ordering at LDS Hospital.

11.1.3 Blood Gases

When blood gases are requested, the computer presents a data entry screen requesting information concerning the amount of oxygen the patient is receiving and by what route. Next, free-text comments may be entered (e.g., reason for the test) followed by the handling priority. Finally, the exact test to be performed (arterial, venous, capillary, or any combination of these) is entered. Requests for specimen pick-up and notifications of specimens to be delivered are similarly entered. The final, verified order is printed in the blood gas laboratory and on the nursing division.

11.1.4 Central Supply

To help manage the ordering of the myriad of items available from the Central Supply Department, the computer presents an initial screen allowing the nurse or clerk to choose a set of items either by category (e.g., dressings, IV supplies, orthopedic supplies, etc.) or alphabetically. The set of items corresponding to the choice is then presented and the request may be made. The request is automatically printed in the Central Supply Department after a handling priority is entered (e.g., routine, stat, etc.) and the order is verified.

11.1.5 Isolation Orders

As an aid to nurses managing patients with infectious diseases, information from the Centers for Disease Control guidelines for isolation procedures [1] has been incorporated into the HELP system. The system currently has information for more than 200 infectious diseases [2]. The nurse may choose a set of diseases either by category (viral, bacterial, gastrointestinal, CNS, etc.) or alphabetically, or the set of the 40 most common infectious diseases may be chosen. The specific infection involved is then indicated.

Once the disease has been indicated, the computer accesses the preprogrammed knowledge to determine which if any isolation procedures are appropriate. These may then be reviewed on a terminal or in a printed report (Figure 11.2). A list of all patients on isolation is also available (Figure 11.3).

11.2 Computerized Dietary System

11.2.1 Overview

Dietary ordering and communications between the nursing unit and the dietary department at LDS Hospital is done through the HELP computer system. Diet-related information is entered in the admitting office when the patient arrives. Additional dietary information is entered by nurses and dietitians on the nursing unit. The computer checks pharmacy information

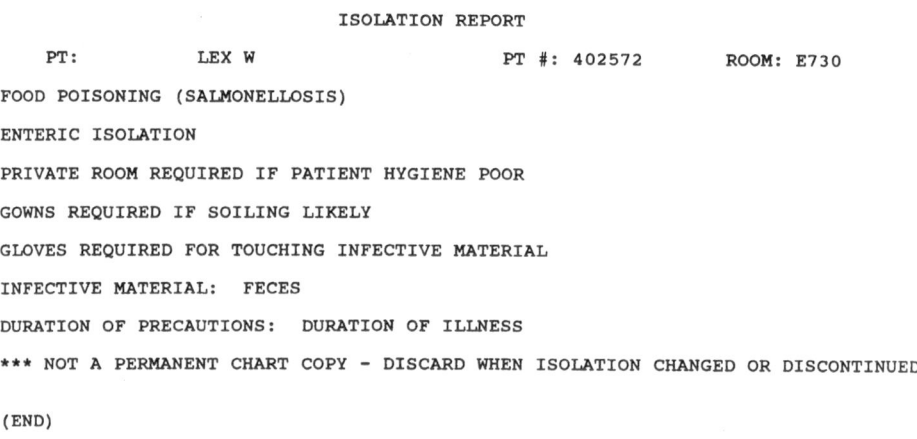

```
                              ISOLATION REPORT
        PT:         LEX W                 PT #:  402572        ROOM: E730
FOOD POISONING (SALMONELLOSIS)

ENTERIC ISOLATION

PRIVATE ROOM REQUIRED IF PATIENT HYGIENE POOR

GOWNS REQUIRED IF SOILING LIKELY

GLOVES REQUIRED FOR TOUCHING INFECTIVE MATERIAL

INFECTIVE MATERIAL:   FECES

DURATION OF PRECAUTIONS:   DURATION OF ILLNESS

*** NOT A PERMANENT CHART COPY - DISCARD WHEN ISOLATION CHANGED OR DISCONTINUED

(END)
```

Figure 11.2. Isolation report for a single patient.

```
                           ISOLATION REPORT

          PATIENT                ROOM      ISOLATION TYPE          DATE
    1          MELANIE ALLEN     W623      CONTACT ISOLATION       07/19/88
    2          DANIEL ALBERT     W742      CONTACT ISOLATION       07/28/88
    3          ROBERT CHARLES    ·E732     PROTECTIVE ISOLATIO     08/09/88
    4      RICHARD LYMAN         W733      CONTACT ISOLATION       08/17/88
    5          CHRISTINE VIRGINIA E701     STRICT ISOLATION        08/17/88
    6              HOWARD A      DSCH      DRAINAGE/SECRETION       06/27/88
    7              RALPH D.      W715      CONTACT ISOLATION        08/19/88
    8      ADRIENNE SMITH        E606      ENTERIC ISOLATION        08/11/88
    9          KAYDEAN WILLIAMS  W325      CONTACT ISOLATION        08/15/88
   10      GORDON GREG           W835      BLOOD/BODY FLUID IS      08/17/88
   11          PAUL ROY          E717      STRICT ISOLATION         08/21/88
   12      RAYMOND LEON          E738      DRAINAGE/SECRETION       07/18/88
   13      JANILEE ROWE          E604      ENTERIC ISOLATION        08/16/88
   14      JANILEE ROWE          E604      BLOOD/BODY FLUID IS      08/04/88
   15      CARMA MAY             W820      DRAINAGE/SECRETION       08/10/88
   16          DAVID EUGENE      E642      BLOOD/BODY FLUID IS      08/11/88
   18          LEX W             E730      ENTERIC ISOLATION        08/21/88
   19          ORRINE NMI        W851      TUBERCULOSIS ISOLAT      08/20/88

 (END)
```

Figure 11.3. Isolation report for all patients at LDS Hospital.

in the electronic patient file and alerts the dietitian when the patient is on certain medications. Alert reports, diet order sheets, and other reports are printed in dietitian's offices, in the kitchen, and on the nursing unit as needed.

11.2.2 General Design of the Dietary System

The computerized dietary system is modular and is menu driven. To enter the system, the nurse, clerk, or dietitian chooses diet orders from the main HELP system menu. He/she then chooses from the following modules:

1. change patient diet
2. diet order sheets
3. dietary communications
4. refrigerator stock orders
5. diet history
6. future diet report

Before each meal, each unit's diet order sheet (choice #2) is automatically printed in the appropriate dietitian's office and reflects the current diet orders for patients on that unit. Dietary assistants compare this to the menu sheets the patients have filled out to make sure there is no conflict. The diet order sheets are then taken to the kitchen in case questions arise while the trays are being assembled. The diet order sheets are also used to update the dietary kardexes, which is the dietitians' means of communicating among themselves.

The computer has logic built in so when a patient is transferred his/her diet orders will automatically appear on the new floor. Also, when a patient is in surgery or recovery, his/her diet orders will not appear on the diet order sheet.

11.2.3 Data Entry

In the admitting office, new patients are asked if they are diabetic or on any special diet. This, or other diet data, may be entered into the computer by the admitting clerk. Once on the floor the nurse or unit clerk can make entries into the system based on the physician's orders. The change patient diet module (choice #1) is used to establish a diet on a new patient, change an existing diet order, or arrange for the diet to be changed at some point in the future (e.g., NPO after midnight for patients undergoing tests in the morning).

Entries into the dietary communications module (choice #3) are printed directly in the kitchen and/or the dietitian's office. This module is used to request a tray to be sent to the patient as soon as possible, to request a specific food item for the patient, to request a guest tray or a father's tray for the maternity ward, or to send a message to the dietitian. Nursing unit staff may request items for the floor refrigerator by entering specific items under the refrigerator stock orders (choice #4) menu.

11.2.4 Implementation of the Dietary System

The dietary assistants request a print out of a diet order sheet for each nursing unit at 5:00 AM, 9:30 AM, and 2:30 PM each day. These reflect the current diet orders and are compared to the menu sheet the patient has completed. Assembly of the tray is from the patient's sheet, with reference to the diet order sheet if necessary. Special orders (ASAP, guest, etc.) are printed in the kitchen or dietitian's office as needed.

11.2.5 Diet-Related Alerts

The computer automatically reviews pharmacy data present in the electronic patient file to see if there are conditions that may be affected by the patient's diet or may influence their ability to eat. These are printed in the kitchen each evening and are brought to the dietitian's offices for review. The dietitian will then follow up and will recommend dietary modification if appropriate. This assists physicians in the care of complicated patients.

Conditions producing alerts are:

1. patients receiving diabetic drugs
2. patients receiving chemotherapy or radiation therapy
3. patients receiving cyclosporine

11.2.6 Dietary Reports

A variety of computer-generated reports are available. As mentioned, diet
order sheets (Figure 11.4) are used by the dietary assistants and kitchen
staff. They may also be used as worksheets on the nursing units. A future
diet report may be generated to see all the patients on the floor with a

```
DIET LIST FOR W7                        05/30/88    TIME OF REPORT:  13:50
=========================================================================

ROOM    PATIENT#    PAT NAME                   DIET
-------------------------------------------------------------------------
W701    25705       DELORES HEAD               SEL WITH CHOC MILK AND
                                          GLASS BID


W702                ....................
W703    25702             JAMES RAY            SELECT


W704    25723            ALLEN GALE<-          NAS


W705    25710            MERLIE INA            1600 ADA


W706                ....................
W707    25728          KENT CHAMBERLAIN        LO CHOL


W708    25707          CHLOE OVIATT            NAS


W709    23259            CHARLES OLIVER        NAS SEL


W710    23225          JESSE RICHARD           SELECT


W711    25700              WESLEY E            NAS


W712                ....................
W714    402161            PAUL DAVID           NAS


W715    402170          DANIEL B               NAS LOW CHOL MECH SOFT
                                      PLEASE SEND GROUND MEAT

W716                ....................
W717    25621            LANE WALTER           SELECT


W718    402178          WESLEY REED            SELECT
```

Figure 11.4. Diet order sheet for a single nursing division.

future diet ordered. This is used to make sure all preoperative patients have
been made nothing by mouth (NPO) after midnight. A diet history (Figure
11.5) will print a report reviewing a patient's complete diet history.

11.2.7 Workload Management Module

The number of dietitians and dietary assistants (DAs) staffing the dietary
department varies from day to day at LDS Hospital. A computer program
has been written to distribute automatically and equitably the staff's daily
duties. After the make-up of that day's available personnel is entered (e.g.,
three dieticians, four DAs, or four dietitians, three DAs, etc.), the computer will automatically assign the staff to the nursing units. In addition,
the computer directs messages from the nursing units to the appropriate
dietitian's office for that day.

11.2.8 Billing

As of this writing, only one of the nursing units (Newborn ICU) bills
separately for itemized dietitian's services (e.g., assessments, re-evalu-
ations). The services performed are entered into the computer and a bill

```
                        PATIENT DIETARY HISTORY
**********************************************************************

 25562            VERA KIRK                      W741    04/08/88



DIET                                 TIME ENTERED              TECH
----                                 ------------              ----

NO ADDED SALT                        05/29/88.14:00             MMM
NO ADDED SALT                        05/29/88.14:00             MMM
FULL LIQUID WITH PLAIN JELLO         05/28/88.08:36             ACW7
FULL LIQUID WITH PLAIN JELLO         05/26/88.16:09             K.O.
FULL LIQ                             05/26/88.09:00             TXS
CL LIQ                               05/25/88.17:53
NPO                                  05/24/88.14:00             CRV
SELECT                               05/16/88.05:11             BJG
SELECT- NO LUNCH OR DINNER 5/15      05/15/88.05:49             BJG
SELECT                               05/14/88.05:07             BJG
SELECT- NO LUNCH OR DINNER 5/13      05/13/88.11:41             BJG
SELECT                               05/09/88.13:41             BJG
SELECT- TO CAFETERIA FOR LUNCH AND DINNE 05/09/88.09:50         BJG
SELECT                               05/07/88.12:51             LW8W
SEL                                  04/25/88.08:35             RW8
SELECT (NO BREADED FOODS)            04/18/88.14:06             EN
SELECT FROM LOW CAL MENU             04/16/88.10:06             MJ
1000 ADA                             04/15/88.18:01             W8SV
SELECT                               04/08/88.19:22             W8SV
                                     04/08/88.18:14             HHA
                                     04/08/88.18:14             HHA

(END)
```

Figure 11.5. Patient diet history. Record of diet changes for a single patient showing new diet, date and time, and technician entering change (or floor where change
was entered).

automatically generated on discharge. This approach accurately reflects the intensity of services rendered for each patient and bills accordingly. The benefits of the individualized billing are being evaluated and it may be extended to the other units in future.

11.3 The HELP System in the Emergency Room

11.3.1 Overview

The HELP system is used in the Emergency Room (ER) for laboratory data retrieval, order entry, ADT functions, blood ordering, respiratory therapy charting, and other services that are available in other locations throughout LDS Hospital. These functions are discussed elsewhere in this book. The HELP system is additionally used to provide functions specific to the ER such as maintenance of a computerized trauma registry and a state-mandated computerized logsheet.

11.3.2 Emergency Room-Specific Programs

Results of all microbiology studies requested from the ER are available through the HELP system. A comprehensive report is printed daily in the ER and reviewed by a nurse to assure appropriate patient follow-up. Information regarding the current price of medications from the hospital pharmacy is available to ER physicians through a program that accesses pharmacy data.

Two HELP programs are used to assist with the gathering and analysis of ER patient data. Information is gathered on all trauma patients presenting to the ER and is entered into a computerized trauma registry. The data gathered includes demographics, details of the mechanism of the injury, transport information, and clinical data such as signs and symptoms, treatment received (including procedures), consultants, complications, and outcome. Half of all data recorded in the trauma registry (e.g., demographics, diagnoses, etc.) are already present in the HELP system from routine computerized ADT and medical records functions. The remaining data are entered manually into the HELP system by hospital personnel.

The HELP system is used to analyze and report data related to trauma patients. At present, the information is used for in-house review purposes only. In the near future, however, the trauma registry data will be made available to State of Utah authorities as part of LDS Hospital's recertification process as a top-level trauma center.

The State of Utah also requires hospitals to keep a record of all patients treated in the ER. The ER log includes information such as demographics, chief complaint, level of urgency, transfer, and disposition data. It is less detailed than the trauma registry and pertains to all patients. The State of Utah mandated in 1988 that all ERs make the log data available in a standardized, computerized format for the purpose of review and analysis. A program was written to allow the LDS Hospital ER log data, which had been kept in a handwritten log, to be entered into the HELP system [3]. The data are easily transferred into the format required by the state and can also be used to generate administrative and clinical data.

As with the trauma registry, part of the patient log data (e.g., demographics, ICD-9 codes, etc.) is entered as part of other routine computerized hospital functions. The remainder of the data are entered by clerks in the ER.

References

[1] Centers for Disease Control. Isolation techniques for use in hospitals. Atlanta: Centers for Disease Control; 1975.
[2] Jacobson JT, Johnson DS, Ross CA, et al. Adapting disease-specific isolation guidelines to a hospital information system. *Infect Cont* 1986;8:411–418.
[3] Russell J. *A Computerized Emergency Room Log*. Salt Lake City, Utah: University of Utah; 1988. Thesis.

III
Use of the HELP System in the Intensive and Acute Care Units

12
The HELP System in the Intensive Care Units

There is not a separate HELP system "ICU Module." Rather, HELP operates in the ICUs as it does throughout LDS Hospital. Certain of the HELP modules (e.g., MIB, hemodynamic monitoring, APACHE scoring) are used almost exclusively within the ICU setting. The benefits of the HELP system's integrated computerized patient database with attendant decision-making capabilities are best realized in the LDS Hospital ICUs with critically ill patients and large volumes of data [1-3]. HELP automatically captures large amounts of diverse patient data (e.g., physiologic patient data automatically captured from patient monitors, computerized nurse charting data, etc.) and reports the data in clinically meaningful ways (e.g., computer-generated reports, automatic computer-generated drug, laboratory, and diagnostic alerts) to assist care-givers in the ICUs. This chapter gives an overview of the impact of the HELP system in the LDS Hospital ICUs.

12.1 The Intensive Care Units at LDS Hospital

Four ICUs at LDS Hospital provide specialized care to critically ill adult patients:

1. Coronary Care Unit (CCU): for patients with nonsurgical cardiac disease (e.g., heart failure, arrhythmias, postmyocardial infarction, etc.).
2. Thoracic Intensive Care Unit (TICU): for postoperative care of critical patients (usually post-openheart surgery cases but occasionally other chest surgeries or complicated vascular cases).
3. The Shock–Trauma–Respiratory Intensive Care Unit (STRICU) provides care for critically ill patients with diseases of all etiologies. Because physician coverage is greater than in the other ICUs, the STRICU admits the most seriously ill patients.
4. The Medical–Surgical Intensive Care Unit (MICU) admits medical and

surgical patients requiring more thorough care than can be provided on a standard acute care nursing division. The MICU also functions as a "stepdown" unit from the other ICUs.

12.2 HELP Applications in the Intensive Care Units

The applications listed below are discussed in detail in the chapters that follow. The applications are introduced here to illustrate the use of the HELP system ICUs.

1. Nursing information system (Chapter 13): With computerized nurse charting, clinical patient data are available for display and computerized decision-making. Computerized nursing reports make use of data from a variety of HELP system modules.

2. Computerized respiratory care charting (Chapter 14): Ventilators are common therapeutic devices in the care of the critically ill. Computerized respiratory care charting provides an easy mechanism for rapid entry of data and for retrieval of legible records. The computerized charting has been shown to be preferred by respiratory therapists. The charting data are used for concurrent quality of care review by the medical director of the Respiratory Care Department and is also available for computer-assisted ventilator management programs.

3. Medication monitoring program (Chapter 15): Automatic screening of computerized pharmacy data for potentially adverse drug-related inter-actions occurs throughout the hospital and specifically in the ICUs where the more acutely ill patients tend to be on more medications. The computer-ized pharmacy data are presented on summary reports.

4. Laboratory information system (Chapter 16): Laboratory data are used with great frequency in medical decision making in the ICUs. The HELP system permits laboratory data to be reviewed quickly and easily using either video terminals or printed reports. Multiple report formats are available for review of the laboratory data. Many of the HELP laboratory reports include other data from the integrated clinical database to facilitate an understanding of the patient's condition.

5. Laboratory alerts (Chapter 17): Studies have shown that man is an imperfect information processor [4]. With large volumes of information, computerized laboratory alerts are helpful in calling physicians' attention to potentially life-threatening events [5].

6. Computerized infectious disease monitor (Chapter 18): Nosocomial infections can be a serious complication for ICU patients. The CIDM auto-matically reviews microbiology data for the purpose of identifying nosoco-mial and other important infections. The alerts generated by the CIDM are reviewed by hospital infectious disease personnel daily.

7. Blood gas (Chapter 19): Computerized decision logic automatically

interprets all blood gases done at LDS Hospital. The blood gas results and the computerized interpretations are printed in a variety of report formats and can also be reviewed on HELP terminals.

8. Medical information bus (Chapter 20): Currently clinically operative in the TICU and still under development, the MIB presents a mechanism for obtaining diverse digital data directly from computerized measuring devices (e.g., pulse oximeters, ventilators, etc.) and allowing computerized control of substance delivery devices (e.g., IV pumps).

9. Hemodynamic monitoring (Chapter 21): Cardiac output, blood pressure, and pulmonary pressures are often critical variables in ICU patients. The HELP system automatically collects data from measuring devices on a regular basis and includes hemodynamic data in summary reports.

10. Total parenteral nutrition (Chapter 22): Critically ill patients commonly require TPN. The HELP system is used by physicians to order the TPN solutions, and by Nutritional Support Services staff to review relevant patient data in concurrent review procedures.

11. Ventilator management protocols (Chapter 23): An expert system is in use in the Shock-Trauma ICU that suggests ventilator adjustments for hypoxemic patients. The system follows ventilator management protocols designed by local experts. The system was originally designed for patients entering a clinical trial of a new treatment for the adult respiratory distress syndrome (ARDS) but is now being used in other hypoxemic and hypoventilatory patients as well.

12. Surgical monitoring (Chapter 24): For patients undergoing open-heart surgery, physiologic data during the surgery are captured automatically by the HELP system. Important events during the surgery are entered into the HELP system by technical personnel. In this way a detailed summary of the surgery can be obtained and personnel in the postoperative ICU can follow the progress of the surgery without contacting the OR directly.

13. APACHE (Chapter 25): The HELP system automatically calculates APACHE severity of illness scores for all patients in the STRICU daily. The scores are calculated from physiological parameters present in the computerized patient database.

Other HELP system modules managing routine hospital services (e.g., order entry, computerized dietary ordering, computerized blood ordering, etc.) are also used in the ICUs.

12.3 Reports

Several computer-generated reports have been designed to present suitably the diversity of data present in the ICUs. A study [6] uncovered the relative frequency with which physicians use data from following different sources

Table 12.1. Data used by physicians in ICU decision making

Data	% of time used in rounds	% of time used at bedside
Bedside monitor	13	22
Direct observation	21	22
Drugs, intake, output, IV	22	13
Blood gas	9	20
Laboratory	33	18
Other	2	5

for clinical decision-making (in rounds and at the bedside) in ICUs (Table 12.1).

To accommodate the diversity of data sources available from the HELP computer system and required by physicians, the ICU Rounds Report (Figure 4.4) was designed. A transparency of the report is created each morning to project the data (with an overhead projector) at the STRICU teaching rounds. The report is temporary and is not entered into the permanent medical record.

A nursing shift report (Figure 13.15) is available at any time to display (graphically) the patient's vital signs and list hemodynamic, intake/output, and medication data. A permanent copy of the report is produced at the end of each shift, initialled by the responsible nurse, and inserted into the patient's permanent medical record.

The 7-day summary report (Figure 4.5) is produced to display, graphically and in tabular form, hemodynamic, pharmacy, fluid balance, and nutritional data from patients in the hospital for an extended period. This report also becomes part of the patient's permanent medical record.

Cardiac output (Figure 21.2) and hemodynamic pressure (Figure 21.3) reports are also available.

References

[1] Gardner RM, West BJ, Pryor TA, et al. Computer-based ICU data acquisition as an aid to clinical decision-making. *Crit Care Med* 1982;10:823–830.
[2] Clemmer TP, Gardner RM. Data gathering, analysis, and display in critical care medicine. *Resp Care* 1985;30:586–598.
[3] Gardner RM. Computerized management of intensive care patients. *MD Comput* 1986;3(1):36–51.
[4] McDonald CJ. Protocol-based computer recorders: the quality of care and the non-perfectibility of man. *New Engl J Med* 1976;295:1351–1355.
[5] Bradshaw KE. *A computerized laboratory alerting system.* Salt Lake City, Utah: University of Utah; 1988. Dissertation.
[6] Bradshaw KE, Gardner RM, Clemmer TP, Orme JF Jr, Thomas F, West BJ. Physician decision-making — evaluation of data used in a computerized ICU. *Intl J Clin Monit Comput* 1984;1:81–91.

13
Nursing Information System

Computerized nurse charting [1] and computerized nursing care plans [2] have been functional on the HELP system since the mid-1980s. Computerized nurse charting allows nurses to chart patient care tasks performed by the nurse and the patient's response to treatment. The charting program at LDS Hospital uses bedside terminals for real-time data entry. (Bedside terminals are currently in place in the four ICUs and on four 48-bed acute care wards and are scheduled to be implemented in the one other acute care nursing division.) The computerized nursing care plan program allows the nurse to identify active patient problems, probable causes for the problems, desired outcomes, and interventions to be taken to achieve the desired outcomes. A nursing care plan history report is produced, which shows the progression of the patient's care plan throughout the hospital stay. The care plan history becomes part of the patient's permanent record. A new version of the computerized care plan program (scheduled for implementation in 1991) will incorporate more patient-specific data and use a nursing diagnosis, instead of a medical diagnosis, approach to patient problems.

Clinical patient data captured through the nursing information system (NIS) are included in a variety of computerized patient reports that become part of the permanent patient medical record and are used for administrative and financial purposes. Patient acuity is calculated either directly from computerized nurse charting data or from manually completed mark sense cards. Patient charges for nursing care are based on acuity data and thus reflect the intensity of nursing services actually delivered. Patient acuity data are also used to determine staffing needs.

The HELP NIS is currently being expanded to include (1) a history module, (2) a physical assessment module, and (3) a computerized nurse planner, which serves as a computerized worksheet for the nurse and also permits access to other HELP system programs such as data review. The history and assessment modules are due to be implemented shortly and the planner is being developed. The NIS is being designed to be a comprehensive system that follows the nursing process (see Table 13.1).

Programs that use HELP system decision logic to help create the computerized nursing care plans are also being developed. Computerized decision logic will automatically examine data entered into the history, assessment, and charting modules and determine which nursing diagnoses may be present. The possible diagnoses will be presented to the nurse when the care plan module is accessed and the nurse will verify or reject the computer's suggestions.

The NIS at LDS Hospital is in a period of transition. Old programs are being upgraded and new programs are being added. To be complete, this chapter describes both the system in current use and the planned modifications.

13.1 Computerized Nurse Charting

The computerized nurse charting module allows nurses to enter patient care tasks, qualitative and quantitative data, and the response to therapy through a set of comprehensive, yet easy-to-use computer screens [1]. The first charting screen is list of 16 charting areas (Figure 13.1). After choosing one of the charting areas, the nurse is directed through a series of screens that allow the tasks performed or the patient's condition to be adequately described. Figures 13.2 through 13.4 show the screens that allow nurses to chart actions related to manipulation of intravenous lines. Figure 13.5 shows a screen where a nurse may indicate hygiene activities with which the patient has been assisted. Currently, many assessment items (e.g., site, status, color, drainage of wounds, vital signs, etc.) are also included in the nurse charting program.

Certain data elements require entry (i.e., they may not be left blank). This ensures compliance with documentation standards. The time the ac-

Table 13.1. Relationships between NIS and the nursing process

Component of NIS	Corresponding aspect of nursing process
Admission history (scheduled for implementation)	Assessment
Physical assessment (scheduled for implementation)	Assessment
Care plan (in current clinical use, new version scheduled for implementation)	Diagnosis and planning
Nurse charting (in current clinical use)	Intervention and evaluation
Planner (under development)	Intervention and evaluation

```
TEST, DIANA FOUNTAIN              50000025  TEST   J 07/11/89 08/18/89
                          CHARTING (FLOORS)
                  ACUTE CARE CHARTING -- MAIN MENU
   0. Select a new patient          9. Teaching
   1. Neurological state/care      10. Emotional support
   2. Cardiovascular/invasive lines 11. Nursing process
   3. Respiratory care/chest tubes 12. Activity/hygiene/isolation
   4. GI/drains                    13. Equipment/procedures/transport
   5. GU (urine)                   14. Admit/discharge/transfer
   6. Wound/skin (dressing changes) 15. Nurse signoff
   7. Antenatal/postpartum         16. IV/Other infusion intake
   8. Vital signs/measurements         (not for W8)
```

Please select 0 to 16 of the above options

Figure 13.1. Main computerized nurse charting screen.

```
TEST, DIANA FOUNTAIN              50000025  TEST   J 07/11/89 08/18/89
                          CHARTING (FLOORS)
     CARDIOVASCULAR/INVASIVE LINES -- SUBMENU

     Invasive Lines (IV's, CVP, Hickman, Etc.)

  1. Invasive lines: checked and patent only
  2. Invasive lines: other management actions

     Cardiovascular

  3. Pacer
  4. Pulses (foot, other peripheral)
  5. Capillary refill
  6. Extremity temp. & color
  7. Peripheral edema
```

Please select 0 to 7 of the above options

Figure 13.2. Cardiovascular/invasive lines charting submenu screen.

```
TEST, DIANA FOUNTAIN              50000025  TEST   J 07/11/89 08/18/89
                          CHARTING (FLOORS)
                  CENTRAL LINE/INTRASIL ACTIONS
     Lumen ## And Actions Are On A   * 7. Manipulated
     Following Screen.                 8. Removed (no lumen # followup)
                                       9. Tip culture
  1. Inspection/care/dressing        10. Pulled out by patient
  2. Inserted == ga.
  3. Pressure dressing               11. By RN
  4. Site held == min.               12. By MD
  5. Sandbag applied                 13. By technician
  6. Sandbag removed
```

Please select 0 to 10 of the above options

Figure 13.3. Invasive lines management screen.

```
TEST, DIANA FOUNTAIN                    50000025   TEST    J 07/11/89 08/18/89
                                CHARTING (FLOORS)
           ──────LUMEN # 1 ACTIONS (CENTRAL/INTRASIL/HICKMAN LINE)──────
  ┌─────────────────────────────────────────────────────────────────────────┐
  │ 1. Checked and patent                                                     │
  │ 2. Heparin/NS lock in place                                               │
  │ 3. Changed to heparin/NS lock                                             │
  │ 4. Solution (FT):                                                         │
  │ 5. Infusion rate (FT): _____                                              │
  │ 6. IV tubing changed                                                      │
  │ 7. Via infusion pump                                                      │
  │ 8. Via rate regulator                                                     │
  │ 9. Lumen action (FT): _____                                               │
  │                                                                           │
  └─────────────────────────────────────────────────────────────────────────┘

Please select 0 to 9 of the above options
```

Figure 13.4. Invasive lines management: manipulation submenu screen.

```
TEST, DIANA FOUNTAIN                    50000025   TEST    J 07/11/89 08/18/89
                                CHARTING (FLOORS)
           ──────────────HYGIENE: BATH & OTHER - DONE BY NURSE──────────────
  ┌─────────────────────────────────────────────────────────────────────────┐
  │  1. Bed bath & linen change          14. Waterpick                        │
  │  2. Partial bath and linen change                                         │
  │  3. Shower and linen change          15. Adult diaper change              │
  │  4. Towel bath and linen change      16. Chux changed                     │
  │  5. Tub bath and linen change        17. Gown changed                     │
  │  6. Chair shower and linen change    18. Other linen change               │
  │  7. Sitz bath and linen change       19. Peri pad change                  │
  │                                                                           │
  │* 8. Hair care                        20. Backrub                          │
  │  9. Shampoo                          21. Nail care                        │
  │*10. Shave                            22. Peri-care                        │
  │                                      23. Routine skin care                │
  │*11. Denture care                     24. Special skin care                │
  │ 12. Routine oral care                                                     │
  │ 13. Teeth brushed                    25. Hygiene (FT): _____              │
  │                                                                           │
  └─────────────────────────────────────────────────────────────────────────┘

Please select 0 to 25 of the above options
```

Figure 13.5. Hygiene submenu charting screen.

tion is (or was) performed is usually required. Nurses are occasionally required to input numerical values directly whereas other choices are made
from lists. Certain categories allow free-text entries. Nurses may go back
and change data using an "Edit" function up to 24 hours after they have
entered it.

The computerized nurse charting data are printed each shift in the computerized nurse comments report (Figure 13.6). The nurse comments report
is signed by the responsible nurse and becomes part of the permanent medical record. The charting data are also used for billing purposes and determining patient acuity and nurse staffing requirements (see below).

13.2 Computerized Nursing Care Plans

The computerized nursing care plan program currently in use at LDS Hospital uses a medical diagnosis approach to patient problems [2]. Because
nurses sought added functionality, a new version of the care plan program

has been designed and is due to be implemented shortly. The new program
will permit patient problems to be described in greater detail and will define
patient problems as active nursing diagnoses rather than medical diagnoses.

13.2.1 Current Nursing Care Plan Program

Currently, nurses identify problems from 1 of the 16 areas listed in Table
13.2. After identifying the problem area (e.g., respiratory function), the
nurse is presented with lists (menus) from which to choose the probable
cause of the abnormality (e.g., pneumonia, asthma, postanesthesia, etc.),
and the expected outcomes (e.g., respirations regular, blood gases normal,
etc.). After this, individual nursing actions that will be taken to achieve the
expected outcome (e.g., record breath sounds q__hrs, incentive spirometry
to __cc, etc.) are entered also from menus. Schedules (e.g., q4h, TID,
etc.) or numerical values (e.g., 500cc) are needed for some items. Free-text
comments may be added. For 15 of the common medical diagnoses encountered at LDS Hospital (e.g., status–postcardiac catheterization, status–postcardiac bypass surgery, acute myocardial infarction, abdominal aortic
aneurysm, thoracotomy, kidney transplant donor, kidney transplant recipi-
ent, etc.), standard care plans have been designed. For patients with one of
these diagnoses, lists of problems, expected outcomes, actions, and standards of care are automatically generated.
   The current care plan may be reviewed from any terminal in the hospital.
Printouts of the nursing care plan (Figure 13.7) are also available. The
nursing care plan is modified throughout the patient's stay. Problems may
be indicated as resolved, causes and outcomes may be modified, and values
may be changed. All changes to the plan, the time of the change, and the
name of the responsible nurse are recorded in the patient's electronic file.
The record of all these changes (the nursing care plan history, Figure 13.8)
may be reviewed on a terminal or printed out at any time. A printed copy
of the care plan history is included in the patient's medical record at the
time of discharge.

13.2.2 New Nursing Care Plan Program

A new nursing care plan program has been designed and will be implemented shortly. The new program was designed because nurses at LDS
Hospital felt the current program did not allow the care plan to be individualized sufficiently for each patient. Specifically, the new program permits
(1) interventions to be described in more detail, (2) goal dates (for resolu-
tion) to be set for each problem, (3) a focus person (e.g., patient, spouse,
child) to be defined for each problem, and (4) collaborative problems,
protocols, and nursing orders to be included in the care plan. The new care
plan program will organize patient problems by nursing diagnosis rather
than by medical diagnosis. Interventions defined in the care plan will be

SEQ #:   6   07/08/90 17:04

NURSE COMMENTS REPORT
SHIFT: 07/08/90 14:01 - 07/08/90 22:01

Name: NMI   Sex: M   Age: 80   No: 211221   Rm: W718
Dr. PREECE, MICHAEL J.   Ht: 173 cm   Admit Wt: 65.00 kg
Nurse: SELLERS, MIKE, LPN   07/08/90 14:01 - 07/08/90 22:00   Current Wt: 65.00 kg   Admit date: 07/06/90 14:43

NEUROLOGICAL
MENTAL STATUS/LOC
  1622 ALERT, ORIENTED X 3, MAE, OBEYS COMMANDS, APPROPRIATE,
SAFETY
  1622 Actions: BED IN LOW POSITION
                 CALL LIGHT WITHIN REACH
        SIDERAILS UP: RIGHT UPPER, LEFT UPPER,

CARDIOVASCULAR/INVASIVE LINES
PULSES
  1622 BILAT. DORSALIS PEDAL 2+   BILAT. RADIAL 2+
EXTREMITY TEMPERATURE AND COLOR
  1622 BILATERAL UPPER EXTREMITIES   Desc: WARM, DRY, PINK,
        BILATERAL LOWER EXTREMITIES
  1622 BILATERAL UPPER EXTREMITIES   Desc: WARM, DRY, PINK,
        BILATERAL LOWER EXTREMITIES

CAPILLARY REFILL
1622 BILAT. UPPER EXTRM., < 3 SECONDS   BILAT. LOWER EXTRM., < 3 SECONDS
PERIPHERAL EDEMA
1622 BILATERAL UE 0 +   BILATERAL LE 1 +
INVASIVE LINES
1622 HEPARIN LOCK: LEFT, FOREARM,
     Management: CHECKED AND PATENT,

RESPIRATORY
CONTINUOUS THERAPY/VENTILATORS
1419 CANNULA: LITERS/MIN: 2
     STO2 (FINGER): 97
1430 CANNULA: ROOM AIR
     STO2 (FINGER): 93

G.U.
G.U. OUT
1624 SPECIPAN URINE 300 ML   Char: CLEAR
                                          YELLOW

VITAL SIGNS/MEASUREMENTS
VITAL SIGNS
1624 HEART RATE 78 PER MINUTE   SYSTOLIC BP 100   Temp: EAR PROBE TEMP.   36. 1 C,
     RESPIRATORY RATE (PER MIN) 18   DIASTOLIC BP 58

Figure 13.6. Nurse comments report.

(Continued.)

MEASUREMENTS
    1419 STO2 (FINGER): 97
    1430 STO2 (FINGER): 93

EQUIPMENT/PROCEDURES/TRANSPORT
SPECIAL EQUIPMENT
    1622 Telemetry
    Management: IN PLACE, IN USE/FUNCTIONING CORRECTLY,

Name: NMI   No: 211221   Room: W718   Shift: 07/08/90 14:01 - 07/08/90 22:01   Sequence #:   6

- 1 -

(END)

Figure 13.6. *Continued.*

Table 13.2. Care plan categories (current program)

---

1. Respiratory function
2. Perfusion/cardiac output
3. Fluid/electrolye balance
4. Neurological status
5. Cardiac rhythm
6. Patient comfort
7. Elimination: bowel, bladder
8. Infection/wound management
9. Physical mobility/skin integrity
10. Psychosocial status
11. Nutrition
12. Impaired coagulation status
13. Knowledge deficit
14. Self-care deficit/rehabilitation
15. Cognitive-perceptual deficit
16. Female reproduction

---

linked to the charting and assessment modules via the computerized planner.

Diagnoses recognized by the North American Nursing Diagnosis Association (NANDA) [3] will be used at LDS Hospital. For each nursing diagnosis identified, the care plan module permits the nurse to enter data regarding (1) related factors, (2) expected outcomes, and (3) interventions that will be performed by the nurses to obtain the expected outcomes. For example, for a patient with the diagnosis of self-care deficit, the nurse may choose any of the 14 related factors shown in Figure 13.9. Option 14 allows the entry of free text. The nurse next chooses the expected outcomes. Expected outcomes for self-care deficit include (1) perform bathing at optimal levels, (2) identify areas of weakness/needs, (3) identify resources that can provide assistance, (4) demonstrate use of adaptive devices, and (5) verbalize/demonstrate understanding of teaching. The nurse also selects whether the outcome applies to the patient, family, or other person. Goal dates are selected for each outcome.

The nurse next identifies the interventions that will be undertaken in order to achieve the desired outcomes. Examples of interventions for self-care deficit include (1) assess for additional related factors, (2) assess current level of functioning, (3) evaluate support systems available, (4) provide assistive devices, (5) teach use of assistive devices, (6) keep call light within reach while bathing, and so forth. The HELP system allows certain interventions for each diagnosis to be designated as "standards of care," meaning that if the diagnosis is indicated as active, the standard of care interventions will be automatically selected. A frequency may be selected for each intervention. The frequency indicates when the intervention should be per-

```
CURRENT NURSING CARE PLAN          E712          VAN                          211150
   07/08/1990 20:02
```

1. ALTERATION IN RESPIRATORY FUNCTION   TIME INITIATED: 07/07/1990 02:52   LAST UPDATED: 07/08/1990 07:02

RELATED TO (CAUSES):   ACTIONS:
  1 ATELECTASIS   1 ASSESS PARADOXICAL CHEST WALL MOVEMENT
                                          2 REPORT ABG VALUES ABNORMAL FOR PATIENT
                                          3 ASSESS CHARACTER/FREQUENCY/EFFECTIVENESS OF COUGH Q SHIFT
                                          4 INCENTIVE SPIROMETRY: Q 2 HR AFTER EXTUBATION
OUTCOMES: 5 ASSESS SECRETIONS FOR CHARACTER & VOLUME Q SHIFT
  1 RESPIRATIONS REGULAR/UNLABORED
  2 RESPIRATORY RATE WNL FOR PATIENT
  3 CHEST CLEAR PER AUSCULTATION   STANDARDS OF CARE :
  4 ABGS WNL FOR PATIENT ON ROOM AIR   1 ASSESS ADEQUACY OF FLUID BALANCE
    OR SUPP OXYGE   2 ASSESS BREATH SOUNDS Q SHIFT & PRN
  5 INDEPENDENTLY HANDLES SECRETIONS   3 EVALUATE RESP PATTERN/EFFORT Q SHIFT & PRN
                                          4 REPORT S/S HYPOXIA

2. ALTERATION IN PERFUSION/CARDIAC OUTPUT   TIME INITIATED: 07/07/1990 02:54   LAST UPDATED: 07/08/1990 07:03

RELATED TO (CAUSES):   ACTIONS:
  1 SURGERY   1 ASSESS HEART SOUNDS Q 4 & PRN
  2 VALVULAR HEART DISEASE   2 EVALUATE HEMODYNAMIC PARAMETERS AS OBTAINED
                                          3 ASSESS CAPILLARY FILLING TIME Q4H & PRN
                                          4 ASSESS PERIPHERAL PULSES Q SHIFT & PRN
OUTCOMES: 5 ASSESS TEMP/COLOR OF EXTREMITIES Q 4 & PRN
  1 BP WNL WITHOUT MEDICATION   6 ASSESS SACRUM & EXTREMITIES FOR EDEMA
  2 BP WNL WITH MEDICATION   7 MONITOR A-V DIFFERENCE FOR ADEQUACY OF CARDIAC OUTPUT
  3 HEART RATE WNL   8 DAILY WEIGHTS
  4 CARDIAC OUTPUT/INDEX WNL
  5 URINE OUTPUT >20-30 CC/HR   STANDARDS OF CARE :
                                          1 ASSESS BREATH SOUNDS Q SHIFT & PRN
                                          2 ASSESS/RECORD LEVEL OF CONSCIOUSNESS
                                          3 EVALUATE FLUID BALANCE Q SHIFT
                                          4 OBSERVE FOR S/S OF FLUID OVERLOAD
                                          5 OBSERVE FOR DYSPNEA
                                          6 OBSERVE FOR WEAKNESS/DIZZINESS/PALPITATIONS
```
```

5. ALTERATION IN CARDIAC RHYTHM TIME INITIATED: 07/07/1990 02:56 LAST UPDATED: 07/08/1990 07:05

RELATED TO (CAUSES):
1 CARDIAC SURGERY

ACTIONS:
1 ASSESS EFFECT OF DYSRHYTHMIA ON CARDIAC OUTPUT
2 TEMPORARY PACER

OUTCOMES:
1 STABLE RHYTHM FOR PATIENT WITH QT
  ADEQUAT
2 HR WNL, NO IRREGULARITIES
3 ELECTROLYTES WNL

STANDARDS OF CARE :
1 ASSESS BP, HR, RR Q SHIFT & PRN
2 ASSESS LEVEL OF CONSCIOUSNESS
3 MONITOR RHYTHM, RATE, INTERVALS, WAVE FORMS Q 4H & PRN
4 DOCUMENT RHYTHM WITH POSTED ECG STRIP Q 4H & PRN
5 OBSERVE FOR WEAKNESS, DIZZINESS, PALPITATIONS
6 CHECK ECG FOR PACER CAPTURE, SENSING & AMPLITUDE

6. ALTERATION IN COMFORT TIME INITIATED: 07/07/1990 02:54 LAST UPDATED: 07/08/1990 07:03

RELATED TO (CAUSES):
1 SURGERY

ACTIONS:

OUTCOMES:
1 VS REFLECT STATE OF COMFORT
2 EXPRESSED RELIEF/ABSENCE OF PAIN

STANDARDS OF CARE :
1 OBSERVE FOR NON-VERBAL INDICATIONS OF PAIN
2 IDENTIFY CARE/ACTIVITIES WHICH INCREASE PAIN
3 MEDICATE PRIOR TO PAINFUL PROCEDURES

LAST NURSE TO UPDATE CARE PLAN: SHINKLE, DEBRA   07/08/1990 07:05   21115092

************************************* TEMPORARY REPORT--DISCARD WHEN UPDATED ***

(END)

Figure 13.7. Nursing care plan printout.

NURSING CARE PLAN HISTORY   VAN   *** FINAL COPY ***   211150   E712   07/08/1990 20:01

1. ALTERATION IN RESPIRATORY FUNCTION   TIME INITIATED: 07/07/1990 02:52   TIME RESOLVED: NOT RESOLVED

A. RELATED TO (CAUSE) (INITIATED - RESOLVED):
   1. ATELECTASIS   (07/07 -   )
   2. POST-ANESTHESIA   (07/07 - 07/08)
   3. COMMENT: HX OF CHF PRE-OP AND PULMONARY EDEMA   (07/07 - 07/08)

B. ACTIONS (INITIATED - RESOLVED):
   1. ASSESS PARADOXICAL CHEST WALL MOVEMENT   (07/07 -   )
   2. ASSESS RETRACTION/ACCESSORY MUSCLE USE   (07/07 - 07/08)
   3. REPORT ABG VALUES ABNORMAL FOR PATIENT   (07/07 -   )
   4. ASSESS CHARACTER/FREQUENCY/EFFECTIVENESS OF COUGH Q SHIFT   (07/07 -   )
   5. VENTILATORY PARAMETERS Q SHIFT & PRN   (07/07 - 07/08)
   6. SV O2 FROM OPTICATH Q SHIFT & PRN   (07/07 - 07/08)
   7. INCENTIVE SPIROMETRY: Q 2 HR AFTER EXTUBATION   (07/07 -   )
   8. ASSESS SECRETIONS FOR CHARACTER & VOLUME Q SHIFT   (07/07 -   )

C. STANDARDS OF CARE (INITIATED - RESOLVED):
   1. ASSESS ADEQUACY OF FLUID BALANCE   (07/07 -   )
   2. ASSESS BREATH SOUNDS Q SHIFT & PRN   (07/07 -   )
   3. EVALUATE RESP PATTERN/EFFORT Q SHIFT & PRN   (07/07 -   )
   4. REPORT S/S HYPOXIA   (07/07 -   )

D. EXPECTED OUTCOMES (INITIATED - RESOLVED):
   1. RESPIRATIONS REGULAR/UNLABORED   (07/07 -   )
   2. RESPIRATORY RATE WNL FOR PATIENT   (07/07 -   )
   3. SPONTANEOUS VENTILATION   (07/07 - 07/08)
   4. CHEST CLEAR PER AUSCULTATION   (07/07 -   )
   5. CHEST EXPANSION SYMMETRICAL   (07/07 - 07/08)
   6. ABGS WNL FOR PATIENT ON ROOM AIR OR SUPP OXYGEN   (07/07 -   )
   7. INDEPENDENTLY HANDLES SECRETIONS   (07/07 -   )

2. ALTERATION IN PERFUSION / CARDIAC OUTPUT   TIME INITIATED: 07/07/1990 02:54   TIME RESOLVED: NOT RESOLVED

A. RELATED TO (CAUSE) (INITIATED - RESOLVED):
1. SURGERY   (07/07 -   )
2. VALVULAR HEART DISEASE   (07/07 -   )

B. ACTIONS (INITIATED - RESOLVED):
1. ASSESS HEART SOUNDS Q— & PRN   (07/07 -   )
2. EVALUATE HEMODYNAMIC PARAMETERS AS OBTAINED   (07/07 -   )
3. ASSESS CAPILLARY FILLING TIME Q4H & PRN   (07/07 -   )
4. ASSESS PERIPHERAL PULSES Q SHIFT & PRN   (07/07 -   )
5. ASSESS TEMP/COLOR OF EXTREMITIES Q— & PRN   (07/07 -   )
6. ASSESS SACRUM & EXTREMITIES FOR EDEMA   (07/07 -   )
7. MONITOR PVO2 FOR ADEQUACY OF TISSUE OXYGENATION   (07/07 - 07/08)
8. MONITOR A-V DIFFERENCE FOR ADEQUACY OF CARDIAC OUTPUT   (07/07 -   )
9. ASCERTAIN FUNCTION/PATENCY OF CHEST TUBES   (07/07 - 07/08)
10. DAILY WEIGHTS   (07/07 -   )
11. MONITOR FOR PULSUS PARADOXUS   (07/07 - 07/08)
12. MONITOR FOR S/S OF CARDIAC TAMPONADE   (07/07 - 07/08)

C. STANDARDS OF CARE (INITIATED - RESOLVED):
1. ASSESS BREATH SOUNDS Q SHIFT & PRN   (07/07 -   )
2. ASSESS/RECORD LEVEL OF CONSCIOUSNESS   (07/07 -   )
3. EVALUATE FLUID BALANCE Q SHIFT   (07/07 -   )
4. OBSERVE FOR S/S OF FLUID OVERLOAD   (07/07 -   )
5. OBSERVE FOR DYSPNEA   (07/07 -   )
6. OBSERVE FOR WEAKNESS/DIZZINESS/PALPITATIONS   (07/07 -   )

D. EXPECTED OUTCOMES (INITIATED - RESOLVED):
1. BP WNL WITHOUT MEDICATION   (07/07 -   )
2. BP WNL WITH MEDICATION   (07/07 -   )
3. HEART RATE WNL   (07/07 -   )
4. CARDIAC OUTPUT/INDEX WNL   (07/07 -   )
5. URINE OUTPUT >20-30 CC/HR   (07/07 -   )
6. STABLE HEMATOCRIT   (07/07 - 07/08)

Figure 13.8. Nursing care plan history printout.

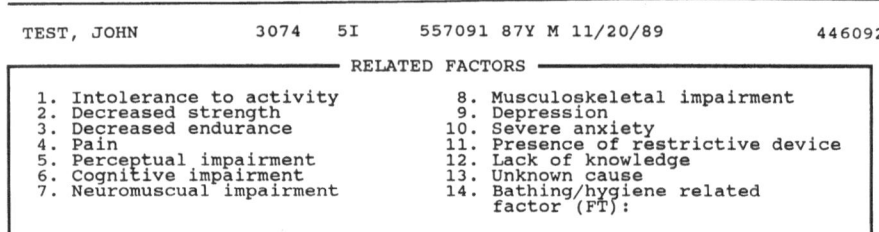

TEST, JOHN   3074   5I   557091 87Y M 11/20/89   446092
━━━━━━━━━━━━━━━━━━━━ RELATED FACTORS ━━━━━━━━━━━━━━━━━━━━

1. Intolerance to activity   8. Musculoskeletal impairment
2. Decreased strength   9. Depression
3. Decreased endurance   10. Severe anxiety
4. Pain   11. Presence of restrictive device
5. Perceptual impairment   12. Lack of knowledge
6. Cognitive impairment   13. Unknown cause
7. Neuromuscual impairment   14. Bathing/hygiene related
                                     factor (FT):

Please select 0 to 14 of the above options

Figure 13.9. Related factors screen for self-care deficit.

formed and determines where the intervention will appear on the planner. Each intervention may be individualized for the patient.

In addition to choosing nursing diagnoses from the new nursing care plan program, the nurse will also select protocols, collaborative problems, and nursing orders. Protocols are generally treatment- or equipment-related standards of care. Collaborative problems are potential or active medical conditions for which the nurse must monitor physical findings and laboratory results so the physician can be notified when necessary. When viewed on the terminal screen, only the name of the protocol or collaborative problem will appear; however, the complete text of the standards of care for protocols and collaborative problems can be reviewed on the terminal or can be printed out. Nursing orders are nurse-to-nurse orders that need to be performed but are not related to any specific active nursing diagnosis. Examples include performing vital signs more often than usual, performing intake/output measurements, or visitation specifications.

Printouts of the current nursing care plan and the care plan history will continue to be available with the new version of the computerized nursing care plan program.

13.3 History Module

The history module of the NIS has been developed and is scheduled for implementation at LDS Hospital shortly. At LDS Hospital, the 11 functional health patterns [4] (FHPs, listed in Table 13.3) are used as a framework for collecting historical data. When a nurse enters the history module, the specific FHP to be assessed is chosen. Next, the nurse may indicate the source of the information (e.g., patient, spouse, child, etc.). For each FHP the nurse is presented with 10 to 20 questions regarding the patient's functioning in that area. For example, in the nutritional—metabolic pattern, the nurse may inquire about recent weight change, appetite, special dietary requirements, allergies, and/or easy bruising. When a question is chosen the nurse is presented with a screen of the most likely answers. For example, under the health perception pattern, if the nurse chooses the question "How has your general health been?," a screen appears with the choices "excellent," "good," "fair," and "poor." This menu-driven approach speeds data entry and also allows the data to be collected in a coded form (see Chapter 3). The nurse is not required to enter data for every question.

The nurse moves through all the appropriate FHPs until the history taking is complete. The history may be updated at any time. History data will be able to be reviewed on any terminal or through a printed report.

A future goal of the NIS is to have decision logic determine which nursing diagnoses may be present based on pertinent historical data. The potential diagnoses will be presented to the nurse when the care plan module is used.

13.4 Assessment Module

The physical assessment module of the NIS has been designed to be used by nurses to capture an assessment of the patient's physical condition upon admission and throughout the hospital stay. The assessment module is due to be implemented in 1991. The assessment categories are physiological systems with an additional category for pain assessment and an option for no abnormalities (Table 13.4). For example, if the nurse chooses the cardiovascular category, the screen in Figure 13.10 will be displayed. The top window shows the normal assessment (regular rhythm, pulses 3+, good capillary refill, and warm and dry extremities). The normal assessment is the default if no modifications are made. The second window shows the findings that can be modified. If "vital signs" is chosen, the nurse will be presented with a screen that permits a full set of vital signs to be entered. If "edema" is chosen, the nurse will be presented with a menu screen that permits a detailed entry (in coded form) of the location and extent of the

Table 13.3. Functional health patterns

1. Health perception — health management
2. Nutritional — metabolic
3. Elimination
4. Activity — exercise
5. Sleep — rest
6. Cognitive — perceptual
7. Self — perception
8. Role — relationship
9. Sexuality — reproductive
10. Coping — stress tolerance
11. Value — belief

Table 13.4. Categories for patient assessment

1. Neurological
2. Cardiovascular
3. Pulmonary
4. Gastrointestinal
5. Genitourinary
6. Wound/skin
7. Perinatal
8. Pain
9. All systems normal

edema. Under GI assessment, normal bowel sounds, no tenderness, and no abdominal discomfort is the normal assessment and detailed data may be entered on bowel sounds, the abdominal examination, abdominal discomfort, bowel movements, tubes, and drains.

Once initial assessment has been entered, the most recent assessment data appear on the screen when the assessment program is accessed. The nurse may then (1) accept the displayed data as current if no change has occurred, (2) select particular items within each category for modification, or (3) modify the assessment of the patient for the entire category.

Data entered into the assessment module will be able to be reviewed on any HELP system terminal and will appear on the nurse comments report (Figure 13.6). Data entered into the assessment module will be examined by decision logic to see if any nursing diagnoses are suggested.

13.5 Planner

The nurse planner module is under development and will serve as a central module integrating the other components of the NIS. The planner will provide an electronic method of tracking nursing interventions required for

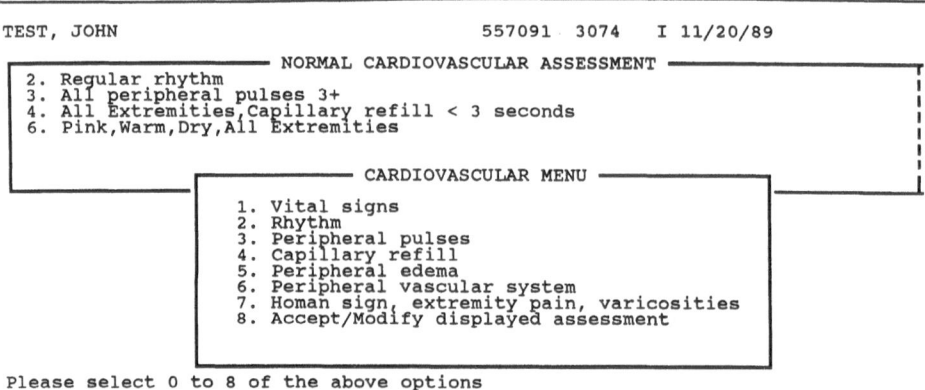

Figure 13.10. Initial cardiovascular assessment screen.

a patient. The nurse will use the planner as an electronic worksheet to assist in planning patient care. The planner will also provide access to data review programs and will keep track of what data items have already been charted.

The planner will display interventions from the care plan and orders from the order entry module (see Chapter 11). The planner will have the configuration shown in Figure 13.11. The two columns separate the orders (meaning interventions and orders) into scheduled (left side) and unscheduled (right side) orders. Scheduled orders are those that have been verified by the nurse and have a time scheduled such as qid or qd. The orders are displayed chronologically. Verification by the nurse is required only for some orders and indicates that the nurse is aware the order has been made and is correct. The right side of the screen displays unverified orders (identified by an asterisk). The asterisk alerts the nurse that there is a new order. Unscheduled orders (those with a frequency of prn, continuous, q3–4 hours prn, etc.) also appear on the right-hand side of the screen.

The planner is designed to facilitate the documentation of care plan interventions. By selecting option C (chart, see bottom of Figure 13.11) and the number of the intervention to be documented, the appropriate charting screen will be displayed. Once the data are entered, the planner is redisplayed with the intervention dimmed to indicate that the item has been documented. For example, "05/09.18:00 Vital Signs" is intervention #2. The user enters C2 and then the vital sign charting screen is displayed. After the vital signs are entered, the user is returned to the planner and "05/09.18:00 Vital Signs" appears in a dim mode. This enables the nurse to review quickly orders that still need to be performed and those that have been completed and documented.

The nurse will be able to access any of the other nursing applications and many other HELP programs from the planner. The assessment, his-

```
TEST, JOHN            3074   5I      557091 87Y M 11/20/89              446092

St Idx   Date/Time Intervention/Order  St  Idx    Date/Time Intervention/Order
   1. 05/09.18:00 Vital Signs          *   1.                Input and Output
   2. 05/10.00:00 Vital Signs          *   2.                24 Hour Urine
                                            3.      PRN       O2 2L








          Options: S,U,#,C#,D#,E#,R#,V#,A,F,K,N,P,T,X,MA,MC
```

Figure 13.11. Planner screen.

tory, and care plan applications will be able to be accessed. Review of essential demographic data and critical clinical information such as allergies, infection precautions, emergency contact, and so forth, and general data review will be able to be performed through the planner as well. The planner has been designed with maximum flexibility as the nurse uses it as the key point in the NIS. Upon entering the patient's room, the nurse will access the planner first.

13.6 Medications

Nurses use the HELP pharmacy program (Chapter 15) to schedule, verify, and record administration of medications to patients. The HELP pharmacy program involves both nurses and pharmacists. At LDS Hospital, either nurses or pharmacists may schedule medications for patients. If the pharmacist enters the medication, the nurse is responsible for verifying that the order is correct. Similarly, if the nurse schedules the medication, the pharmacist verifies the order. The HELP medication monitoring program described in Chapter 15 presents medication alerts to whomever is scheduling the drug. (The responsibility for follow-up of the alerts ultimately falls to the pharmacist, however, and complete lists of alerts are printed in the hospital pharmacy to facilitate follow-up.) A schedule of medications to be administered is printed to assist the nurses (Figure 13.12). The schedule

```
                ***** MEDICATION ADMINISTRATION RECORD *****
                          07/08/90    17:54
   18:00               CALVIN                    111194
   W841    38. KCL 20 MEQ, POWDER, ORAL, BID 0900 1800,
            LAST GIVEN 07/08/90    09:30

   18:00               WIDTSOE                   111070
   W845    61. BACTRIM 14 ML, INJ, IVPB, Q 12 H 1800,
            LAST GIVEN 07/08/90    05:50
            D5W 250 ML

   19:00               CALVIN                    111194
   W841    20. ERYTHROMYCIN 1000 MGM, INJ, IVPB, Q 6 H 1900,
            LAST GIVEN 07/08/90    13:30
            NORMAL SALINE 250 ML

   21:01               CALVIN                    111194
   W841    42. OMEPRAZOLE (LOSEC) 20 MGM, CAP, ORAL, QHS

   00:00               CALVIN                    111194
   W841    19. CEFTRIAXONE (ROCEPHIN) 2000 MGM, INJ, IVPB, Q 24 H 0000,
            LAST GIVEN 07/08/90    00:05
            NORMAL SALINE 150 ML

   01:00               CALVIN                    111194
   W841    20. ERYTHROMYCIN 1000 MGM, INJ, IVPB, Q 6 H 1900,
            LAST GIVEN 07/08/90    13:30
            NORMAL SALINE 250 ML

   06:00               WIDTSOE                   111070
   W845    61. BACTRIM 14 ML, INJ, IVPB, Q 12 H 1800,
            LAST GIVEN 07/08/90    05:50
            D5W 250 ML

   07:00               CALVIN                    111194
   W841    20. ERYTHROMYCIN 1000 MGM, INJ, IVPB, Q 6 H 1900,
            LAST GIVEN 07/08/90    13:30
            NORMAL SALINE 250 ML
```

Figure 13.12. Medication administration record.

lists the medications the patients are due to receive, the times the medications were last given, and the times the medications should be administered during the current shift. Nurses chart administered medications using a module of the pharmacy program. When the planner is implemented, scheduled medications will appear as tasks to be performed and charting of administered medications will be performed directly through the planner. A patient care flow sheet (Figure 13.13) lists the current IVs the patient is receiving and the amount of fluid remaining.

13.7 Acuity and Management Functions

Billing for nurses' services and determination of staffing needs are done from computerized calculation of patient acuity [5]. Calculation of patient acuity may be done automatically directly from the computerized nurse charting data and, in fact, this is done on some wards at LDS Hospital. However, because of the data-intensive nature of the acuity calculation program (i.e., all computerized nurse charting data items must be examined for all patients and numerous calculations and look-ups made on the basis of the found data), acuity on most wards is measured from mark-sense cards completed manually by the nurses to indicate services the patient received. Data from the mark-sense card reader is transferred to the Tandem computer. Specific nursing actions have time values (minutes) associated with them (Figure 13.14). More acutely ill patients will have more nursing activities recorded. A fixed amount is added for noncharted functions (e.g., charting, communications, etc.).

Patients are billed for nurses' services based on care they actually received. Each nursing action has a charge associated with it. When a patient is discharged, the computerized acuity data are used to generate an itemized bill. The bill accurately reflects the intensity of the service rendered. An automated patient acuity report is generated for each unit at the end of each shift based on the amount of care that the patient required. The acuity report is used to determine staffing needs for the following shift.

13.8 Evaluations: Bedside Terminals, Charting, and Care Plans

The NIS at LDS Hospital utilizes 80286-based PC terminals at the bedside. Bedside terminals were tested in the four ICUs and on one acute care ward at LDS Hospital for the latter half of the 1980s. Hospital-wide installation of bedside terminals began in 1989. Evaluation studies at LDS Hospital [6] have measured the impact of bedside terminals and the NIS on the structure, process, and outcome of medical care. Questionnaires revealed nurses had a marked preference for the bedside terminals over nursing station-

PATIENT CARE FLOW SHEET   07/08/1990 22:02

E   AGE: 70 SEX:M   DR. WALSH, KEVIN J   # 111195   E734

ADMIT DATE: 07/04/90 DIAGNOSIS: ATRIAL FLUTTER
ALLERGIES:
SURGICAL PROCEDURE & DATE : _____
CURRENT DIAGNOSIS : _____

                                              2. DIET: _____

1. ACTIVITY: _____
3. TREATMENTS: _____
4. TESTS: _____
5. _____

IVS REMAINING:

TIME	CONCEN	REMAIN																			
D5W 200 ML, INJ		200																			
AMINOPHYLLINE 200 MGM, INJ	1000	200																			
D5W 100 ML, INJ	500	100																			
50.0 MGM/HR		50.000																			

Figure 13.13. Patient care flow sheet.

```
Codes for time spent:
    A  = Activity (ambulation, up in chair, etc.)
    D  = Dressing change (suture care, pin care, etc.)
    T  = Teaching            ES = Emotional support
    TR = Transport           P  = Special procedure
```

```
AUTOMATED ACUITY DATA:      TICU         07/08/90   06:01
            111142              PHYLLIS
```

		(CHARTED)	COUNTED
PA MEAN PRESSURE		(1)	1
PORTABLE X-RAY X ==		(1)	1
#1		(2)	1
FOLEY CATH URINE ==== ML		(5)	1
BLOODGAS, BLOOD TYPE =		(1)	1
RESPIRATORY RATE (PER MIN) ==		(5)	5
TIME SPENT ==== MIN. (ES)		(1)	20
TIME SPENT ==== MIN. (T)		(1)	20
CPAP	OR (	0)	
FACE MASK	OR (	0)	
FACE TENT	OR (	0)	
HEATED NEBULIZER	OR (	2)	
CANNULA	OR (	1)	
NON-REBREATHER	OR (	0)	
T-PIECE	OR (	0)	
TRACH MASK	OR (	0)	
VENTI-MASK		(0)	1
PERI-CARE		(1)	1
TIME SPENT ==== MIN. (A)		(2)	25
AMBULATED	OR (	0)	
BED BICYCLE	OR (	0)	
CHAIR	OR (	2)	
DANGLED	OR (	0)	
STATIONARY BICYCLE	OR (	0)	
STOOD AT BEDSIDE	OR (	0)	
STROKE CHAIR		(0)	2
INCENTIVE SPIROMETER;		(3)	3
COUGH AND DEEP BREATHE		(3)	3
OTHER LINEN CHANGE		(1)	1
TURNED & POSITIONED		(1)	1
BACKRUB		(1)	1
SVO2: ==		(1)	1
IV - NO CHARGE			1
INTRAVENOUS			4
ORAL			1
INTRAMUSCULAR			1

```
06.4 HR,   53 % RN      full shift                       SORTG
(END)
```

Figure 13.14. Acuity report for a single patient. Statistics for full shift shown at bottom. CHARTED = number of times this item was charted during the shift, COUNTED = total number of minutes for this activity, OR = this item may be required because of orders.

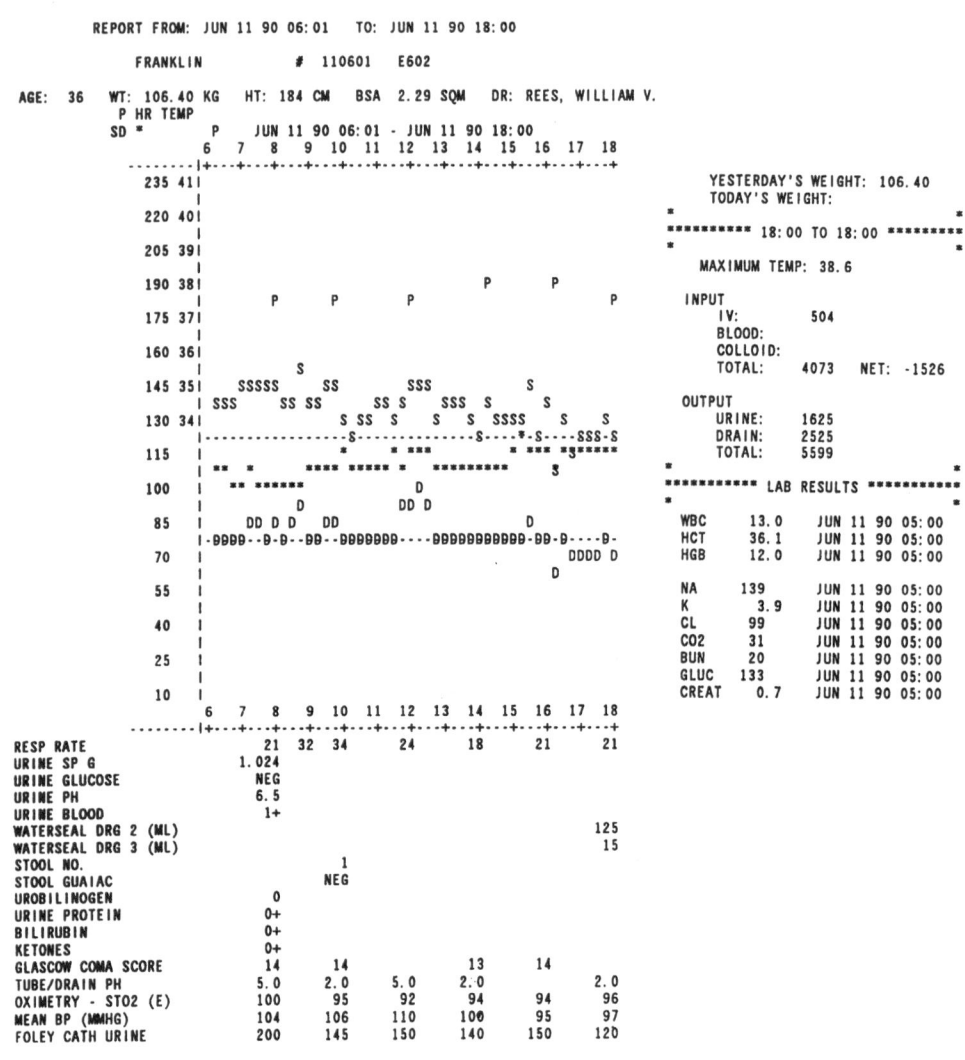

Figure 13.15. Nursing shift report (see text for details).

based terminals for computerized nurse charting. Nurses stated they welcomed the extra opportunity to be in the patient's room. Patients did not mind the presence of the terminal, reacting to it as another piece of equipment. Use of the bedside terminals showed less redundant charting (i.e., writing data on a piece of paper for entry into the computer at a later time) and higher (80% vs. 75%) real-time data entry (i.e., entry within 30 minutes of gathering the data) when compared with the use of nursing station-based

```
                                                        JUN 11 90 06:01 - JUN 11 90 18:00
                                                     6    7    8    9   10   11   12   13   14   15   16   17   18
                                                 -------------|+---+---+---+---+---+---+---+---+---+---+---+---+
NYSTATIN, POWDER                               1 APPLICTOPIC            1                        1
ALBUTEROL (PROVENTIL), INHALATION SOLUTION  2.50 MGM     INHAL               1              1              1
MAALOX EXTRA STRENGTH, LIQUID                 30 ML      NG                       1                                   1
METOCLOPRAMIDE (REGLAN), INJ                  10 MGM     IV                                 1                         1
SUCRALFATE (CARAFATE), TAB                  1000 MGM     NG              1                        1
TAP WATER, LIQUID                             10 ML      NG        2    3    2                        3              2
NORMAL SALINE, INJ                             3 ML      INHAL                    1         1              1
POTASSIUM CHLORIDE, LIQUID                    40 MEQ     NG              1
OSMOLITE, LIQUID                             200 ML      NG D    ******
  BLUE FOOD COLORING, ORAL                   0.4 ML
MIDAZOLAM (VERSED), INJ                     68.0 MGM     IV      ***************************
  INFUSION RATE                            10.00 MGM/HR         1111111111111111111111111
  NORMAL SALINE, INJ                          68 ML
POTASSIUM CHLORIDE, INJ                      1.2 MEQ     IV      *******************************************************

  INFUSION RATE                            5.000 CC/HOUR        1111111111111111111111111111111111111111111111111111111
  D5 IN 0.45 NACL, INJ                        59 ML
MAGNACAL, LIQUID                             300 ML      NG D                     ********************************************
MIDAZOLAM (VERSED), INJ                     61.0 MGM     IV                       ****************************
  INFUSION RATE                            10.00 MGM/HR                           1111111111111111111111111111
  NORMAL SALINE, INJ                          61 ML
                                                     6    7    8    9   10   11   12   13   14   15   16   17   18
                                                 -------------|+---+---+---+---+---+---+---+---+---+---+---+---+

INTAKE (ML):   NON-BLOOD IV        188    OUTPUT (ML):   INSENSIBLE LOSS       702
               ENTERAL FEEDING     620                   FOLEY CATH URINE      905
               NG DRUG              60                    WATERSEAL DRG, 2      125
               OTHER DRUG            9                    WATERSEAL DRG, 3       15
               OTHER IRRIGATION    540                   DRAINAGE
                                                         STOOL                   1
               TOTAL              1417                   TOTAL                 2553    NET BALANCE      -1136


JUN 11 08:00 GLASGOW COMA SCORE  14
JUN 11 09:45 GLASGOW COMA SCORE  14
JUN 11 14:15 GLASGOW COMA SCORE  13
JUN 11 16:10 GLASGOW COMA SCORE  14

        FRANKLIN          #  110601    E602

TIME OUT JUN 12 90 15:34
(END)
```

Figure 13.15. *Continued.*

terminals. An average waiting time of approximately 8 minutes per nurse per shift to use a terminal at the nursing station was absent when nurses used bedside terminals. When computerized charting was implemented, charting on actions specified by the nursing care plan increased from 30% to 60%. Additionally, legibility increased 26% and the frequency of notes being timed and dated increased 23%. Factors that were measured and were not found to be influenced by computerization were the average patient length of stay and the frequency of incidence reports.

The effect of implementing computerized nursing care plans was also evaluated. The presence of nursing care plans increased from 40% to 90% after computerization. The fraction of nursing care plans that were current (i.e., that had been updated within the past 48 hours) dropped slightly initially, but after 6 months had increased to 69%, which was 22% above preimplementation levels.

13.9 Computerized Nursing Reports

The nurse comments report (Figure 13.6) displays the results of the computerized nurse charting for the shift. The comments report becomes part of the permanent medical record. The computerized nursing shift report (Figure 13.15) contains a wide variety of objective patient data and also becomes part of the patient's permanent record. Patient identifying data are seen at the top with age, weight, height, body surface area, and the physician's name displayed next. On the upper left of the report is a graphical representation of the patient's temperature from a pulmonary artery catheter (P), heart rate (*), and systolic (S) and diastolic (D) blood pressures for the shift. To the right are data for the patient's weight, maximum temperature, inputs and outputs, and laboratory results. Below the graph are timed data for respiratory rate, urine testing, chest drainage, stool testing, Glasgow coma scores, oximetry, mean blood pressure, and urine output. Medication data are displayed next indicating the name of the drug, dosage, route, and time administered. Intravenous drips are displayed too, the number indicating the multiple of the dosage being administered. Intake and output data complete the report.

Figure 13.15 is for an ICU patient. A shift report for a patient on an acute care ward contains less information but has the same general layout. Shift reports become part of the permanent medical record but temporary copies may be printed at any time. They may be displayed on video terminals as well. The shift report is only possible because of HELP's integrated database, which permits data from a variety of sources to be viewed together and thus give a comprehensive view of the patient's overall current status.

References

[1] Pryor TA. Computerized nurse charting. *Intl J Clin Monit & Comput* 1989;6: 173–179.
[2] Cengiz M, Ranzenberger J, Johnson DS, Killpack AK, Lumpkin RW, Pryor TA. Design and implementation of computerized nursing care plans. *Symposium on computer applications in medical care (SCAMC)* 1983;7:561–565.
[3] *Taxonomy 1 – Revised 1989 with official diagnostic categories.* St. Louis, Mo; North American Nursing Diagnosis Association (NANDA).
[4] Gordon M. *Nursing Diagnosis: Process and Application. 2nd ed.* New York: McGraw-Hill; 1987.
[5] Killpack AK, Johnson DS, Pryor TA, Chapman RH, Ranzenberger J. Automating patient acuity from nursing documentation. *Symposium on computer applications in medical care (SCAMC)* 1984;8:709–711.
[6] Halford G, Pryor TA, Burkes M. Measuring the impact of bedside terminals. *Symposium on computer applications in medical care (SCAMC)* 1987;11:359–362.

14
Computerized Respiratory Care Charting

14.1 Overview

Respiratory care charting at LDS Hospital is done through the computer system. Respiratory therapists enter patient treatment data into the HELP system through a series of menu-driven screens. From these data, a variety of reports for medical, administrative, and QA purposes are generated. Billing is calculated directly from the computerized charting and accurately reflects services rendered. The therapists' notes are printed by the computer and become part of the permanent medical record.

Using preprogrammed logic, alerts are generated daily for the department's medical director concerning potentially dangerous medical conditions. Also, situations that violate the department's quality standards are determined by the computer and printed in a weekly report for the QA coordinator. The value of the computer lies in its ability to review large amounts of data (most of which is uninteresting) to find the few cases that may need medical attention. The system is well received and seems to improve productivity.

14.2 Data Stored

The respiratory therapists enter data into the computerized system through a menu-driven format [1]. After the therapists enter the patient's name and their own identifying number, the initial screen asks for the treatment rendered. The choices are:

1. pulmonary exercise
2. medication nebulizer
3. IPPB
4. chest physical therapy
5. oxygen

14. Computerized Respiratory Care Charting

6. nebulizer monitoring
7. nebulizer treatment
8. patient assessment
9. treatment not done
10. ICU treatment
11. nursery treatment
12. miscellaneous treatment
13. other (comments only)
14. home care

Depending on their response to the above screen, the therapists are presented with a series of screens that allow them to enter the data relevant to the treatment rendered. With most treatments they will be asked to enter data such as:

1. date and time of treatment
2. patient position during treatment
3. type of patient interface to respiratory equipment
4. medication received
5. patient condition during treatment
6. cough quality
7. sputum quality and quantity
8. pretreatment lung findings
9. posttreatment lung findings
10. heart rate (before, during, and after treatment)
11. respiratory rate (before, during, and after treatment)
12. comments
13. duration of treatment

If the choice on the initial screen is "Treatment not done" the computer will ask for a reason. If the initial choice is "Oxygen" the computer will ask for the data related to initiating or altering O_2 therapy. Similarly, various initial choices prompt questions specific to that area.

If a data item is not relevant to the particular treatment, it may be left blank. Free text may be entered in the "Comments" section for clarifying information. Therapists are encouraged to document physicians' verbal orders in the "Comments" section in case the written orders are not updated after a requested change.

14.3 Data Integration

In implementing its decision logic and generating reports for the respiratory care module, the HELP system uses computerized data from the patient's admission record, blood gas results, data from the CIDM (computerized infectious disease monitor), and isolation information from nursing charting in addition to the respiratory care charting data.

14.4 Decision Logic

A program has been implemented that monitors the quality of respiratory care at LDS Hospital [2,3]. The computer triggers an alert if it finds certain potentially adverse conditions.

The Medical Director's Alert Report (Figure 9.5) contains alerts related to physiologic results that are out of range (or not recorded) and may result in a poor outcome if not corrected. The logic here, although effective, may not be complex. Often, certain values are checked against a specified range.

Examples include PEEP/CPAP pressures greater than 10 cm H_2O, cuff pressures greater than 27 cm H_2O, humidification temperatures greater than 39°C, arterial blood gas with pH > 7.54, FiO_2 values greater than 0.60 for at least 16 hours, and PaO_2 values less than 55 mmHg without a change in oxygen therapy or repeat blood gases.

The computer also automatically screens incoming respiratory care data and generates an alert report for the quality care coordinator (Figure 9.6). This report identifies deviations from accepted policies and procedures on aspects of care such as ventilator management (documentation of set-up, tracheal cuff pressure monitoring, daily equipment changes, FiO_2 analysis, proximal airway temperatures), aerosolized bronchodilator treatments (documentation of heart rates, adverse reactions), and extubations (documentation of spontaneous ventilation parameters before extubation). Similar monitors are in place for other aspects of respiratory care.

14.5 Reports

A rich variety of respiratory therapy reports is produced for medical, administrative, and QA purposes. The key to the ease of generating these reports is that all the information is taken from the computerized charting and no re-entry of data or retrospective gathering has to be done.

1. The Medical Director's Report (Figure 9.5) is produced daily at 7:30 AM and is reviewed on morning rounds at 8 AM. This allows prompt attention to possibly adverse situations.

2. The QA Report (Figure 9.6) is produced weekly for the Respiratory Care Department's QA coordinator. This allows identification of errors within the department and appropriate counseling. A monthly summary (Figure 14.1) is also produced for review by the medical director and respiratory care QA committee.

3. Management reports are produced that chart individual therapist (Figure 14.2) and departmental (Figure 14.3) productivity. These reports detail activity by therapist, service rendered, and by clinical service. A report is also produced listing all scheduled treatments not administered and the reasons.

4. Physicians, nurses, or respiratory therapists may print or display on

```
M O N T H L Y    A L E R T    T O T A L S    F O R    M A Y

MAY 19, 1988

BRONCHODILATOR GIVEN WITH PRE HR > 130:                  69

BRONCHODILATOR GIVEN WITHOUT PRE, DUR, OR POST HR:       27

ADVERSE REACTION DURING BRONCHODILATOR THERAPY:           9

FIO2 (HPN) NOT ANALYZED WITHIN FIRST 4 HOURS OF SHIFT:    7

FIO2 (VENT) NOT ANALYZED WITHIN FIRST 4 HOURS OF SHIFT:  12

FIO2 (CPAP) NOT ANALYZED WITHIN FIRST 4 HOURS OF SHIFT:  13

PARAMETERS NOT DONE 3 HOURS PRIOR TO EXTUBATION:         14

ABNORMAL HUMIDIFICATION TEMPERATURE:                     44

USN EQUIPMENT NOT CHANGED AFTER 24 HOURS OF USE:         15

HPN EQUIPMENT NOT CHANGED AFTER 24 HOURS OF USE:          0

VENT EQUIPMENT NOT CHANGED AFTER 24 HOURS OF USE:        23

CPAP EQUIPMENT NOT CHANGED AFTER 24 HOURS OF USE:         3

CUFF PRESSURE NOT CHECKED WITHIN FIRST 4 HOURS OF SHIFT: 23

VITAL CAPACITY PERFORMED BY PATIENT WITH GLASCOW < 9:     1

CUFF PRESSURE GREATER THAN 27 CM H2O:                    16

TRACH CARE CATHETERS NOT CHANGED:                        22
```

Figure 14.1. Monthly summary of respiratory therapy quality assurance alerts.

the terminal a ventilator patient's latest spontaneous respiratory parameters (Figure 14.4). This report is generated with only a single keystroke at the appropriate menu.

5. A Corrective Action Alert Report (Figure 14.5) is generated to monitor initial equipment orders and supposed interruptions in oxygen therapy. This helps to ensure that all oxygen delivered is billed for. Also patient charges for O_2 of more than 24 hours in 1 day appear in this report so double billing is eliminated.

6. Part of the function of the Infectious Disease Monitor Report (Figure 9.4) is to produce a report of all nosocomial pneumonias (see Chapter 18 for further details). If a common organism is found in two patients fewer than 3 days apart the CIDM searches for a common respiratory therapist between them. The CIDM also identifies respiratory infections of concern to respiratory therapists such as tuberculosis, staphylococci, and gram-negative bacilli. These reports are presented to the medical director on morning rounds.

RESPIRATORY CARE THERAPIST REPORT   18 MAY 1988

24 HOUR MANAGEMENT REPORT

THERAPIST: DAVE   MAY 18, 1988   00:00   -   MAY 18, 1988   23:59

DATE	TIME	ROOM	PATIENT	TREATMENTS	RVU'S	UNITS	DURATION	TA	CT	ENTRY
05-18	00:35	N683	JOHN	O2	8.04	12	5	-5	.5	00:35
05-18	00:36	E637	GEORGE	O2	8.04	12	5	-5	.5	00:36
05-18	22:38	E433	PAUL	BLOOD GAS - CATH	5.47	1	10	467	1.5	06:34
05-18	22:10	E433	PAUL	VENTILATOR MONITORING	26.10	2	8	24	2.5	22:40
				ON O2	1.34					
05-18	21:21	E433	PAUL	VENTILATOR MONITORING	.00	0	0	0	.5	21:21
05-18	20:40	E433	PAUL	CPT (INFANT)	12.30	1	20	2	2.5	21:00
05-18	20:22	E433	PAUL	VENTILATOR MONITORING	26.10	2	8	-7	1.5	20:22
				ON O2	1.34					
05-18	22:20	E433	FREDERICKA	VENTILATOR MONITORING	26.10	2	8	16	2.5	22:42
				ON O2	1.34					
05-18	21:16	E433	FREDERICKA	BLOOD GAS - CATH	5.47	1	10	547	1.5	06:32
05-18	20:24	E433	FREDERICKA	VENTILATOR MONITORING	26.10	2	10	-9	1.5	20:24
				ON O2	1.34					
05-18	21:30	E432	SAMUEL	CPT (INFANT)	12.30	1	20	-1	2.5	21:47
05-18	06:21	E636	IMA	CPAP MONITORING	26.10	2	8	-7	1.5	06:21
				ON O2	1.34					
05-18	04:16	E636	IMA	CPAP MONITORING	26.10	2	8	-8	.5	04:16
				ON O2	1.34					
05-18	02:32	E636	IMA	CPAP MONITORING	26.10	2	8	-7	1.5	02:32
				ON O2	1.34					
05-18	00:08	E636	IMA	CPAP MONITORING	26.10	2	8	1	1.5	00:16
				ON O2	1.34					
05-18	23:40	E432	IMA	DELIVERY CALL	10.00	10	10	8	1.5	23:57
05-18	06:24	W620	FRANK	HPN MONITORING	11.95	5	20	-20	.5	06:24
				ON O2	3.35					
05-18	04:08	W620	FRANK	O2	2.01	3	5	-4	1.5	04:08
				O2 EQUIPMENT SET UP	13.67					
05-18	21:50	E433	WHITNEY	CPT (INFANT)	12.30	1	20	2	2.5	22:10
05-18	00:33	E630	ELISE	O2	8.04	12	5	-5	.5	00:33
05-18	01:30	DSCH*	SUSAN *DONOR*	OTHER	.00	1	2	52	.5	02:24
05-18	03:30	C459A	KIM	SVN	23.24	1	15	1	2.5	03:44
05-18	06:36	DSCH	JERRY	OTHER	.00	1	5	-4	1.5	06:36
05-18	06:25	DSCH	JERRY	HPN MONITORING	28.68	12	20	-19	1.5	06:25
				ON O2	8.04					
05-18	03:59	DSCH	JERRY	OTHER	.00	1	5	-5	.5	03:59
05-18	06:26	E632	PEGGY	HPN MONITORING	28.68	12	45	-45	.5	06:26
				ON O2	8.04					

Figure 14.2. Twenty four-hour management report for a single respiratory therapist. RVU's = relative value unit for procedures performed, UNITS = number of procedures performed, DURATION = actual length of time to perform procedure (as estimated by therapist), TA = turnaround time (time from performance of procedure to charting), CT = charting (computer) time, ENTRY = time of data entry.

7. The computer automatically prints a list of all patients in the hospital receiving oxygen therapy and notes which have not been assessed in the past 3 days (Figure 14.6). It is the responsibility of the home care coordinator (in our hospital this person is responsible for this duty, discharge planning, and other functions) to review the cases needing assessment, make sure

SHIFT TOTALS

	RVUs	POINTS	% OF TOTAL	CHARGES	% OF TOTAL	% OF RVUs COMPLETED	% CHARGES COMPLETED	RVU:TX MIN	TOTAL TX DUR	CHART DUR	ENTERIES
ADULT:	15323.20	510.8	78.6	15838.37	73.1	96.1	97.3	1.78	8589		
NBICU:	3366.52	112.2	17.3	3318.30	15.3	100.0	100.0	1.90	1769		
HOME:	171.07	5.7	.9	100.65	.5	100.0	100.0	1.14	150		
HBO:	642.55	21.4	3.3	2410.00	11.1	100.0	100.0	1.19	540		
TOTALS:	19503.34	650.1	100.0	21667.32	100.0	96.9	98.0	1.77	11048	1391	813
AVERAGES:	591.01	19.7							335	2	25

THERAPISTS: 33   AVG. TURNAROUND TIME: 93

TREATMENT TOTALS

TREATMENTS COMPLETED	TOTAL (INITIAL)		RVUs	CHARGE

ADULT THERAPY

I.S.:	116	(24)	2138.24	1592.00
SVN:	126	(3)	2977.44	1750.50
CPT:	58	(4)	1492.98	1180.40
O2:	2388 HRS		1599.96	6137.16
SUCTION:	2		27.34	13.00
SPUTUM INDUCTION:	1		27.34	20.15
INHALER:	11	(0)	150.37	110.00
MEMBRANE LUNG:	2		.00	.00
INTERHOSPITAL TRANSPORT:	90 MIN		90.00	108.00
APNEA ALARM MONITORING:	4 HRS		16.00	7.20
EXTUBATION:	3		41.01	18.00
MEDS - IN LINE:	12		98.40	165.00
INTRAHOSPITAL TRANSPORT:	40 MIN		40.00	48.00
THORACIC DEMO:	3		131.22	105.00
HPN MONITORING:	297 HRS		709.83	484.11
USN MONITORING:	250 HRS		597.50	407.50
CPAP CONTINUOUS:	30 HRS		391.50	295.50
VENTILATOR MONITORING:	196 HRS		2557.80	1930.60
RESP PARAMETERS:	8		360.88	96.00
I.S. ASSESSMENT:	30		738.30	304.50
CPT ASSESSMENT:	3		73.83	30.45
USN EQUIPMENT SET UP:	2		27.34	20.00
HPN EQUIPMENT SET UP:	7		143.57	70.00
VENTILATOR EQUIPMENT SET UP:	4		87.48	248.80
O2 EQUIPMENT SET UP:	41		560.47	492.00
VENTILATOR EQUIPMENT CHANGE:	6		90.00	.00
CPAP EQUIPMENT CHANGE:	2		30.00	.00
OTHER:	7		.00	.00
TRACH LAVAGE:	1		34.18	.00
TUBE/TRACH CARE:	5		68.35	.00
TRACH CARE CATHETER:	8		.00	192.00
			15323.20	15838.37

Figure 14.3. Respiratory therapy management report for a single shift. RVUs = relative value unit, POINTS = RVUs/30, DUR = duration, TX = treatment,

```
                NBICU THERAPY
                -------------------------
                CPT (INFANT):                 21          258.30    315.00
                AIRWAY INTUBATION:             1           20.51     10.00
                O2:                          102 HRS       68.34    262.14
                VENTILATOR MONITORING:       104 HRS     1357.20   1024.40
                TcmCO2 MONITORING:           120 HRS      452.40    750.00
                TcmO2 MONITORING:             60 HRS      226.20    360.00
                HEADBOX MONITORING:           12 HRS       82.08     27.60
                PULSE OXIMETER MONITORING:   138 HRS      471.96    333.96
                VENTILATOR EQUIPMENT SET UP:   1           21.87     62.20
                TcmCO2 EQUIPMENT SET UP:       2           27.34     40.00
                VENTILATOR EQUIPMENT CHANGE:   4           60.00      .00
                DELIVERY CALL:               190 MIN      190.00    133.00

                STANDBY:                      10 MIN       10.00      .00
                TUBE/TRACH CARE:               2           27.34      .00
                BLOOD GAS, CATH:               6           32.82      .00
                BLOOD GAS, PUNCTURE:           4           60.16      .00
                                                        -----------------------
                                                         3366.52   3318.30
                HOME CARE
                -------------------------
                O2 (IN-TOWN):                  1          141.00     70.50
                INHALER INSTRUCTION:           1           30.07     17.50

                OXYMISER:                      1            .00      12.65
                                                        -----------------------
                                                          171.07    100.65
                HBO THERAPY
                -------------------------
                HYPERBARIC CHAMBER:          540 MIN      540.00   2160.00
                HBO EQUIPMENT SET UP:          5          102.55    250.00
                HBO EQUIPMENT CHANGE:          5            .00       .00
                                                        -----------------------
                                                          642.55   2410.00
                                                        =======================
                TOTALS                                 19503.34  21667.32

                REASONS TREATMENTS NOT DONE
                -------------------------------------------------------------------

                ADULT THERAPY
                -------------------------
                NOT ON UNIT:                   6          166.79    115.00
                ASLEEP:                        1           15.04     10.00
                EATING:                        1           23.24     13.75
                RECEIVING OTHER CARE:          4          135.35     97.50
                NAUSEATED:                     2           38.28     23.75
                REFUSED CARE:                  4           69.73     50.00
                ADVISED NOT TO GIVE:           1           24.61     20.00
                OTHER:                         5          149.02    101.25
                                                        =======================
                TOTALS                        24          622.06    431.25
```

Figure 14.3. *Continued.*

I.S. = incentive spirometry, SVN = small volume nebulizer, CPT = chest physical therapy, HPN = heated pneumatic nebulizer, USN = ultrasonic nebulizer, TcmCO2 = transcutaneous CO_2 monitoring.

```
                      SPONTANEOUS RESPIRATORY PARAMETERS
                                 MAY 20, 1988

        JOHN                          E609                  PATIENT ID#: 402071
   DR. STEVENS, MARK H          SEX: M   AGE: 72            ADMITTED: MAY 05, 1988

   DATE     TIME    HR   VR   VT   VE    VC    MIP              COMMENTS
   ===========================================================================

   05/20/88 08:30   92   29  438  12.7  1400   -60  ALERT, COOPERATIVE, PT
                                                    OBSERVED FOR 10 MIN PRIOR
                                                    TO DATA COLLECTION, STERNAL
                                                    RETRACTIONS NOTED DURING
                                                    SPONTANEOUS BREATHING.   -
                                                    GORDON, STEVE
   05/18/88 09:50  107   33  436  14.4  1100   -38  SITTING, ALERT, FATIGUED,
                                                    COOPERATIVE, PT HAS BEEN UP
                                                    IN CHAIR FOR 1 HOUR,
                                                    APPREATS FATIGUED AND MAY
                                                    HAVE DONE BETTER HAD HE NOT
                                                    BEEN TIRED.  -
                                                    HAIDENTHALLER, BRENDA
   05/17/88 07:37   89   23  390   9.0  1800   -48  SEMI-FOWLER, ALERT,
                                                    COOPERATIVE, - CARPENTER,
                                                    LORI
   05/16/88 23:50   95   16  420   6.7  1500   -60  SEMI-FOWLER, CALM,
                                                    COOPERATIVE, GOOD EFFORT.
                                                    PT PUT BACK ON VENT TO REST
                                                    FOR NIGHT.^@  - WARNOCK,
                                                    ROSALEE
   05/16/88 07:42   83   25  340   8.5         -60  FOWLER, NOT RESPONDING, -
                                                    CARPENTER, LORI
   05/15/88 16:22   97   32  510  16.3  1700   -58  SEMI-FOWLER, ALERT,
                                                    COOPERATIVE, - MYERS, LARRY
   05/15/88 07:10  103   40  405  16.2         -62  SEMI-FOWLER, NOT
                                                    RESPONDING, WOULD NOT DO
                                                    VC  - MYERS, LARRY
   05/14/88 14:15  124   40  405  16.2         -60  FOWLER, RESTLESS, - LEWIS,
                                                    ROGER
   05/14/88 08:13  104   28  455  12.7         -60  FOWLER, NOT RESPONDING, -
                                                    LEWIS, ROGER
   05/09/88 07:19   90   20  590  11.8  1260   -50  FOWLER, ALERT, COOPERATIVE,
                                                    - LEWIS, ROGER
```

Figure 14.4. Example of printout of spontaneous respiratory parameters.

oxygen is still clinically indicated, and ensure the patient is actually receiving the prescribed therapy.

8. Similarly, patients receiving incentive spirometry (IS) are evaluated every 3 days to see whether their regimen is still appropriate. The computer prints the list of patients to be evaluated on a given day (Figure 14.7).

9. While the patient is in the hospital, the respiratory care charting data are kept in "electronic" form with printed review reports available at any time (Figure 14.8). At discharge, a copy of the respiratory care charting is printed in the Medical Records Department for inclusion in the final chart.

10. A Patient Locator (Figure 14.9) is generated each shift for each unit and allows the supervisor to be aware of all the patients active on the service, their diagnosis, and the therapy they are due to receive.

11. A program is under development (Figure 14.10) that will make the

```
                              RESPIRATORY   CARE
                           CORRECTIVE ACTION ALERT

                            RUN: 10:09  MAY 20 1988
                       REVIEWS INFORMATION BEGINNING MAY 19

                                        CPAP  VENT  NEB   O2    O2    O2
       ROOM        NAME                  HRS   HRS   HRS   HRS  CONT  TOTAL
       ------------------------------------------------------------------

       N683        HOWARD                                 24    48    48


       E433        MELVIN              4    16           20    56    56
              CONT O2 DC'D OR INTERRUPTED YESTERDAY
              VENT AND CPAP CONT MONITORING NOT CONTINUOUS YESTERDAY


       E608        ASHTABULA                 36          36    84    84
              > 24 HRS O2 CHARGES / DAY
              > 24 HRS VENT CHARGES / DAY
              > 24 HRS O2 CHARGES / DAY  2 DAYS AGO


       E716        FREDERICK           4    28           32    68    68
              > 24 HRS O2 CHARGES / DAY
              > 24 HRS VENT CHARGES / DAY
              INITIAL CPAP EQUIP SET-UP NOT DOCUMENTED
              - CPAP FIRST DOCUMENTED: 05/19 20:25
                THERAPIST: WHITE, LAWRENCE
              INITIAL SVN TX & EDUCATION NOT DOCUMENTED
              - SVN FIRST DOCUMENTED: 05/19 20:38
                THERAPIST: WHITE, LAWRENCE


       E707        ASHTON                    8           20    20    20
              VENT MONITORING NOT CONTINUOUS YESTERDAY
              INITIAL O2 EQUIP SET-UP NOT DOCUMENTED
              - O2 FIRST DOCUMENTED: 05/19 12:15
                THERAPIST: WILKERSON, LYNN


       E601        MICHAEL                   2            2     2     2
              VENT MONITORING NOT CONTINUOUS YESTERDAY
              INITIAL VENT EQUIP SET-UP NOT DOCUMENTED
              - VENT FIRST DOCUMENTED: 05/19 22:20
                THERAPIST: HOLVERSON, JEFF
```

Figure 14.5. Corrective action alert report indicating irregularities in oxygen therapy. CPAP HRS = hours on continuous positive airways pressure, VENT HRS = hours on ventilator, NEB HRS = hours on nebulizers, O2 HRS = hours on oxygen in past 24 hours, O2 CONT = continuous hours on oxygen, O2 TOTAL = total hours on O_2 for this hospital stay.

therapist's assignment schedule and beeper numbers available via the computer. The goal is to reduce the problem of identifying and contacting the therapist assigned to a specific area, in either routine circumstances or, more importantly, in emergency situations.

Additionally, all routine billing is done from computerized charting. As mentioned in Chapter 3, billable procedures are recorded in a transaction file that is transferred nightly to the IBM financial computer.

```
                        RESPIRATORY  CARE
                        O2  ASSESSMENTS

                     RUN: 10:09  MAY 20 1988
                 REVIEWS INFORMATION BEGINNING MAY 19
```

ROOM	NAME	LAST ABG	PAO2	SAO2	FIO2
N690	FREDERICK				

NO ABG'S HAVE BEEN DRAWN

ROOM	NAME	LAST ABG	PAO2	SAO2	FIO2
N477	DOE, GEORGE	14.13:40	98.3	97.9	40
N683	EGBERT	20.08:00	85.6	96.9	3

*** O2 ASSESSMENT NOT DONE IN 3 DAYS ***

ROOM	NAME	LAST ABG	PAO2	SAO2	FIO2
W633	FRANKLIN	18.06:10	60.2	92.9	50
E741	DOE, HARRIET	15.18:50	60.7	92.9	21

CHECK O2 MEDICAL NECESSITIES
NO ABG'S IN THE LAST 4 DAYS

ROOM	NAME	LAST ABG	PAO2	SAO2	FIO2
E837	JACQUELINE	17.07:40	52.2	89.0	21

NO ABG'S IN THE LAST 2 DAYS

ROOM	NAME	LAST ABG	PAO2	SAO2	FIO2
W840	OUTTA	15.18:35	94.4	98.1	4

NO ABG'S IN THE LAST 4 DAYS
*** O2 ASSESSMENT NOT DONE IN 3 DAYS ***

NO ABG'S IN THE LAST 4 DAYS
*** O2 ASSESSMENT NOT DONE IN 3 DAYS ***

ROOM	NAME	LAST ABG	PAO2	SAO2	FIO2
N679	LOTSA				

*** O2 ASSESSMENT NOT DONE IN 3 DAYS ***

ROOM	NAME	LAST ABG	PAO2	SAO2	FIO2
W818	HENRY	15.21:19	42.3	82.4	21

NO ABG'S IN THE LAST 4 DAYS
*** O2 ASSESSMENT NOT DONE IN 3 DAYS ***

Figure 14.6. List of patients requiring assessment of their oxygen therapy.

14.6 Evaluations

A study [1] was done before and after implementation of the system at LDS Hospital to evaluate therapists attitudes and the effect on certain aspects of care.

Therapists were generally happier with computer charting than with the manual method (64% vs. 20%). More than three quarters of those responding felt it would make their job easier, whereas 74% felt their productivity was better after implementation. The quality of documentation with the

```
                         LDS HOSPITAL
                       RESPIRATORY THERAPY
                  I.S. EVALS DUE MAY 21, 1988
```

ROOM	PATIENT	PATIENT #	PRED. IC	LAST EVAL
E736	ALAN	23223		NO ASSESSMENT
E812	JERRY	25658	2210	05/18 20:45
N674	KIM	25606	2500	05/18 16:45
N675	SUSAN	25577	2440	05/18 19:30
N679	KEOKI	40211	2400	05/18 13:00
N684	WHITNEY	25606	4000	05/18 17:10
N685	ELISE	25584	2220	05/18 19:00
N692	GEORGE	23212	2290	05/18 10:45
N694	BARBIE	25577	3000	05/18 11:21
W319	JOE	25645	2160	05/18 08:53
W321	TERRI	23213	3960	05/18 20:30
W325	FRED	402130	3730	05/18 21:15
W333	ELLEN	23152	3720	05/18 11:05
W338	REPHAEL	23206	1600	05/18 16:30
W343	JENNIFER	23171	3960	05/18 17:15
W605	SUKI	4021274	3300	05/18 20:10
W610	IMA	23197	2300	05/18 07:50
W619	HIROSHI	23213	3150	05/18 11:00
W620	DOSIE	23234	3150	05/18 13:20
W704	HODIE	25580		NO ASSESSMENT
W718	WORTHSY	25616		NO ASSESSMENT

Figure 14.7. Incentive spirometry (IS) report. List of all patients requiring IS evaluation. PRED. IC = predicted inspiratory capacity.

computer system equalled or exceeded manual charting. Legibility was greatly improved.

Recorded productivity of the Respiratory Care Department increased 18.2% with the system, despite the need for therapists to enter the data into the computer. It is unclear whether this was due to increased charge capture or actual increased efficiency. Audit of only one aspect of the system, the Corrective Action Alert (described above), was shown to account for $80,000 in what would have been lost but deserved revenues.

Another study [2] has documented the frequency of medical alerts to

LDS HOSPITAL   RESPIRATORY CARE CHARTING
MAY 20, 1988

FRED
TRAUMA

AGE: 72   SEX: M   E609
DR. STEVENS, MARK H

PATIENT ID#: 402071
ADMITTED: MAY 05, 1988

05/20/88

VENTILATOR MONITORING

	VENT MODE	VR	Vt	O2	PF	IP	TEMP	IE RATIO	PK	PL	MAP	PP	m-Vt	c-Vt	s-Vt	MR	SR	TR	m-VE	s-VE	t-VE	Cth	aOX	vOX	Pc	CF
20 12:54	B-I A/C	12	800	50	40		36.0	1: 4.2	20	17		5	860	796		13			10.3			66.3	94			4.3
20 09:34	B-I A/C	12	750	50	40		37.0	1: 3.1	21	16		5	800	731		14			10.2			66.5	95			4.3
20 08:00	B-I A/C	12	750	.50	40		35.0	1: 4.0	23	17		5	860	783		13			10.2			65.2	97		20	4.3
20 05:57	B-I A/C	12	750	40	50		37.0	1: 3.8	21	17		5	810	741		13			9.6			61.8	95			4.3
20 02:10	B-I A/C	12	750	50	40		37.0	1: 3.9	23	18		5	780	703		14			9.8			54.0	92			4.3
20 00:15	B-I A/C	12	750	50	40		37.0	1: 4.0	21	16		5	820	751		13			9.8			68.3	92			4.3

VENTILATOR OBSERVATIONS

05/20/88 DUR/ENTRY

20 12:54 10/12:56   INTERFACE: OROTRACH; ALARMS CHECKED; TEMP SETTING: 5.5; POSITION: FOWLER; PATIENT CONDITION: QUIET;
THERAPIST: GORDON, STEVE

20 09:34 10/09:37   INTERFACE: OROTRACH; ALARMS CHECKED; TEMP SETTING: 5.5; POSITION: SUPINE; PATIENT CONDITION: QUIET;
THERAPIST: GORDON, STEVE

20 08:30 15/08:33   -RESPIRATORY PARAMETERS-

HR	RR	VC	VT	VE	MIP	MEP	MVV	PK FLOW
92	29	1400	438	12.7	-60			

PATIENT CONDITION: ALERT, COOPERATIVE; COMMENT: PT OBSERVED FOR 10 MIN PRIOR TO DATA COLLECTION, STERNAL RETRACTIONS
NOTED DURING SPONTANEOUS BREATHING. THERAPIST: GORDON, STEVE

20 08:00 10/08:02   INTERFACE: OROTRACH; ALARMS CHECKED; TEMP SETTING: 5.5; POSITION: FOWLER; PATIENT CONDITION: QUIET;
THERAPIST: GORDON, STEVE

```
20 05:57  20/05:59  INTERFACE: OROTRACH;  ALARMS CHECKED;  TEMP SETTING:  5.5;  POSITION: FOWLER;  PATIENT CONDITION: CALM;
                    THERAPIST: HOLVERSON, JEFF
20 02:10  20/02:16  INTERFACE: OROTRACH;  ALARMS CHECKED;  TEMP SETTING:  5.5;  POSITION: FOWLER;  PATIENT CONDITION: CALM;
          SUCTIONED, 2 CC, YELLOW, MUCOID, VISCID;  THERAPIST: HOLVERSON, JEFF
20 00:15  20/00:16  INTERFACE: OROTRACH;  ALARMS CHECKED;  TEMP SETTING:  5.5;  POSITION: FOWLER;  PATIENT CONDITION: CALM;
                    THERAPIST: HOLVERSON, JEFF

05/20/88.04:10      -MEDICATION NEBULIZER (IN-LINE) (Q4 HRS)-  -CPT (Q4 HRS)-
PATIENT INTERFACE: IN LINE WITH VENT;  POSITION: FOWLER;  MEDICATION: ALBUTEROL, 2.5 MG, DILUENT: NORMAL SALINE;  PRE BREATH
SOUNDS: COARSE CRACKLES, RHONCHI;  MECHANICAL PERCUSSION: FOR 15 MIN, LEFT/FLAT, RIGHT/FLAT, BLL;  COUGH: WHEN ENCOURAGED,
MODERATE, PRODUCTIVE;  SPUTUM: SUCTIONED, 3 CC, CLOUDY, YELLOW, THICK;  POST BREATH SOUNDS: IMPROVED;  PATIENT CONDITION: CALM,
COOPERATIVE;  COMMENT: RX TOL WELL;
          HR: 89/ 90/ 91                         DUR/ENTRY: 30/04:38      THERAPIST: HOLVERSON, JEFF

05/20/88.00:15      -MEDICATION NEBULIZER (IN-LINE) (Q4 HRS)-  -CPT (Q4 HRS)-
PATIENT INTERFACE: IN LINE WITH VENT;  POSITION: FOWLER;  MEDICATION: ALBUTEROL, 2.5 MG, DILUENT: NORMAL SALINE;  PRE BREATH
SOUNDS: COARSE CRACKLES;  MECHANICAL PERCUSSION: FOR 15 MIN, LEFT/FLAT, RIGHT/FLAT, BLL;  COUGH: FREQUENT, STRONG, PRODUCTIVE;
SPUTUM: SUCTIONED, 3 CC, MUCOID, YELLOW, VISCID;  POST BREATH SOUNDS: IMPROVED;  PATIENT CONDITION: CALM, COOPERATIVE;  COMMENT: RX
TOL WELL;
          HR: 92/ 95/ 93                         DUR/ENTRY: 30/01:01      THERAPIST: HOLVERSON, JEFF

05/20/88.00:00      -OTHER-
COMMENT: EQ NOT CH DUE TO NEW THIS PM.;          DUR/ENTRY: 5/04:25       THERAPIST: HOLVERSON, JEFF
```

Figure 14.8. Sample of respiratory care charting printout. VR = ventilatory rate, Vt = tidal volume, O2 = fraction inspired oxygen, PF = peak flow, IP = inspiratory pressure, TEMP = airways temperature, I:E = length of inspiratory phase to expiratory phase ratio, PK = peak pressure, PL = plateau pressure, MAP = mean airways pressure, PP = positive end expiratory pressure, m-Vt = machine tidal volume, c-Vt = corrected tidal volume, s-Vt = spontaneous tidal volume, MR = machine rate, SR = spontaneous rate, TR = total rate, m-VE = machine minute ventilation, s-VE = spontaneous ventilation, t-VE = total minute ventilation, Cth = thoracic compression, aOX = arterial oxygen, vOX = venous oxygen, Pc = cuff pressure, CF = compliance factor of tubing.

```
                    SHOCK-TRAUMA ICU
                    MAY 20, 1988
                        13:14
   E603    . ,FRED              402075    SEX: M  AGE: 47
      ADMIT DATE: MAY 05, 1988  DR. SWENSEN, SWEN
      ADMIT DIAGNOSIS: TRAUMA, HEAD INJURY
            -CPT-               Q8 HRS

   E604     GEORGE             402124    SEX: M  AGE: 44
      ADMIT DATE: MAY 16, 1988  DR. RICHARDS, KENT F.
      ADMIT DIAGNOSIS: TRUAMA; MULTIPLE INJURIES
            O2 CANNULA @ 3 L/M             05/20 06:09

   E605     PETER              23194.    SEX: M  AGE: 60
      ADMIT DATE: MAY 06, 1988  DR. SORENSON, CHARLES W.
      ADMIT DIAGNOSIS: URETERAL LEAK
            -CPT-               Q4 HRS

   E606     ROGER              23211    SEX: F  AGE: 50
      ADMIT DATE: MAY 10, 1988  DR. RICHARDS, L. STEPHEN JR.
      ADMIT DIAGNOSIS: CORONARY ARTERY DISEASE
            BEAR I: A/C  12,  600,  60%,
            PP:  8.  OROTRACH            05/20 11:56

   E607     KEITH              23248    SEX: F  AGE: 40
      ADMIT DATE: MAY 19, 1988  DR. PEARL, JAMES E.
      ADMIT DIAGNOSIS: ACUTE ARDS
            900c: VOLUME CONTROL  10,   50%,
            PP: 12.  NASOTRACH           05/20 13:05

   E608     JOHN               23202.   SEX: F  AGE: 26
      ADMIT DATE: MAY 08, 1988  DR. PEARL, JAMES E.
      ADMIT DIAGNOSIS: ADULT RESPIRATORY DISTRESS SYNDROME
            -MEDS-             Q4 HRS
            900c: VOLUME CONTROL  30,   80%,
            PP: 14.  NASOTRACH           05/20 12:58

   E609     PAUL     402071    SEX: M  AGE: 72
      ADMIT DATE: MAY 05, 1988  DR. STEVENS, MARK H
      ADMIT DIAGNOSIS: TRAUMA
            BEAR I: A/C  12,  800,  50%,
            PP:  5.  OROTRACH            05/20 12:54
            -CPT-               Q4 HRS
            -MEDS-              Q4 HRS
```

Figure 14.9. Sample of patient locator report.

potentially dangerous conditions in our 475-bed, tertiary care hospital. An average of 26 patients per month were identified as having $PaO_2 <$ 55mmHg without a follow-up blood gas or an increase in their FiO_2. Thirty one patients per month were documented as having an FiO_2 of greater than 0.60 for longer than 16 hours.

14.7 Future Directions

At this time the Respiratory Care Department is examining the possibility of having respiratory care orders entered directly into the HELP system. Currently, orders reach the Respiratory Therapy Department through a

```
              R E S P I R A T O R Y    C A R E
              BEEPER / AREA ASSIGNMENTS

      ADMINISTRATIVE CALL    GREENWAY, LOREN        4822293
      SUPERVISOR             AUSTIN, ANITA              452
      LIFE FLIGHT           LEWIS, ROGER              1070
      HYPERBARIC OXYGEN     ENGER, KIP             4822382
      STRICU                MYERS, LARRY               802
                            GORDON, STEVE              562
      MSICU                 ESKELSON, MAX              522
      TICU                  HOSS, NEAL                 594
                            ZIMMER, WILLIE             592
      CCU                   WILKERSON, LYNN            460
      NBICU                 STOUT, BECKY               472
                            DAVIS, TERIANNE            472
      E8, W8                STAPLEY, TERI              542
      W7                    WHITING, SHARON            582
      E4, W4                DAVIS, TERIANNE            472
      ER, W3                MCKEAN, JOAN               462
      N4, N5, N6            WONG, AMY
                            WOLDEN, JONI               742

   <RETURN> TO CONTINUE, P TO PRINT
```

Figure 14.10. Beeper/area assignment list.

paper-based system. With computerization, not only could the services that are not being rendered be determined, but also the orders to which those services corresponded.

References

[1] Andrews RD, Gardner RM, Metcalf SM, Simmons D. Computer charting: an evaluation of a respiratory care computer system. *Resp Care* 1985;30:695–707.

[2] Elliot CG, Simmons D, Schmidt CD, et al. Computer-assisted medical direction of respiratory care: computer-assisted medical direction of respiratory care. *Resp Manage* 1989;19:31–35.

[3] Elliot CG. Computer-assisted quality assurance: Development and performance of a respiratory care program. *QRB* 1991;17:85–90.

15
The Pharmacy Application
of the HELP System

15.1 Overview

The pharmacy application of the HELP system serves two main functions. First, the application manages medication data for inpatients at LDS Hospital. Pharmacists and nurses enter medication scheduling data into HELP and the data automatically appear on the medication administration records (Figure 13.12) and will appear on the nurses' computerized work planner when development of the planner is complete (see Chapter 13 for further details about the planner). Medication administration data are also entered into HELP and these data are used for billing purposes. The record of administered medications is available to other HELP decision-making programs. The second major function of the pharmacy module is a computerized medication monitoring system that automatically screens the medication scheduling data and other computerized patient data for possible adverse reactions. The program screens for drug–drug, drug–lab, drug–allergy, drug–diet, and other types of interactions. Certain alerts are displayed on the video terminal at the time of medication scheduling, whereas others are displayed in reports generated at a given time each day. The hospital pharmacists are responsible for informing the physician about the alert. The pharmacy module also provides assistance with ordering TPN and calculating appropriate aminoglycoside dosages.

Research is under way to examine the possibility of having physicians enter medication orders directly into the computer. This system will be intelligent in that it will have rules that will assist physicians with the ordering process. A knowledge frame editor that allows pharmacists to write the system's logic has been developed.

15.2 Data Flow

Physicians at LDS Hospital currently handwrite medication orders onto standard order sheets that are part of the patient's medical record. The medication orders are entered into HELP through a scheduling program by either the clinical pharmacist assigned to that division or by the patient's nurse. The scheduling program has a verification feature so that if the clinical pharmacist schedules the drug, the nurses perform the verification and vice versa. Medication administration data are entered into the HELP system by the patient's nurse.

To schedule a medication, the patient's hospital number is entered into HELP and the computer displays the patient's name and asks for confirmation. The drug name (trade or generic, or code number), dose, route, and schedule of administration are then entered. The entire prescription is redisplayed for confirmation. When drugs are discontinued they are also entered into the computer to keep the record current. Review of the chart from any terminal will show discontinued drugs in addition to current ones.

Allergies written on admitting orders are entered into the system by the pharmacist. They are also available to the system from other sources (e.g., the long-term patient file described in Chapter 3).

15.3 Decision Logic and Data Integration

Every time a drug is scheduled, the order is subjected to the logic criteria of the HELP system to see if an alert is warranted [1,2]. Data are used from the drug entry as well as from the rest of the patient file. The knowledge base that provides the logic is maintained by physicians, pharmacists, and pharmacologists and is based on their experience and review of the literature. The hospital Pharmacy and Therapeutics Committee is taking an increasingly active role for review and approval of the decision logic.

Interactions that generate alerts are drug–drug, drug–allergy, drug–lab, drug–disease, drug–diet, drug–dose, and drug–interval. Examples of the types of alerts generated are shown in Figure 9.3.

In addition to the pharmacy data, the system accesses information from the entire computerized patient file to generate its alerts. Specifically, the medication monitoring system uses: (1) height, weight, sex, and age, (2) blood pressure and temperature, (3) laboratory, blood gas, and ECG results, and (4) surgical procedures.

Certain alerts (or messages) are presented to the nurse or pharmacist at the time the medication order is entered into the system. Others (such as those that require laboratory values for full evaluation) are produced in report form at a given time each day for review by the pharmacist. The decision-making capabilities of the HELP system are automatic and activating the system requires no additional effort on the part of hospital personnel after orders are entered.

Examples of messages are: (1) "Suggest that serum creatinine be checked every 3 days when using an aminoglycoside" if an aminoglycoside is ordered and a creatinine level has not been checked recently, or (2) "Suggest potassium supplement, as low serum potassium concentration may result in digitalis toxicity" if the potassium is low and digitalis is ordered.

Ultimately, it is the responsibility of the clinical pharmacist to follow up on alerts. He/she may contact the physician directly, leave a copy of the alert with a note in the patient's chart, contact the medication nurse, or refer the alert to another pharmacist. The pharmacist determines whether any given alert requires action on the physician's part or if it is just informational. This determination depends on the patient's clinical state. In 1987, 52% of all alerts were felt to require action. The rate of physician compliance to the actional alerts is shown in Chapter 9 in Figure 9.2 and is now greater than 97%. A graph of the number of alerts per month from March, 1987 to March, 1988 is shown in Figure 9.1. A report of the total number of alerts, and the breakdown by category, is compiled in a monthly log shown in Table 15.1.

Additionally, copies of the alerts generated by the CIDM (discussed in Chapter 18) [3,4] regarding appropriate use of therapeutic and prophylactic antibiotics are sent to the pharmacy. It is the responsibility of the pharmacist to review and take action on these alerts as well.

15.4 Results of Studies

In a 16-month study [1], 88,505 medication orders were entered on 13,727 patients. Of these, 690 orders (0.8%) generated alerts involving 5.0% of the patients. The causes of the alerts were distributed as follows: (1) drug–drug interactions, 30%, (2) drug–lab, 45%, (3) drug–allergy, 16%, (4) drug–dose, 7%. Data in this study are from 1975 but analysis of recent data show the relative frequencies have not changed. In a separate 12-month study [5], alerts were generated on 8.45% of the patients (227/2684). Forty nine patients (1.8%) received an alert that was considered life threatening.

In another study [2], therapeutic antibiotic alerts were studied separately. In a 12-month period HELP automatically detected mismatches between antibiotic sensitivity studies and the therapeutic regimen 696 (32%) out of

Table 15.1. Pharmacy Medication Alert Report for March 1988

	March	Y.T.D.
Total number of valid alerts	183	774
Total number of action alerts	113	399
Total number of information alerts	70	375
Compliance on action alerts	99.1%	99.2%
Total number of drug–allergy alerts	14	67
Total number of drug–drug interactions	79	356
Total number of drug–lab alerts	90	351
Total number of HELP Sectors activated	45	—

2157 times. Four hundred twenty of the 696 alerts were judged to be true alerts. (Colonization and contamination of specimens accounted for the majority of the false alerts.) When physicians were contacted directly regarding the alerts, therapy was changed in 125 (30%) of the 420 cases. In the 70% of cases where therapy was not changed, the physician was either using an antibiotic not tested for on the routine susceptibility studies or the physician felt that the clinical response justified continuation of the current regimen.

The system has received a positive response from physicians, nurses, and pharmacists.

15.5 Nonalert-Related Functions

The HELP system allows physicians to enter TPN orders from the terminals. It provides a menu from which physicians can order components while simultaneously displaying current laboratory values for easy reference. TPN ordering is discussed fully in Chapter 22.

The system also provides a function to assist with calculating aminoglycoside dosages. Using patient data (already in the system) and knowledge of drug kinetics, it calculates expected peak and trough values for various dosage regimens. This allows the physician to consider multiple options rapidly while choosing the most efficacious regimen.

15.6 Future Directions

A module is being developed that will allow physicians to order medications through the computer [6]. Under this system, the gathering of medication orders from physicians will be "intelligent" in a number of ways. When the physician enters the name of the drug, the computer examines a knowledge base to find the suggested dosage, route, and schedule for the administration of that drug and will display it to the physician. These parameters are based on patient-specific data such as sex, age, weight, laboratory values, and others. Comments regarding use of the drug (e.g., loading doses, special uses of the drug, etc.) will also be displayed.

The module will use a knowledge base to examine the incoming medication order, compare it to other data present in the computerized patient file, and, if warranted, generate an alert for the physician. These alerts may refer to abnormal laboratory values (that may be affected with use of the drug), other currently ordered drugs (that may interact with the one being considered), or allergies (that are listed on the patient record). The computer will ask the physician if he/she wants to continue with the order in view of the information.

If the displayed prescription is acceptable to the physician, he/she need only press the enter key to order the medication. If an alteration is required, a simple method has been devised to allow the physician to change any parameter easily (e.g., route, dosage, schedule, starting time, etc.) before entering the order. Only changes that conform to a specified range or set of values (that may depend on patient-specific data) will be allowed. This will prevent gross medication errors. In special circumstances physicians will be allowed to override the limits.

To help keep the pharmacy knowledge base up to date, the Pharmacy Medication Ordering Knowledge Frame Editor (PMOKFE) has been designed and is being tested. The program will allow pharmacy personnel to enter the data for the suggested routes, schedules, and doses into the computer's knowledge base without the help of a programmer. The pharmacists will also be able to input the information needed for the computer to generate alerts. The goal is for pharmacists to be able to create and maintain their own departmental knowledge base. PMOKFE functions on a personal computer.

References

[1] Hulse RK, Clark SJ, Jackson JC, Warner HR, Gardner RM. Computerized medication monitoring system. *Am J Hosp Pharm* 1976;33:1061–1064.
[2] Pestotnick SL, Evans RS, Burke JP, et al. Therapeutic antibiotic monitoring: surveillance using a computerized expert system. *Am J Med* 1990;88:43–48.
[3] Evans RS, Hulse RK, Gardner RM, et al. Computer surveillance of hospital-acquired infections and antibiotic use. *JAMA* 1986;256:1007–1011.
[4] Evans RS, Gardner RM, Bush AR, et al. Development of a computerized infectious disease Monitor (CIDM). *Comput Biomed Res* 1985;18:103–113.
[5] Pryor TA, Gardner RM, Clayton PD, et al. The HELP system. *J Med Systems* 1983;7:87–102.
[6] Prokosch HU, Hulse RK, Wall M, Wong TW, Pryor TA. New decision support concepts for the pharmacy application in the HELP system. EFMI Special Topic Congress, Hannover, West Germany, September, 1988.

16
The LDS Hospital
Laboratory System[1]

LDS Hospital and its managing corporation, IHC, have adopted a unique strategy to solve the problem of laboratory computerization. A scheme has been devised where a single laboratory computer system and sophisticated telecommunications techniques serve the needs of multiple IHC hospitals in the intermountain region. The laboratory computer communicates with HELP systems installed at each of the hospitals to provide the laboratory functions required of a hospital information system. Sections 1 to 6 of this chapter are reprinted from an article appearing in the 1988 SCAMC proceedings that detailed the architecture of the link between the laboratory system and HELP and elaborated on issues related to implementation. A system that alerts for life-threatening laboratory results is in place at LDS Hospital and is discussed in Chapter 17.

16.1 Overview

In order to provide cost effective patient care and to provide better information access and exchange capabilities for health-care providers, a regional healthcare network has been created. The goal of the network is to provide the decision support capabilities of the HELP system to LDS Hospital and a group of affiliated hospitals in the intermountain region and to allow access and exchange of clinical patient data among the various institutions. The structure and design of the network are explained below and a discussion of issues and problems related to the implementation follows thereafter.

---

[1]Reprinted with permission from "Proceedings of the Symposium on Computer Applications in Medical Care" by Huff and Pryor, *SCAMC* 11, 1988.

16.2 Network Overview

The current network consists of three geographically distinct locations. The primary node of the network is in downtown Salt Lake City, Utah, at the corporate offices of IHC. The network node at corporate headquarters consists of a cluster of Tandem central processors and associated disks that hold the MPI, the master doctor file, and a master copy of the PTXT file. (The MPI and the PTXT file are discussed in Chapter 3.) The other two sites involved in the network are LDS Hospital and McKay–Dee Hospital in Ogden, Utah. McKay–Dee Hospital is a primary–secondary care hospital located 35 miles north of Salt Lake City.

As shown in Figure 16.1 the health-care network consists of six main components at LDS and McKay-Dee hospitals:

1. the HELP system operating at each site on Tandem hardware
2. a laboratory information system (LIS) physically located at LDS Hospital operating on Prime hardware
3. a series of LANs connected to the LIS
4. financial and billing systems operating at each site on IBM hardware

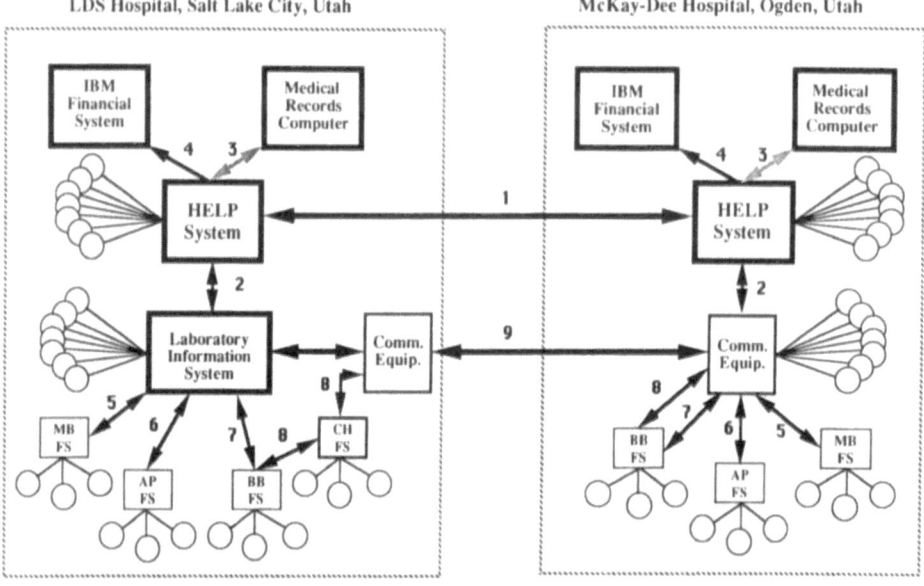

Figure 16.1. Simplified schema of the regional health care network. The two facilities are separated by 35 miles. AP = anatomic pathology, BB = blood bank, CH = Clearing House, FS = file server, MB = microbiology.

5. communication equipment (statistical multiplexors and modems) that connects the geographically distant sites
6. medical records software operating at each site on Data General hardware (see Chapter 8)

16.3 Information Flow Within the System

Information flow within the network is dictated by the needs of health-care providers within the system. The first use of the network takes place at the time of registration of inpatients or outpatients at LDS and McKay–Dee Hospitals. After the admitting clerk has entered demographic and historical information from the patient into the HELP system, a comparison is made across the Tandem link with information at corporate headquarters in the MPI.

As care is provided for the patient, laboratory studies and therapeutic procedures are requested. This process is currently paper-based but is in the process of being incorporated into the HELP system's order communications program (Chapter 11). The order entry process on the HELP system will be driven by knowledge frames that examine the orders in the context of the patient clinical findings and administrative policies and standards of care. Pharmacy orders and radiology orders are handled directly by the HELP system itself. However, orders for clinical laboratory tests, blood bank ordering, or services in anatomic pathology will be passed from the HELP system to the laboratory information system as shown in Figure 16.1. Tests ordered at LDS Hospital will be transferred across a direct line linking the HELP system at LDS Hospital with the LIS. Orders from McKay–Dee hospital will be transferred from the McKay–Dee HELP system by a separate communication line that links to the LIS located at LDS Hospital. In both cases, orders for general laboratory tests (chemistry, hematology, serology) will be processed and sent directly to the LIS minicomputer.

An innovative part of the LIS is that it uses LANs for processing and recording results for microbiology, blood bank, and anatomic pathology. Thus, the LIS forwards all orders from the LDS Hospital HELP system to the three LANs at LDS Hospital. Orders originating from the HELP system at McKay–Dee Hospital will be sent to the LIS, and then forwarded from the LIS back across the communications lines to the blood bank, anatomic pathology, and microbiology LANs at McKay–Dee Hospital.

The blood bank component of the system has some special features. A central blood bank file server (called the Clearing House) at LDS Hospital contains a central database of all blood donors that have donated blood at any of the IHC facilities, as well as a tabulation of all units of blood available in inventory at the various institutions. Each time a new donor is registered at any site, an entry is also made in the master donor file at the Clearing House. The central donor file allows exclusion of permanently deferred donors from making inappropriate donations and allows centralized administration of credits to blood donor groups. The Clearing House inventory files are updated automatically by a user-initiated process. At intervals dictated by clinical need, software at the Clearing House queries the blood bank LANs at the various facilities and uploads data about units available, units cross matched, and units in processing. With this information the Clearing House can quickly respond to blood shortages by requesting transfer of blood from institutions with a surplus.

Test results from the laboratory follow the reverse route of the orders. That is, the LANs communicate results back to the LIS, which then forwards the results to the HELP system at the appropriate hospital. Once the results are back in the HELP system they can be examined by HELP decision frames and are available for physician and nurse review. Stat or urgent results can be automatically printed in the ER or ICUs. Results can be viewed by terminals directly connected to the HELP system or from physicians' offices and homes using dial-up lines connected by modems to the HELP system (Chapter 31). Additionally, the LIS has the ability to do auto-dial download of formatted reports to printers or PCs for outpatient work performed for clinics or other institutions that are not a part of IHC.

Finally, the HELP systems pass needed information to the financial systems. Using a combination of real time and batch processing, charge capture takes place on the HELP system and these data are forwarded to the financial system. The financial system then produces patient and third-party bills and handles the other aspects of medical accounting. The connection between the HELP system and the Code 3 Data General computer is less direct, as indicated by the lighter arrow in Figure 16.1. Medical records personnel use a combination of manual chart review and on-line data review to determine the correct DRG, CPT-4, and ICD-9 codes for the patient's stay.

16.4 Interfacing Strategy and Details

Nine interfaces or communication links are currently involved in the network, with most interfaces operating at more than one site. The details of each link will be outlined briefly below and are labeled in Figure 16.1. The current links include:

1. HELP to HELP data link
2. HELP to LIS interface

3. HELP to medical records interface
4. HELP to financial system interface
5. LIS to microbiology LAN interface
6. LIS to anatomic pathology interface
7. LIS to blood bank interface
8. Blood bank to clearing house communication link
9. Data link between LIS and remote LIS terminals and LANS

The link between HELP at LDS Hospital and HELP at McKay–Dee Hospital is provided using standard Tandem "Expand" software. This is a proprietary protocol that allows remote data files and processes to be accessed as if they were local to the user. The same type of software is also used to link the hospitals to the MPI and master doctor index files at corporate headquarters in Salt Lake.

The HELP to LIS interface adheres to the seven layer International Standardization Organization (ISO) open systems interconnection (OSI) model as much as possible. It uses a vendor-provided X.25 protocol for the physical link, data link, and network control layers. The transport and session control layers were created locally, but conform to IEEE standards. The presentation and application layers have also been created locally, but no applicable standards are yet available.

The transactions supported on the HELP-LIS interface are summarized in Table 16.1. A simplified diagram of the interface data flow is shown in Figure 16.2. A clinical user on the HELP system interacts via a video terminal with a laboratory ordering process or with an ADT process. These processes then write appropriately formatted data to a circular queue that holds data that are to be sent to the LIS. A requester process (using X.25 protocol) then sends data to the LIS and receives an acknowledgment from the LIS that the data were received. As the server on the LIS side receives the data it writes it to a circular queue of data, where it awaits processing by the LIS. After order processing and actual testing has occurred, the results are entered into a circular queue that holds data awaiting transfer to the HELP system. A requester on the LIS then reads the data and sends it (using X.25 protocol) to the HELP system and receives acknowledgment that the data arrived. The server on the HELP system immediately writes the transaction to a circular queue, which is processed by the result storing process, and which translates the incoming data from the network format into PTXT codes that are then stored in the patient file. The translation process used by the result storing process is table driven. The same translation tables are also used for formatting ADT and order transactions before they are written to the "out to lab queue."

The interface between the HELP system and the medical records system is structurally the same as the LIS to HELP interface, and uses the same software on the HELP system side.

Table 16.1. HELP-LIS transactions

---

1. ADT (HELP to LIS)
    Inpatient registration
    ER registration
    Outpatient registration
    Pre-admit
    Inpatient transfer
    Demographic update
    Inpatient discharge/death notification
    Outpatient discharge
    Change medical records number
    Change visit number
    Cancel admission
2. Order processing
    Order entry and add to accession
    Order cancel (LIS to HELP)
    Order status (LIS to HELP)
3. Results return
    Structured and encoded results
    Corrected results
4. Configuration maintenance
    Locations
    Doctors
    Test codes
    Result codes
    Microbiology codes

The interface between the HELP system and the financial system uses the same architecture but different (IBM) protocols for the communications. The lower levels of the protocol use SNA (system network architecture), whereas the upper levels are LU 6.2 using locally defined transaction types and message contents. The SNA and LU 6.2 software is vendor supplied and supported.

The interfaces between the microbiology, blood bank, and anatomic pathology LANs and the LIS have the same functionality as described above, but are vendor provided and use proprietary protocols at the discretion of the LIS vendor. The communication links between the Clearing House LAN and the other blood bank LANs are standard vendor-supplied "bridge" software.

Communications equipment between the LIS at LDS hospital and McKay–Dee Hospital for both the LANs and the remote LIS terminals is vendor supplied and supported, and consists of a bank of statistical multiplexors and modems.

Laboratory Information System   HELP System

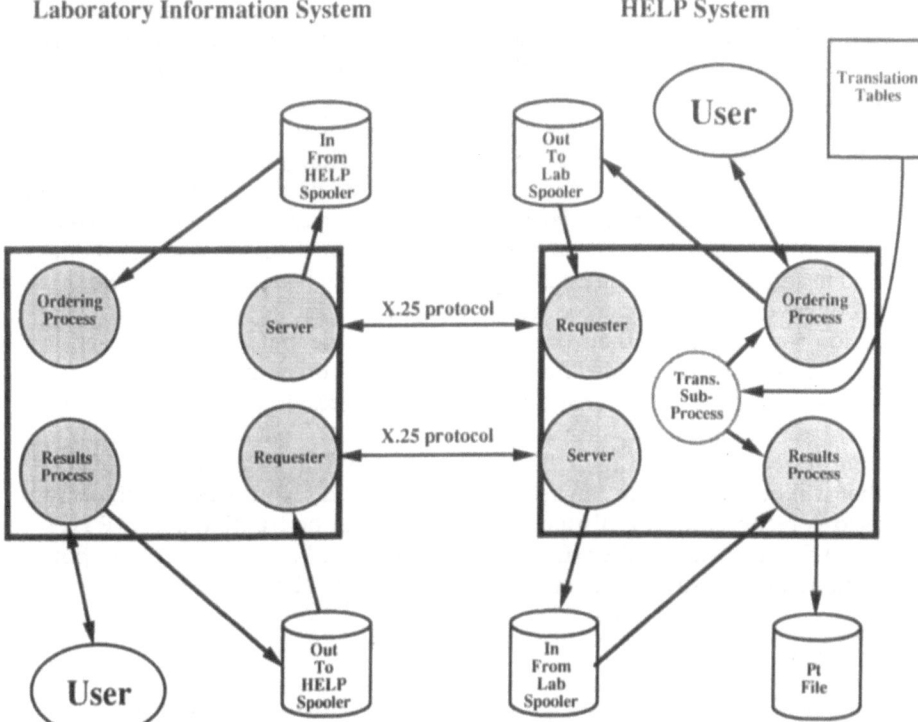

Figure 16.2. Processes and files used in the HELP to LIS interface.

16.5 Discussion

16.5.1 Standard Protocols

The design philosophy for implementing the network is to adhere to national or international standards whenever possible. But as can be seen from above, "the beauty of standards is that there are so many to choose from." Thus, in interfacing to non-IBM systems, X.25 is the low-level protocol of choice, but for the HELP to financial system interface, SNA coupled with LU 6.2 was the most cost effective choice. The point is that the standards are not really standards yet. Second, at the presentation and application levels only partial standards exist. The work of Clement McDonald and subcommittee E-31.11 of ASTM [1] and of the HL-7 group is excellent for defining the domain of content for LIS to HIS interactions, but the particular syntax using delimited fields is restrictive (binary or graphic data cannot be transmitted, difficulty in translation of data items that can be alpha or numeric, exclusion of the delimiter from use as data in a record) and difficult to maintain (addition of fields, subfields or sub-

subfields will cause program rather than table changes). There are currently no standards for transactions like ADT, anatomic pathology orders, and blood bank orders.

The work of subcommittee P1157 MEDIX of IEEE is encouraging [2]. This subcommittee has taken on the task of defining the presentation and application level protocols for the transfer of all medical information. The direction is to establish standard transaction and message contents as well as syntax for the general needs of medical information transfer. A major reduction in software development and maintenance can be realized if this work is productive. Our experience indicates that the standard of tagged field transaction formats rather than fixed field or simple delimited field transaction formats will provide needed flexibility and greater breadth of expression.

An important characteristic of all the interfaces described above is that they transfer information in a structured format. That is, result information passed back from the LIS to the HELP system is not a block of text that will be displayed to the user, but a structured record where each result field is recognized as a distinct variable that can be translated into PTXT codes for inclusion in the HELP patient database. From a decision support standpoint only structured data are of value to the HELP system.

16.5.2 Translation Issues

Because the HELP database is encoded, some difficulties were encountered in translating data received from the LIS into the HELP system format. For instance, a single term in the LIS occasionally maps to more than one term in the HELP system. An example is microbiology data, where the value of the "specimen site" field is expressed by a single code representing "right arm," whereas the HELP system has two separate codes, "right" and "arm," for representing the same information. This required that the tables used to translate incoming data from the LIS support one to many mappings of the primitive field values.

A related but distinct problem is presented by some clinical chemistry tests like T4, creatinine kinase, and many other enzyme measurements. These tests can be performed by different methods such as radioimmunoassay or ELISA methods, or can be measured at different temperatures. In the LIS this means that distinct tests are created for CK measured at 37°C and CK measured at 30°C, and that separate tests are created for T4 by RIA and T4 by ELISA. However, in the HELP system each analyte has a single code and the methodology used to measure the analyte is expressed in a modifier field that is attached to the analyte result. The implication for the translation process on the HELP system is that when a T4 by RIA result is received from the LIS it maps that single value to two fields, one field with a value and one field that is a modifier that shows the methodology. These kinds of complexities have caused us to begin investigating ways of using the knowledge-driven frames of the HELP system to drive the translation process.

16.5.3 Queuing Versus Real Time

As depicted in Figure 16.2, it is apparent that the interface between HELP and the LIS is a queuing interface. This is essentially true of all of the interfaces to the HELP. This mode of interfacing was selected as superior to real-time interfaces (where the user process waits on the completion of a transaction with a remote system before proceeding to the next task) for several reasons. The first is that a queuing interface allows work to continue on the HELP system regardless of whether or not the LIS is up. This means that order entry can continue on the HELP system during scheduled or unscheduled down time on the LIS. Furthermore, a queuing interface provides insulation of the HELP response time from LIS response time and vice versa. The time that it takes a user to order a laboratory test depends only on the time it takes the HELP system to store the order in the queue, not the time it takes for the LIS to receive and process the order. Another advantage is that since the queues used are circular, there is a period of time during which retransmission of data can be initiated if for some reason (hardware failure, software bugs, or operator error) data are lost either in the LIS or on HELP. A final advantage is that queuing allows the incorporation of LANs into the LIS. Without queuing it would really not be feasible for a blood bank order to be passed from HELP to the LIS and then to the blood bank LAN.

These advantages were weighed against some disadvantages, however, and some compromises had to be made. The first problem with a queued interface is with error recovery. If a user on the HELP system creates a duplicate order or an order that is incomplete, an error condition will be created in the LIS. Since the order was processed through two queues, it is likely that the user placing the order will no longer be an active user on the HELP system. To communicate the error to the user requires either electronic mail or manual methods such as telephone calls. If the interface was real time the user would still be at the terminal and the error message would be returned directly to the user. The strategy used by the HELP system is to do extensive order verification on the HELP system so that errors are infrequent, but when they occur they are handled by manual methods.

A second situation that is not handled well by a queuing interface is new orders created in the laboratory. It should be noted that "new orders" are not merely additions to an existing order or accession, but a new test that requires a new specimen. This situation arises when the results of a previously ordered test indicate further work-up is needed and the laboratory technologist orders the needed tests on the LIS. The most common areas of the laboratory affected are anatomic pathology, blood bank, and microbiology. The conflict arises because the LIS requires a valid HELP order number before it can create a valid order on the system. This problem was solved by changing the LIS software so that orders can be created with temporary HELP order numbers. It could also have been solved by allowing the requesters and servers to process transactions real-time rather than from the queue. This second alternative would imply that laboratory users would be waiting for real-time response from the HELP system to do the new test order function, and that if the HELP system was down then tests could not be ordered.

A final problem relates to tests canceled by clinicians. By policy, laboratory personnel actually control the cancellation of a test. This means that a procedure on the HELP system to cancel a test is actually a request of the laboratory people to cancel a test. The laboratory people base their actions on how far work has proceeded as to whether the test can be canceled. This situation is handled poorly by a queuing interface unless several transactions are created. The needed transactions would be "request cancellation" from HELP, "cancellation denied" from LIS, and "cancellation completed" from the LIS. Although this is clearly feasible, it was determined that cancellations were relatively infrequent and that it was easier to have the clinician call the laboratory on the phone and verbally request cancellation. The laboratory person can respond immediately and either cancel the test or deny cancellation. If the test is canceled, the only transaction needed is a "cancel test" transaction from the LIS to HELP.

16.5.4 Functional Overlap

Some earlier points of discussion have pointed to another problem in interfacing systems, that of functional overlap. An example previously mentioned involved checking for duplicate orders. Duplicate orders frequently occur, especially in teaching hospitals where several physicians are responsible for a patient's care. Most LIS systems provide for duplicate checking. In order to prevent an inordinate number of bad orders from being transmitted from the HELP system to the LIS, duplicate checking is also done on the HELP system. Other functions between systems can overlap as well. Many LIS vendors now offer various forms of interpretive reporting, which is redundant with the alert and decision support logic of HELP. Most vendor-supplied LIS systems provide billing capabilities, but charge capture and billing are also provided as features of the HELP system. Other examples can be cited involving medical records. Redundant features create unwanted system overhead. The question arises as to what responsibilities should be handled by which systems.

The following guidelines were adopted. All functions of the system that require knowledge-based behaviors will be accomplished on the HELP system. Since order entry is a knowledge-driven process, it is done on the HELP system. It can be argued that duplicate checking could be provided at the laboratory more efficiently than on the HELP system. This may be true for simple duplicate checking but for other knowledge-driven behaviors like checking for cost effectiveness of the test, checking for drug-laboratory interactions, and verifying clinical need for blood orders, the

HELP system has the required clinical information to make a decision whereas the laboratory does not. For the same reasons, interpretive reporting is a function of the HELP system and not of the LIS. Also, the decision to provide access to clinical laboratory data through the HELP system rather than through the LIS is based on a desire to provide decision support information to the clinicians that cannot be provided through the LIS.

However, many functions remain for the other systems in the network. For instance, it is appropriate for the LIS to handle workload capture, machine maintenance records, delta checking, and all other functions that do not require direct correlation with clinical patient data. The same is true for the billing system and the medical records systems. Although billing and account management could be accomplished by the HELP system, it seems a nice division of labor to let the financial system take that responsibility.

One concern that is raised because the HELP system is the interface to the patient clinical data is a feeling of a loss of control by pathologists. The real "product" of the clinical laboratory is information and until the results of laboratory tests are communicated to the physician the job of the clinical laboratory is not complete. Whereas pathologists have had exclusive control of laboratory information systems, the HELP system must serve the needs of pharmacists, radiologists, nurses, and physicians. A particular problem has been the format of laboratory data review screens with concerns being raised about how reference ranges should be displayed and updated and about specimen collection dates and times and about horizontal versus vertical formats. The loss of control is more apparent than real, however. What is needed is for pathologists to assert themselves as the supervisors of pathology data on the HELP system, which is a natural extension of their current role that is caused by the expanding health-care network.

16.5.5 Local Area Network Integration

The integration of LANs into the LIS has both advantages and disadvantages. The advantages are in the independence and decentralization allowed by the LAN architecture. For text- and database-intensive activities like blood banking, microbiology, and anatomic pathology decentralization means that greater processing power is available to individual users. Another distinct advantage is that the system can grow modularly with the addition of new work stations as needed with little effect on the configuration of the rest of the system. A third advantage is that software development tools in the PC environment are ahead of the minicomputer environment because of the large potential market for PC software.

The disadvantages of LAN integration also results from the fact that they are independent systems. Static and configuration files must be maintained across all systems to maintain a high level of performance. Additionally, since order entry and charge capture take place on the HELP system but workload and other reports are generated centrally from the LIS minicomputer, most blood bank, anatomic pathology, and microbiology data passes through the LIS anyway. The combination of static file updates and results transfer imply a sophisticated interface between the LANs and the LIS minicomputer. However, since these interfaces are vendor supplied this is not a problem for the end-user. The performance implications of these interfaces will require further evaluation in the future. If performance is a problem then interfaces will be created directly from the LANs to the HELP system.

16.6 Future Directions

The purpose of the direct link between HELP systems at various institutions is to provide access and electronic transfer or copying of patient data files. Work is under way to define the policies and procedures to allow this kind of data sharing, but it is clear that the MPI will play a major role in coordinating this activity. The number of nodes on the network is also expected to increase, with eight hospitals scheduled for addition to the network over the next few years. Each new node will have its own version of the HELP system, but all laboratory work will be coordinated by the single minicomputer at LDS Hospital.

The next phase planned for the HELP to LIS interface is to implement presentation and application protocols that are consistent with the developing IEEE MEDIX standards. Specifically, transaction messages will be converted to a tagged (name–value) format from the current fixed-field format. With this more flexible type of record structure there is the opportunity to exchange patient data between the HELP system and physician or clinic database systems using structured and encoded data formats.

```
          SU KRISTINE                      402210              E601

    LAB DATA

        0   SELECT A NEW PATIENT
        1   CBC                    22   URINE ELECTROLYTES
        2   PROTIME                24   URINE UREA NITROGEN
        3   PTT                    32   AMIKACIN
        5   DIFFERENTIAL           41   GENTAMICIN
        7   SMA-7-CHANNEL          74   FIBRIN SPLIT PRODUCTS
        8   SMAC                   78   IRON/IBC
        9   LACTIC  ACID           83   IBC
       10   OSMOLARITY             94   AMYLASE
       11   MAGNESIUM              97   CORTISOL
       15   CPK                   110   LIPASE
       20   URINALYSIS            155   FIBRINOGEN
       21   URINE CREATININE

    CHOOSE ONE OR MORE OF THE ABOVE
    OR TYPE <shift> ? FOR INSTRUCTIONS --
```

Figure 16.3. Menu screen for choosing laboratory data for review.

```
          Jane                              X02210          E601
LAB DATA - SMA-7

   DATA  TIME             NA+      K+     CL-     CO2    BUN    GLUC   CREAT
   17AUG 05:10    B       138     4.1      97      33      8     129     .5
   16AUG 05:00    B       137     4.5      99      29      7     120     .4
   15AUG 05:10    B       138     4.2     103      31      8     114     .4
   14AUG 05:05    B       137     4.6     102      31      9     123     .5
   13AUG 05:05    B       138     4.2     100      31      9     137     .5
   12AUG 05:10    B       136     4.2     100      35      6     115     .5
   11AUG 05:15    B       138     4.3      99      29      7     126     .6
   10AUG 05:25    B       140     4.2     101      35      7     122     .5
   09AUG 05:15    B       138     4.2     100      34      9     113     .4
   08AUG 05:10    B       136     4.3     102      33     11     141     .4
   07AUG 05:20    B       135     4.4      97      31      9     126     .4
   06AUG 05:45    B       138     4.6     104      34     10     120     .4
   05AUG 05:30    B       138     4.4     104      35     10     125     .4
   04AUG 05:15    B       138     4.0     102      35     12     124     .4

LAB DATA - CBC

   DATE  TIME          WBC     RBC     HGB     HCT     MCV     MCH    MCHC    RDW    PLAT
   17AUG 05:10 B      12.2    4.12    13.0    37.8    91.6    31.4    34.3   17.2    294
   16AUG 05:00 B      11.2    4.32    13.1    39.8    92.1    30.3    32.9   16.8    188
   15AUG 05:10 B      10.5    4.09    12.8    37.8    92.4    31.3    33.9   17.3    345
   14AUG 05:05 B      10.5    3.97    12.5    36.6    92.1    31.5    34.2   16.9    366
   13AUG 05:05 B       9.0    4.02    12.1    37.0    91.9    30.2    32.8   17.7    354
   12AUG 05:10 B       9.9    4.04    12.4    37.3    92.3    30.6    33.2   17.7    428
   11AUG 05:15 B       9.3    3.75    11.4    34.5    92.1    30.4    33.0   17.3    361

LAB DATA - DIFFERENTIAL

DATE  TIME      VALUES
17AUG 05:10      META      BAND     POLYS    LYMPH      MONO     BASO    EOSIN   PLASMA
            B               2        77       10          5                6
          PLATELETS:                       ADEQUATE
16AUG 05:00      META      BAND     POLYS    LYMPH      MONO     BASO    EOSIN   PLASMA
            B     1         5        81       11          1                1
          ANISOCYTOSIS:                      1+
15AUG 05:10      META      BAND     POLYS    LYMPH      MONO     BASO    EOSIN   PLASMA
            B               2        79        6          4                9
          PLATELETS:                       ADEQUATE
14AUG 05:05      META      BAND     POLYS    LYMPH      MONO     BASO    EOSIN   PLASMA
            B               1        77       15          2                5
          PLATELETS:                       ADEQUATE
13AUG 05:05      META      BAND     POLYS    LYMPH      MONO     BASO    EOSIN   PLASMA
            B               4        78       12          6
          PLATELETS:                       ADEQUATE
12AUG 05:10      META      BAND     POLYS    LYMPH      MONO     BASO    EOSIN   PLASMA
```

Figure 16.4. Sample of laboratory data display.

16.7 Reports

Laboratory data are available for review in a variety of different formats on the HELP system. Because of HELP's integrated database, laboratory data are also often displayed in conjunction with other clinical data to give a comprehensive view of the patient's status. Some comprehensive reports that include laboratory data are described in other chapters. Examples include the ICU Rounds Report (Figure 4.4), the 7-day report (Figure 4.5), and the nursing shift report (Figure 13.11). All of the laboratory reports

```
************************************************
*   ROUNDS REPORT FOR PAST 72 HOURS ON JUL 02 90 AT 21:52   *
************************************************
```

                         WILLIAM   AGE: 82 ADMITTED: 07/01/90 22:57   PAT NUM: 113370 SEX: M ROOM: E606
ATTENDING: SANDERS, JOHN M.   RAD NUM: 4505300   HEIGHT: 183 cms   WEIGHT: 78.00 kgs
ADMITTING DX: EPIDURAL HEMATOMA   BIRTHDATE: 06/01/08

```
************************************************
HEMATOLOGY SECTION
************************************************
```

	WBC	RBC	HGB	HCT	MCV	MCH	MCHC	PLATE
02 05:55	15.0	4.7	14.8	42.9	90.9	31.4	34.6	218
01 20:21	10.5	4.6	15.5	42.6	92.6	33.6	36.3	204

	META	BAND	POLYS	LYMPH	MONO	BASO	EOSIN
02 05:55		4	89	19	3	4	
01 20:21			67		9	9	5

```
************************************************
COAGULATION SECTION
************************************************
```

PROTIME:
PT		PTT
01 23:19 11.3		01 23:19 31

```
************************************************
ELECTROLYTES SECTION
************************************************
```

	NA	K	CL	CO2	BUN	GLUC	CREAT
02 16:45 SE	134	3.9	104	23	15	194	1.9
02 11:00 SE	133	4.3	108	16	14	251	1.8
02 05:55 SE	135	4.4	109	15	16	253	1.8
01 20:21 SE	145	3.4	113	22	17	125	2.1

OSMOLARITY
02 16:45 SE	293
02 15:00 SE	288
02 11:00 SE	294
02 05:55 SE	301
01 23:19 SE	294

LACTIC ACID
02 15:00 PL 2.4

MAGNESIUM
01 23:19 SE 2.2

	NA	K	CL	CO2	CA++	PO4	GGT	GLUC
01 23:19 SE	141	4.0	105	20	8.7	2.6	13	253

```
************************************************
SMAC-20 AND LIVER FUNCTION SECTION
************************************************
```

	CREAT	BUN	URIC	CHOL	PROT	ALB	T.BIL	D.BIL	ALK	LDH	SGOT	SGPT
01 23:19 SE	2.1	19	6.8	276	7.0	4.0	.6	.1	96	289	19	11

```
************************************************
CARDIAC SECTION
************************************************
```

BB(IU/L)	MB(IU/L)	MM(IU/L)	BB(%)	MB(%)	MM(%)
0	7	458	.0	2.0	98.0

CREAT KINASE(CPK)
	CPK	ISOENZYME TOTAL
02 08:20	465	
02 08:20		465.0

```
********************************************************        ************************************        ********************************
01 23:19  COLOR: STRAW       DENSITY: CLEAR             URINE ANALYSIS SECTION          HEME: 2+            BILE: NEGATIVE
          KETONES: 1+        GLUCOSE: TRACE             SPEC. GRAVITY: 1.013            UROBILINOGEN:       pH : 7.0
          RBC (/HPF): 0-2    WBC: 0-2                   PROTEIN: TRACE                  BACT:
                                                        EPITH: OCCASIONAL
```

BLOOD GAS SECTION

JUL 02 90	pH	PCO2	HCO3	BE	HB	CO/MT	PO2	SO2	O2CT	%O2	AVO2	VO2	C.O.	A-a	Qs/Qt	PK/	PL/PP	MR/SR
NORMAL HI	7.45	43.3	27.7	2.5	17.7	2/ 1	- -	- -			5.5	300	7.30	29	5			
NORMAL LOW	7.35	29.9	17.3	-2.5	13.7	0/ 1	53	86	17.9		3.0	200	2.90		0			
02 17:50 A	7.45	33.2	22.9	.5	15.1	0/ 1	91	96	20.4	5								

SAMPLE # 3. TEMP 38.4. BREATHING STATUS : NASAL CANNULA
NORMAL ARTERIAL ACID-BASE CHEMISTRY
PULSE OXIMETER SO2 96.0

	pH	PCO2	HCO3	BE	HB	CO/MT	PO2	SO2	O2CT	%O2	AVO2	VO2	C.O.	A-a	Qs/Qt
02 10:00 A	7.41	30.5	19.1	-3.6	15.0	0/ 1	86	95	20.1	4					
01 20:40 A	7.43	30.7	20.2	-2.2	14.9	1/ 1	69	93	19.5	21				19	

SAMPLE # 1. TEMP 37.0. BREATHING STATUS : ROOM AIR
NORMAL ARTERIAL ACID-BASE CHEMISTRY

```
********************************************        CURRENT MEDICATIONS        ********************************
```

	DATE ORDERED
ACETAMINOPHEN 650 MGM, ELIXIR, NASOGASTRIC Q 4, PRN,	02JUL
MAALOX PLUS 30 ML, LIQUID, NASOGASTRIC Q 4, PRN,	02JUL
SUCRALFATE (CARAFATE) 1000 MGM, TAB, NASOGASTRIC Q 4, PRN,	02JUL
PHENYTOIN (DILANTIN) 200 MGM, INJ, INTRAVENOUS Q 4, AS DIRECTED,	02JUL
POTASSIUM CHLORIDE 40.0 MEQ, INJ, INTRAVENOUS DRIP Q 4, AS DIRECTED,	02JUL
TOBRAMYCIN 0.3% (TOBREX) 1 GTTS, OPHTH S, OPHTHALMIC QD,	02JUL
ARTIFICAL TEARS (LACRIL) 1 GTTS, SOLUTIO, OPHTHALMIC Q 4, AS DIRECTED,	02JUL
POTASSIUM CHLORIDE 20.0 MEQ, INJ, INTRAVENOUS DRIP Q 4, AS DIRECTED,	02JUL
PHENYTOIN (DILANTIN) 100 MGM, INJ, INTRAVENOUS Q 6,	02JUL
POTASSIUM CHLORIDE 40.0 MEQ, INJ, INTRAVENOUS DRIP Q 6, AS DIRECTED,	02JUL

(END)

Figure 16.5. 72-hour rounds report showing demographic, laboratory, blood gas, and pharmacy data.

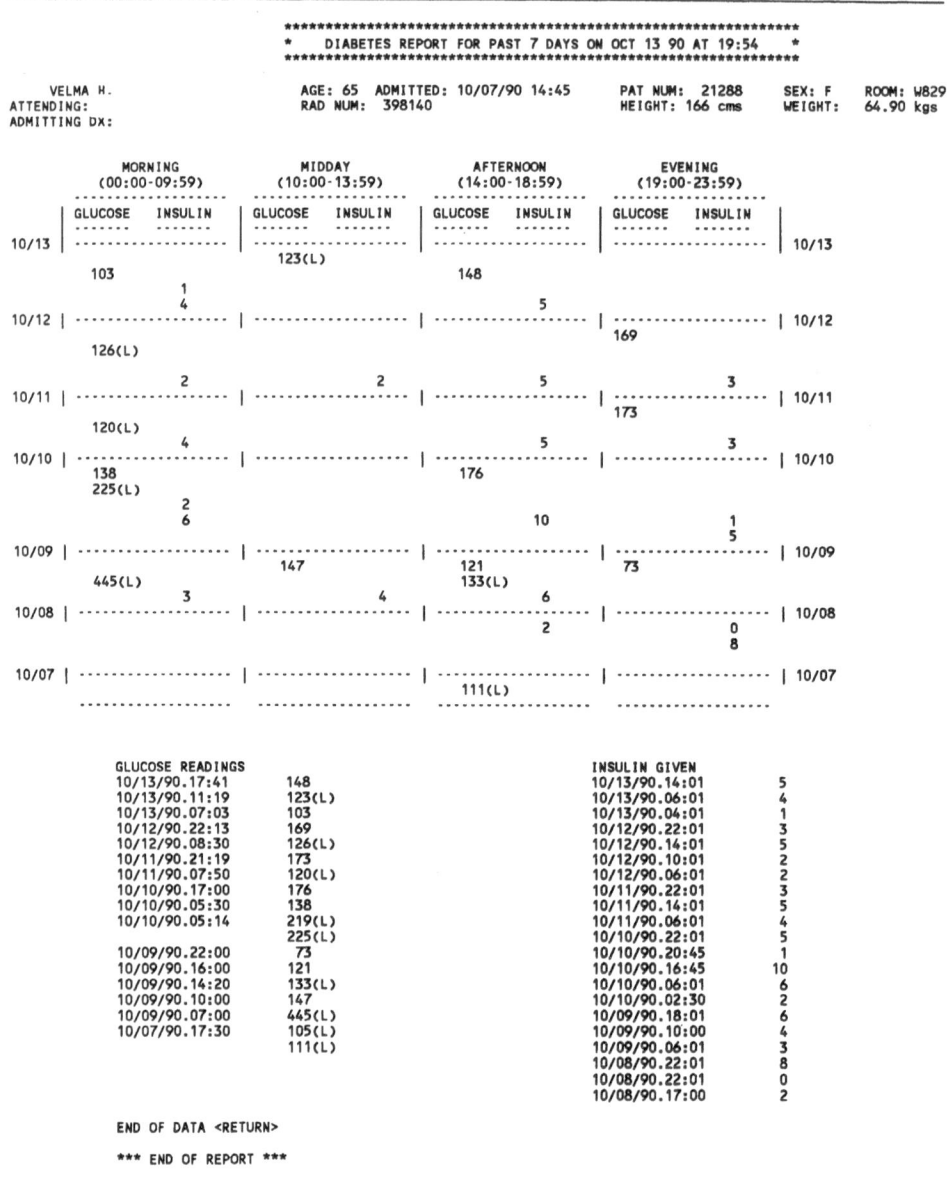

Figure 16.6. Diabetes report showing glucose values and doses of regular administered. The report is organized by time of day so that glucose readings and insulin doses for a given time period can be examined easily. (L) indicates a laboratory glucose measurement. The other measurements are glucometer readings and are entered into the HELP system through the computerized nurse charting module.

may be reviewed on a terminal screen or from a printed copy. Descriptions of other routine laboratory reports follow.

16.7.1 Routine Laboratory Data Review

The HELP laboratory data review program permits the results of each set of tests (e.g., CBC, SMA-7, etc.) to be reviewed in reverse chronological order. Tests may be reviewed individually or in groups. The HELP system displays a menu of tests for personnel to choose from (Figure 16.3). The menu displays only the tests for which results have been recorded in the database for the patient under consideration. A sample of the display is shown in Figure 16.4.

16.7.2 Seventy Two-Hour Report

The 72-hour report (Figure 16.5) is intended to be a summary of laboratory and other data for physicians to use to review the status of acutely ill patients. Data for the past 72 hours for hematology, electrolytes, urinalysis, blood gases, and medications are displayed.

16.7.3 Diabetes Report

The diabetes report (Figure 16.6) displays measured glucose values and amounts of administered insulin (from computerized pharmacy data) for a patient's stay. The data are organized into four time intervals for each day, thus facilitating regulation of insulin dosages. A time sequential display of glucose measurements is also displayed.

References

[1] McDonald CJ, Wiederhold G, Simborg DW. A discussion of the draft proposal for data exchange standards for clinical laboratory results. *Annu Symp Comp Appl Med Care* IEEE: 1984.
[2] Rutt T. Minutes of IEEE P1157 MEDIX Organizational Meeting. IEEE: Nov. 1987.

17
Computerized Laboratory Alerting System

17.1 Overview

A program that uses the HELP system to screen laboratory data automatically for potentially life-threatening situations has been developed, implemented, and evaluated at LDS Hospital. The CLAS program uses decision logic to screen automatically laboratory data being stored to the patient database for the presence of life-threatening conditions. If such a condition is detected, a message is stored in a computerized file. The message is displayed the next time the HELP system is used to review the alerting patient's laboratory data.

Implementation of the program led to (1) an increase in the frequency of appropriate treatment, (2) less time spent in life-threatening situations, and (3) shorter hospital stays for general ward patients for certain of the alerts. No measurable differences were noted for ICU patients where the routine level of care is higher than on the general wards. A questionnaire revealed that users felt the system was generally useful.

17.2 Background

In the late 1970s a pilot system for generating alerts and communicating them to physicians was designed, implemented, and evaluated at LDS Hospital [1]. This system transmitted the alerts to nurse clinicians who evaluated their appropriateness and, if necessary, informed the physician. This system required the nurse clinician to check the computer frequently (approximately every 15 minutes) and the nurse clinicians were only on duty 16 hours per day. Additionally, staff nurses felt their normal duties, that is, monitoring of lab results and conveying them to physicians, were being subsumed by the nurse clinicians.

A system was sought that would make monitoring and follow-up of the alerts a more routine part of the nursing unit function and would not require additional personnel.

17.3 Design

Whereas the earlier system screened for 60 different alerting conditions, the new CLAS [2,3,4] focused on only 11. The criteria to generate the alerts were developed in a painstaking, multistage process that involved a substantial amount of expert physician input. The goal was to maximize the reliability of any given alert.

Because of the life-threatening nature of the alerts being generated, notifying hospital personnel of the presence of an alert in a rapid manner was considered an important component of the project. In one experiment, a flashing yellow light was installed on one nursing unit. The light (similar to that found on an emergency vehicle) was set off when an alert was posted for that division. Although response time to this notification method was quite rapid, the hospital personnel felt the method was too obnoxious and its use was abandoned. Displaying a message passively on a video terminal was considered but discarded because the terminal might not be examined for a long period of time and also because the patient's family members might become alarmed seeing a "life-threatening alert" message posted.

After a variety of notification methods were studied, it was decided to display the alert automatically when a hospital staff member (nurse, physician, ward clerk, or other) accessed an alerting patient's laboratory data through the HELP system. Reviewing laboratory data through HELP is the routine method of laboratory data retrieval at LDS Hospital and, indeed, the laboratory data review program is the single most accessed program at LDS Hospital.

An example of the alert report is shown in Figure 17.1. After the patient's identifying data are displayed, the time the alert was generated, the type of alert, and the specific values involved are displayed. The alert is considered to be acknowledged when it is reviewed. The alerts implemented and the criteria for their generation are shown in Table 17.1.

```
W815      042356788    SMITH, DONALD NMI

CURRENT LAB ALERTS:

27JUN  07:15            HYPOKALEMIA (K+ IS 3.1, K+ HAS FALLEN 1.3
                        MEQ/L IN 24 HRS)

27JUN  07:15            SEVERE HYPERGLYCEMIA (GLUCOSE IS 735)


********************END OF LAB ALERT REPORT********************
```

Figure 17.1. Sample laboratory alert.

Table 17.1. Alerts and their criteria

Alerting condition	Criteria
Hyponatremia (NAL)	$Na^+ < 120$ meq/l
Falling sodium (NAF)	Na^+ fallen 15+ meq/l in 24 hr and $Na^+ < 130$ meq/l
Hypernatremia (NAH)	$Na^+ > 155$ meq/l
Hypokalemia (KL)	$K^+ < 2.7$ meq/l
Falling potassium (KLF)	K^+ fallen 1+ meq/l in 24 hr and $K^+ < 3.2$ meq/l
Hypokalemia, patient on digoxin (KLD)	$K^+ < 3.3$ meq/l and patient on digoxin
Hyperkalemia (KH)	$K^+ > 6.0$ meq/l
Metabolic acidosis (CO_2L)	$CO_2 < 15$ and BUN > 50 or $CO_2 < 18$ and BUN < 50 or $CO_2 < 18$ or CO_2 fallen 10+ in 24 hr and $CO_2 < 25$
Hypoglycemia (GL)	Glucose < 45 mg%
Hyperglycemia (GH)	Glucose > 500 mg%
Falling hematocrit	Hct fallen 10+ since last HCT and Hct < 35 or Hct fallen 5+ and (HCT > time Hct/2.5 or HCT ≥ 15)

17.4 Data Sources and Reports

Alerts are generated based on laboratory values (either one-time values or time rate of change of laboratory values), and in one instance (hypokalemia — patient on digoxin), pharmacy data is used as well.

All the alerts for a given unit may be reviewed at any terminal. Printouts can also be obtained (Figure 17.2). The alerts that appear on the screen contain the same information as the printed reports.

17.5 Evaluation

The CLAS was evaluated [2] for its effects on clinical parameters such as length of stay, frequency of appropriate treatment, and length of time in a life-threatening state. Program-related parameters were also measured, such as time from posting of the alert until acknowledgment. A user survey was also conducted.

A control group for the evaluation was obtained by retrospectively reviewing computerized data to see when an alert would have been generated, and then reviewing the patient's chart to see if appropriate action was taken. The same method was used after the system was implemented to collect information on the study group. Computerized data were examined to find the time of the alerting laboratory test and the time the alert was acknowledged, and the corresponding chart was examined for the times of treatments, follow-up tests, and other data.

```
                    SUMMARY OF PATIENT LAB ALERTS FOR 8E/N

TIME OF REPORT: 22 APR 87 11:12
--------------------------------------------------------------

2284129              COLE, LONA PERREY           E802
10APR 18:52          HYPOKALEMIA--K+ IS 3.1, K+ HAS FALLEN 1.4
                     MEQ/L IN 24 HRS
PAST ALERT           TIME OF REVIEW: 11 APR 07:00

2284129              COLE, LONA PERREY           E802
11APR 00:37          HYPOKALEMIA--K+ IS 2.9, K+ HAS FALLEN 1.6
                     MEQ/L IN 24 HRS
PAST ALERT           TIME OF REVIEW: 11 APR 07:00

2437960              QUIRK, DIANA NMI            E815
21APR 11:38          HCT FALLING--HCT HAS FALLEN 10.5 SINCE LAST
                     HCT, HCT IS NOW 29.0
CURRENT ALERT        TIME TIL REVIEW: 7.2 HOURS

2437692              SMITH, KRISTI MORTENSEN     E824
21APR 16:08          HCT FALLING--HCT HAS FALLEN 10.6 SINCE LAST
                     HCT, HCT IS NOW 27.5
CURRENT ALERT        TIME TIL REVIEW: 2.7 HOURS

                    *****END OF REPORT*****
```

Figure 17.2. Summary laboratory alert report for a single nursing division. Past alert means alert has already been acknowledged. Time of review is given. Current alert means alert is only now being acknowledged. Time since alert was generated is given. Names have been falsified.

Metabolic acidosis and falling hematocrit were the most frequent alerts. The remaining nine (having to do with sodium, potassium, and glucose, called Group 1 in the study) were less frequent and were considered as a single group. Alerting for falling hematocrit was shown not to have an impact on outcome and was not extensively studied.

For the Group 1 alerts, the frequency of appropriate treatment on general nursing floors increased significantly ($p = 0.017$) from 68.1% to 83.8% after the program was implemented. The length of time patients were in the "life-threatening" condition decreased from an average of 30.4 hours to 15.7 hours ($p = 0.0125$). The average total length of stay of patients generating an alert significantly from 350.6 hours to 211.9 hours.

Patients in the study group generating a metabolic acidosis alert spent significantly less time in the life-threatening condition (26.5 hours vs. 44.3 hours, $p = 0.033$). There was also a decrease in total length of stay (269.4 hours vs. 356.9 hours, $p < 0.2$) for the study group but the decrease was not statistically significant. The frequency of appropriate actions was about 33% for both study and control groups.

Alerts were acknowledged an average of 3.6 hours after they were posted on the computer. This is felt to be acceptable. On West 8, where the flashing light was tried, acknowledgment occurred 0.02 hours (< 2 minutes) after posting. However, as mentioned, the method was too annoying to be continued.

Personnel using the system were grouped as follows: Nurses, 53.0%; physicians, 28.5%; ward clerks, 11.5%; others, 7.0%. An opinion survey showed user's feelings toward the system were generally positive, with 25.6% of users classifying the system as "very useful," 24.4% as "frequently useful," 34.6% as "somewhat useful," 7.7% as "seldom useful," and 7.7% as "not useful."

17.6 Current Status

The program is currently a functional aspect of the HELP system. Future plans involve expansion of the CLAS to make metabolic acidosis and hematocrit alerts more accurate and useful and evaluation of the CLAS in another hospital setting.

References

[1] Gardner RM, Clemmer TP, Larsen KG, Johnson DS. Computerized alert system use in clinical medicine. *Symposium on computer applications in medical care (SCAMC)* 1979;3:136–140.
[2] Bradshaw KE. *A Computerized Laboratory Alerting System to Warn of Life-Threatening Events.* Salt Lake City, Utah: University of Utah; 1988. Dissertation.
[3] Bradshaw KE, Gardner RM, Pryor TA. Development of a computerized laboratory alerting system. *Comput Biomed Res* 1989;22:575–583.
[4] Tate KE, Gardner RM, Weaver LK. A computer laboratory alerting system. *MD Comput* 1990;7:296–301.

18
Computerized Infectious Disease Monitor

18.1 Overview

The microbiology module of the HELP system analyzes microbiology data and, with information from other modules in the patient file, indicates when infections are nosocomial, reportable, or otherwise unusual. It alerts pharmacists when antibiotic use is inappropriate with respect to adequate coverage of the involved organism, cost, timing of perioperative administration, and length of use for postoperative prophylaxis [1].

The microbiology module consists of two components. One is the clinical microbiology laboratory system, which produces the routine microbiology results for patients at LDS Hospital. The other component is the computerized infectious disease monitor (CIDM), which generates reports for the Department of Infectious Diseases and for the pharmacy based on analysis of microbiology and other data. The clinical microbiology component is part of the laboratory information system (LIS, described in detail in Chapter 16), which uses Prime hardware and LabForce software to manage laboratory workload and data entry. Data from the LIS are transferred to the HELP system for storage in the computerized database. The CIDM is a HELP system program that uses the microbiology data and other data in the patient database to generate its alerts.

18.2 Data Stored

The following data from microbiology test results are entered by laboratory technicians into the clinical microbiology laboratory system:

1. type of test
2. stage of test (e.g., preliminary, interim, final)
3. time and date of collection
4. date of completion

5. specimen accession number
6. specimen source and body site
7. stain results
8. culture results
9. antibiotic sensitivities
10. comments

18.3 Decision Logic

When data are entered into the clinical microbiology laboratory computer system, the information is also sent electronically to the HELP system. Here, the CIDM uses the microbiology data, data from the other departments (already in the HELP system), and a knowledge base created with the help of infectious disease specialists to generate alerts.

The system generates an alert if it identifies [2]:

1. a nosocomial infection
2. infection at a normally sterile body site
3. an unusual sensitivity pattern
4. an infection where the patient is not receiving antibiotics, or is receiving antibiotics to which the pathogen is resistant
5. an antibiotic that is not the least expensive suitable one
6. a reportable disease
7. prophylactic antibiotics given longer than indicated
8. patients whose preoperative antibiotic order timing deviates from accepted practice

The alerts are printed daily in a report described below.

To generate its alerts, the CIDM uses computerized data from the following departments:

1. chemistry results (renal and hepatic values, urinalysis)
2. long-term patient file (admission date, previous admission date, etc.)
3. surgery (date of surgery, type of surgery)
4. pharmacy (antibiotic usage, costs, allergies)
5. radiology (chest x-ray results)
6. respiratory therapy charting
7. nursing (isolation information)

18.4 Reports

The laboratory information system described in Chapter 16 produces routine microbiology results. An example of a routine microbiology report is included in Figure 18.1.

```
SMITH, JOHN NMI                     6001200              4S30

MICROBIOLOGY/INFECTIOUS DISEASE

-ROUTINE CULT-  ** PRELIMINARY REPORT **    25OCT 22:30
SOURCE: SPUTUM
STAIN:  MODERATE NUMBER OF EPITHELIAL CELLS
        MODERATE NUMBER OF WBCS
        NO BACTRIA OBSERVED

COMMENT: SPECIMEN MAY CONTAIN OROPHARYNGEAL MATERIAL


-BLD CULT-     ** PRELIMINARY REPORT **    24OCT 13:23
SOURCE: BLOOD
RESULT: STAPHYLOCOCCUS AUREUS   MODERATE GROWTH
        -SENSITIVE TO: Cephalosporin, Chloramphenicol, Erythromycin
                       Clindamycin, Nafcillin, Tetracycline
        -RESISTANT TO: Ampicillin, Penicillin-G
COMMENT: GROWTH IN 2 OUT OF 2 BOTTLES


-ROUTINE CULT-   ** FINAL REPORT **    22OCT 8:45     COMPLETED 24OCT
SOURCE: OF SPECIMEN, CATHETER, CVP TIP
STAIN:  NUMEROUS WBCS
        NUMEROUS GRAM POSITIVE COCCI
RESULT: STAPHYLOCOCCUS AUREUS   50,000 - 100,000 ORGANISMS PER ML
        -SENSITIVE TO: Ampicillin, Cephalosporin, Chloramphinicol,
                       Erythromycin, Gentamicin, Clindamycin,
        -RESISTANT TO: Penicillin-G, Nafcillin, Tetracyline,
                       Nitrofurantoin, Vancomycin
RESULT: STREPTOCOCCUS NON HEMOLYTIC   MODERATE GROWTH
RESULT: MICROCOCCI   LIGHT GROWTH
```

Figure 18.1. Example of output form clinical microbiology laboratory system. Name has been falsified. © 1985 Academic Press, New York. Reprinted with permission from "Development of a Computerized Infection Disease Monitor" by Evans, Gardner, Bush et al., *Computers and Biomedical Research*, 18: 103–113, 1985.

The alerts generated by the CIDM are printed in the Infectious Disease Monitor Report (IDMR). A sample of the IDMR is included as Figure 9.4 and is described in detail in the text. The IDMR is generated daily automatically at 12:30 PM and is reviewed by infectious disease specialists at their daily rounds.

Subsets of the IDMR are produced for two departments. The Respiratory Therapy Department receives a report of all patients with a nosocomial pneumonia. If two patients have nosocomial pneumonias with the same organism within 3 days of each other this will be identified. If they had a common respiratory therapist, that person will be identified. Additionally, the system lists daily patients who have pending or positive tests for infectious diseases that concern respiratory therapists (e.g. hepatitis, legionella, tuberculosis, mumps, rubella).

The pharmacy receives a report of all the antibiotic alerts. It is the responsibility of the clinical pharmacist to follow up on all drug related alerts (see below and Chapter 15 for further discussion).

18.5 Results of Studies

Studies [3,4,5] were done to examine patterns of antibiotic use in the hospital and to measure the effects of reminders to physicians when their practice differed from accepted standards.

Using HELP, it was easy to determine that only 40% of surgical patients requiring preoperative antibiotics were receiving them "on time" (i.e., within 2 hours before the start of surgery). A program (approved by the Pharmacy and Therapeutics Committee) was then implemented, whereby if the computer identified a patient that did not have an appropriate antibiotic order at 3:30 PM on the day before surgery, a clinical pharmacist would leave a note reminding the physician of that fact on the patient's chart. After implementation of this program, the number of cases with appropriate timing increased to 40% from 58% and cases with late administration dropped from 27% to 14%. More importantly, postoperative wound infection rates dropped significantly from 1.9% to 0.9% ($p < 0.02$) [4]. From further examination of the data, it seems that the wound infection rate increases the longer the perioperative dose is delayed.

It was also determined with the computer [6] that patients were receiving prophylactic antibiotics longer than was clinically indicated in many cases. Here, if the computer identified a patient receiving postoperative antibiotics without evidence of infection more than 48 hours after surgery, the pharmacist would place a "stop order" in the chart. This would terminate the antibiotics in 24 hours unless they were reordered by the physician. This reduced the percentage of patients receiving antibiotics more than 48 hours after surgery, as shown in Figure 18.2.

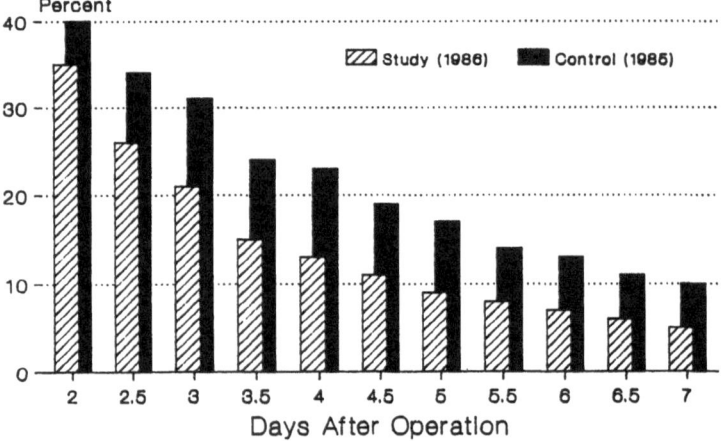

Figure 18.2. Percent of patients receiving antibiotics more than 48 hours after surgery before (control) and after (study) implementation of computer-generated alerts (used with the permission of Dr. Scott Evans).

The ability of the computer to identify nosocomial infections was examined and was found to have the same sensitivity and specificity as the traditional method of using infection control practitioners while requiring only one third the time [2]. This is the method now used for identifying nosocomial infections at LDS Hospital.

In a 2-month study [2], the computer generated 108 alerts regarding the use of therapeutic antibiotics. Fifty percent of these were because patients were on antibiotics to which the organism was resistant. Thirty six, or 67%, of these were felt to be clinically relevant by infectious disease specialists reviewing the alerts. The computer generated 20 false-positive alerts and had difficulty determining appropriate therapy for patients with multiple organisms.

References

[1] Evans RS, Hulse RK, Gardner RM, et al. Computer surveillance of hospital-acquired infections and antibiotic use. *JAMA* 1986;256:1007–1011.
[2] Evans RS, Gardner RM, Bush AR, et al. Development of a computerized infectious disease monitor (CIDM). *Comput Biomed Res* 1985;18:103–113.
[3] Evans RS, Gardner RM, Burke JP, et al. A computerized approach to monitor prophylactic antibiotics. *Symposium for computer applications in medical care (SCAMC)* 1987;11:241–245.
[4] Larsen RA, Evans RS, Burke JP, et al. Improved peri-operative antibiotic use and reduced surgical wound infections through the use of computer decision analysis. *Infect Control Hosp Epidemiol* 1989;10(7):316–320.
[5] Pestotnik SL, Evans RS, Burke JP, et al. Therapeutic antibiotic monitoring: surveillance using a computerized expert system. *Am J Med* 1990;88:43–48.
[6] Evans RS, Pestotnik SL, Burke JP, et al. Reducing the duration of prophylactic antibiotic use through computer monitoring of surgical patients. *DICP Ann Pharmacother* 1990;24:351–354.

19
Computerized Blood Gas Interpretation and Reporting

19.1 Overview

At LDS Hospital the HELP system is used to order blood gases and to calculate, interpret, and report the results. Orders are entered into the system on the nursing division and are automatically transmitted to the blood gas laboratory. When technicians enter the blood gas results into the HELP system, HELP's decision logic present is activated and interpretations of the test are generated. The computerized interpretation requires previous blood gas results and other data present in the computerized patient file. The raw blood gas results, along with the computerized interpretation, are reported automatically and promptly to the nursing unit and are included in a variety of summary reports.

19.2 Order Entry

Nurses or clerks request blood gases from the acute care or intensive care units by entering the order into the computer system (see Order Entry, Chapter 11). They specify whether the request is for arterial or venous gases (or both) and the time the test is to be done. The order information is automatically printed out in the blood gas laboratory (and on the nursing unit as well) and a blood gas technician goes to the nursing unit at the appropriate time.

19.3 Measurements

The pH, pCO_2, pO_2 (corrected for temperature), and hemoglobin are measured routinely in our laboratory. Oxygen saturation, carboxyhemoglobin, and methemoglobin are also measured in most patients. Techniques of measurements, calculations, and interpretations of results are based on published regional standards [1]. Frequently, oxygen saturation from a pulse oximeter is also entered.

19.4 Calculations

From the measured data and the recorded FiO_2, the HCO_3-, base excess (BE), oxygen content, and A-a gradient are calculated. Arterial and venous samples are compared and the right-to-left shunt is calculated [2]. Oxygen saturation can be calculated if it is not measured directly but this practice is decreasing at our hospital. Oxygen content is calculated from the hemoglobin, oxygen saturation, and pO_2 values.

19.5 Data Integration

In forming its reports and interpretations, the blood gas module makes excellent use of other patient data available in the HELP database [3]. Respiratory therapy data are examined to see if the patient is on oxygen therapy. If the patient is hypoxemic and receiving oxygen therapy, this fact is indicated when the interpretation is printed. The computer examines urinalysis data in patients with metabolic acidosis. If glucosuria is present it suggests that diabetic ketoacidosis may be an etiology for the acid-base disturbance. Often ICU patients have cardiac outputs measured through thermodilution techniques. These data, with simultaneous arterial and venous blood gases, can be used to calculate oxygen consumption.

19.6 Decision Logic

The computer uses data from published standards [1] to determine if PO_2, oxygen saturation ($\%SO_2$), or acid-base measurements are normal or not. Abnormalities of PO_2 and $\%SO_2$ are characterized as moderate or severe. An acid-base disturbance (acidosis, alkalosis) may be characterized as mild, although if it is moderate or severe it will be identified as acute or chronic, respiratory or metabolic, or a mixture of these.

From the PCO_2 data the computer will comment on the quality of ventilation. From the PO_2 it will indicate hypoxemia if present, with an additional comment if the patient is already receiving oxygen therapy. If the PO_2 is in a precarious range, an alert statement will be printed for the technician to contact the nurse or physician.

The computer also performs trend analyses and indicates when previously normal values are abnormal and when abnormal ones have improved.

L D S H O S P I T A L B L O O D G A S R E P O R T

MARIE   NO. 034196   DR. NATHAN C.   SEX F AGE 53 ROOM E608

JUL 02 90

	pH	PCO2	HCO3	BE	HB	CO/MT	PO2	SO2	O2CT	%O2	AVO2	VO2	C.O.	A-a	Qs/Qt	PK/ PL/PP	MR/SR
	--	----					---										
NORMAL HI	7.45	41.3	26.4	2.5	15.9	2/1	61	90	16.0		5.5	300	7.30	23	5		
NORMAL LOW	7.35	27.9	16.1	-2.5	11.9	0/1					3.0	200	2.90		0		
02 09:05 A	7.31	67.9	33.5	5.9	11.1	1/0	57	88	13.7	50				163		58/ 47/10	38/
02 05:15 A	7.29	69.6	32.7	4.9	10.5	1/1	70	91	13.5	50				148		/ /12	26/
01 20:40 A	7.36	59.1	32.9	5.9	14.9	1/1	52	93	19.5	40				117		/ /12	26/
01 20:15 A	7.33	65.0	33.6	6.6	10.5	1/1	75	91	13.5	50				148		62/ 44/12	26/
01 15:20 A	7.32	65.9	33.3	6.0	10.7	2/1	64	89	13.4	50				158		/ /10	25/

02 09:05 A
SAMPLE # 302, TEMP 37.0, BREATHING STATUS : ASSIST/CONTROL
SEVERE CHRONIC RESPIRATORY ACIDOSIS
MODERATE HYPOXEMIA
MODERATELY REDUCED O2 CONTENT
HYPOVENTILATION NOT IMPROVED
PULSE OXIMETER SO2 89.0

02 05:15 A
SAMPLE # 301, TEMP 37.5, BREATHING STATUS : ASSIST/CONTROL
SEVERE MIXED CHRONIC AND ACUTE RESPIRATORY ACIDOSIS
MODERATELY REDUCED O2 CONTENT
HYPOVENTILATION MUCH WORSE
PULSE OXIMETER SO2 92.0

01 20:40 A
SAMPLE # 300, TEMP 37.7, BREATHING STATUS : ASSIST/CONTROL
SEVERE CHRONIC RESPIRATORY ACIDOSIS
MODERATE HYPOXEMIA
HYPOVENTILATION IMPROVED
PULSE OXIMETER SO2 88.0

01 20:15 A
SAMPLE # 299, TEMP 37.7, BREATHING STATUS : ASSIST/CONTROL
SEVERE CHRONIC RESPIRATORY ACIDOSIS
MODERATELY REDUCED O2 CONTENT
HYPOVENTILATION NOT IMPROVED
PULSE OXIMETER SO2 93.0

01 15:20 A
SAMPLE # 298, TEMP 37.9, BREATHING STATUS : ASSIST/CONTROL

SEVERE CHRONIC RESPIRATORY ACIDOSIS
MILD HYPOXEMIA
MODERATELY REDUCED O2 CONTENT
HYPOVENTILATION NOT IMPROVED
PULSE OXIMETER SO2 91.0

01 11:20 A 7.34 64.1 34.0 7.0 11.3 2/1 56 87 13.8 50   168   63/ 50/10 33/
SAMPLE # 297, TEMP 37.5, BREATHING STATUS : ASSIST/CONTROL
SEVERE CHRONIC RESPIRATORY ACIDOSIS
MODERATE HYPOXEMIA
MODERATELY REDUCED O2 CONTENT
HYPOVENTILATION IMPROVED
PULSE OXIMETER SO2 91.0

01 05:35 A 7.30 72.7 35.0 6.8 11.5 2/1 81 92 15.0 70   256   39/ 51/12 35/
SAMPLE # 296, TEMP 37.9, BREATHING STATUS : ASSIST/CONTROL
SEVERE CHRONIC RESPIRATORY ACIDOSIS
MILDLY REDUCED O2 CONTENT
HYPOVENTILATION WORSE
PULSE OXIMETER SO2 93.0

01 02:36 A 7.32 66.5 33.6 6.3 10.8 1/1 54 85 12.9 60   229   55/ 41/10 32/
SAMPLE # 295, TEMP 37.8, BREATHING STATUS : ASSIST/CONTROL
SEVERE CHRONIC RESPIRATORY ACIDOSIS
MODERATE HYPOXEMIA
MODERATELY REDUCED O2 CONTENT
HYPOVENTILATION WORSE
PULSE OXIMETER SO2 86.0

PRELIMINARY INTERPRETATION -- BASED ONLY ON BLOOD GAS DATA. ***(FINAL DIAGNOSIS REQUIRES CLINICAL CORRELATION)***
KEY: CO=CARBOXY HB, MT=MET HB, O2CT=O2 CONTENT, AVO2=ART VENOUS CONTENT DIFFERENCE (CALCULATED WITH AVERAGE OF A &V HB VALUES),
VO2=OXYGEN CONSUMPTION, C.O.=CARDIAC OUTPUT, A-a=ALVEOLAR arterial O2 DIFFERENCE, Qs/Qt=SHUNT, PK=PEAK, PL=PLATEAU, PP=PEEP
MR=MACHINE RATE, SR=SPONTANEOUS RATE.
*** SPECIMEN IDENTIFICATION: BLOOD (A=ARTERIAL, V=VENOUS, C=CAPILLARY, W=WEDGE);
FLUIDS (P=PLEURAL, J=JOINT, B=ABDOMINAL, S=ABSCESS); E=EXPIRED AIR;
ECCO2R (I=INFLOW, M=MIDFLOW, O=OUTFLOW)

KEEP FULL PAGE FOR RECORDS

Figure 19.1. Blood gas report (see text for further details).

19.7 Reports

The blood gas results are entered into the central computer system by the technicians [4]. Within minutes a report is automatically printed on the unit where the test was ordered. A complete page is printed which contains the five most recent blood gas results. A copy of this page, or a complete list of the patient's blood gas results, is available on the nursing unit at any time. Results may also be viewed on any computer terminal.

A sample blood gas report is shown in Figure 19.1. Blood gas data for the previous 12 hours appear on the computerized nursing shift report, which is shown in Figure 13.10. The ICU rounds report (Figure 4.4) contains blood gas data for the previous 24-hour period.

In Figure 19.1, patient identifying data appear first followed by normal values (based on sex and age) for all measured and calculated values. Next, all the blood gas results for the patient are listed, the most recent one first. (The abbreviation for the headings are explained at the bottom of the figure.) For each test (1) the number of the test is listed (e.g., 13 means the 13th blood gas analysis on this patient on this hospital stay), (2) an "A" or "V" appears indicating whether the test was arterial or venous, (3) the time the sample was drawn, and (4) the results. If arterial and venous samples were done simultaneously the difference in oxygen content is listed and if a cardiac output was measured, this is listed with the calculated oxygen consumption. If the patient is receiving oxygen the FiO_2 is listed and if the patient is on a ventilator the settings are noted.

These results are followed by the computer's interpretations as well as a disclaimer stating that final interpretations require clinical correlation. When the patient is discharged, a copy of the entire blood gas record is printed in the medical record department for inclusion in the permanent medical record.

19.8 Evaluation

Shortly after the system was implemented, a study [5] was done to examine how the computer's interpretation compared with seven pulmonary specialists. Because the physicians' interpretations of individual blood gas measurements varied greatly, the statistic measured was agreement of the computer with most of the physicians. The specialists agreed with the majority 61.1% of the time (range, 52.5–74.5%). The computer agreed with the majority 74.0% of the time, performing as well as the highest scoring physician.

The same study also evaluated the physicians' attitudes regarding the helpfulness and accuracy of the computer's interpretations. Evaluation forms for physicians to complete were placed in the chart with the computer interpretation. Out of 68 responses, 54 (80%) said they felt the interpretations were helpful. In 19 cases (28%) physicians said the interpretation led to a change in therapy. Only once did a physician state that he disagreed with the computer interpretation.

A later, independent evaluation of the HELP system said that the physicians, despite high expectations from the computerized blood gas laboratory, felt their expectations were fulfilled 81% of the time.

References

[1] Morris AH, Kanner RE, Crapo RO, Gardner RM. *Clinical Pulmonary Function Testing: A Manual of Uniform Laboratory Procedures.* 2nd ed. Intermountain Thoracic Society, Salt Lake City, Utah. 1984.

[2] Clemmer TP, Gardner RM, Orme JF Jr. Computer support in critical care medicine. *Symposium on computer applications in medical care (SCAMC)* 1980; 4:1557–1561.

[3] Gardner RM, Crapo RO, Morris AH, Beus ML. Computerized decision-making in the pulmonary function laboratory. *Resp Care* 1982;27:799–808.

[4] Clemmer TP, Gardner RM. Data gathering, analysis, and display in critical care medicine. *Resp Care* 1985;30:586–598.

[5] Gardner RM, Cannon GH, Morris AH, Olsen KR, Price GA. Computerized blood gas interpretation and reporting system. *IEEE Comput* 1975;8(1):39–45.

20
Medical Information Bus

20.1 Overview

What began as a joint effort between LDS Hospital and Phoenix Baptist Hospital in Phoenix, Arizona, has led to an approach to the systematic gathering of computerized patient physiological data and the management of therapeutic devices in the intensive care and surgery settings [1–3]. The Medical Information Bus (MIB) is a network that enables up to 250 micro-processor-based patient devices (either infusion pumps or monitoring devices) to communicate with a central computer through the use of program-mable device communications controllers, or DCCs. Data from medical devices are converted into a standardized format through the use of the DCCs. The DCCs from the different devices transmit the data across a network bus to a master communications controller (MCC). In the HELP system the data are transferred to a CRDS (Charles River Data Systems; see Chapter 2) minicomputer which in turn transmits it to the Tandem computer for permanent storage. Data may flow in the reverse direction as well and instructions may be sent to the devices by health-care personnel from any of the nursing division's computer terminals.

The cable backbone required by the bus is in place in all the LDS Hospi-tal ICUs and surgical suites. Electronic transmission of data from, and instructions to, microprocessor-based infusion pumps is currently opera-tional in the adult ICUs. Device communications controllers are being de-veloped for other monitoring and therapeutic devices and implementation-related issues are being examined and resolved. Since clinical monitoring devices can generate large amounts of data that can overwhelm even power-ful computers, current efforts are focusing on how to gather data from the monitoring devices selectively, yet still meaningfully [4].

The early work done at LDS Hospital has contributed to the develop-ment of a proposed IEEE standard (P1073) for the medical information bus. Since some of the early and difficult lessons regarding the implementa-

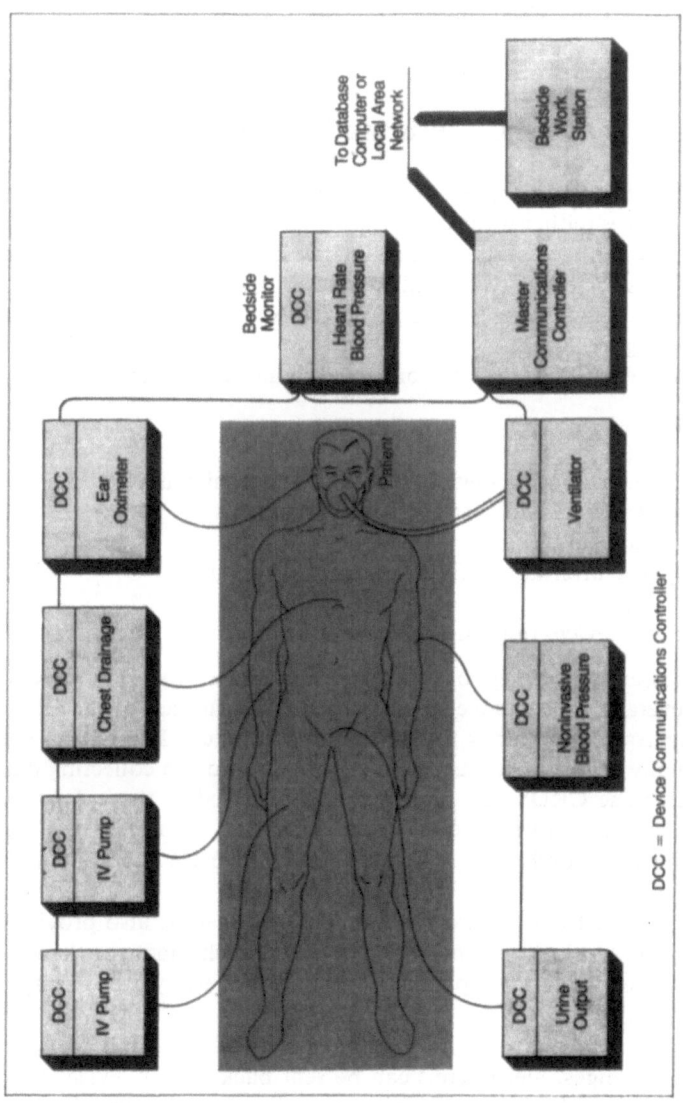

Figure 20.1. Organization of the medical information bus (MIB) in the intensive care setting (see text for further details). © 1986 Springer-Verlag, New York. Reprinted with permission from "Computerized Management of Intensive Care Patients" by Gardner, *MD Computing* 44, March 1986.

tion of such a system have been learned at LDS Hospital, differences exist between some of the proposed standards [3] and what has actually been implemented here. This chapter describes what has been implemented and will point out when the local implementation differs from the proposed standards.

20.2 Design of the Medical Information Bus

As mentioned, the MIB is a network that enables several microprocessor-based patient devices (either infusion pumps or monitoring devices) to communicate with the central computer through the use of DCCs. The motivation for the MIB is to be able to centralize data collection and device control in intensive care settings where multiple devices may be involved in a patient's care (see Figure 20.1).

Examples of devices that may be connected to an MIB are [5]:

1. infusion pumps
2. ventilators
3. blood pressure monitoring devices (invasive and noninvasive)
4. ECG monitors
5. temperature monitors
6. oximeters (pulse/arterial and fiberoptic mixed venous)
7. gastric pH meters
8. fluid collection devices (e.g., urine, chest tube drainage)

An MCC (essentially a network controller) manages the flow of data between the several DCCs and a central computer. In the case of the HELP system, the central computer is a CRDS minicomputer. The CRDS machines serve many other functions in the ICU in addition to collecting data from the MIB. (The CRDS machines are also involved in the capture of data from bedside monitoring devices to the Tandem computer and also serve as multiplexors for HELP system terminals. See Chapters 2 and 21 for details.)

In the proposed IEEE standard for the MIB [3], there is also provision for a bedside communications controller (BCC), which manages the data flow for all devices for a given patient. The BCC has not yet been implemented at LDS Hospital.

Communication flows along the MIB in two directions, and messages (e.g., flow rate changes, alerts, etc.) can be sent back to the devices from the central computer or the MCC. This capacity theoretically gives the computer the ability to make a decision (e.g., increase the level of an infused medication to correct a physiological deficit) based on monitored data and automatically transmit an instruction to the appropriate device (i.e., "closed loop" treatments). There are no plans, however, to exploit this capacity at LDS Hospital in the foreseeable future.

20.3 Device Communications Controller

A DCC consists of:

1. a central microprocessor
2. computer memory
3. a real-time clock
4. a serial interface unit (SIU), which connects the device to the bus
5. two other input/output ports (one for communications with the device, and one for another instrument, e.g., a bar code reader, alarm, etc.)
6. an erasable programmable read-only-memory (EPROM) unit

The memory and the clock can be used if the patient is moved away from the bedside but the device (e.g., infusion pump) is still operative. During that time data would be stored in memory and transmitted to the MIB when the patient returns and the DCC is reconnected to the bus.

The EPROM contains the software necessary to translate the output from the unit into a common (standard) language, the Medical Device Data Language (MDDL). The EPROM also contains parameters for collection of data (e.g. transmit a blood pressure reading every 1 minute). The EPROM can also screen data and eliminate obviously erroneous readings according to specifications (e.g., ignore values for respiratory rate > 200). These parameters can be changed by health-care personnel using menus on the computer terminals.

20.4 Standards

Monitoring devices and infusion pumps are made by several different manufacturers and generate electronic digital output data in a variety of formats. One purpose of the DCCs is to convert the data into a standard language (MDDL). The MDDL provides for device addresses and for error checking as it is transmitted along the bus [6]. The DCCs are connected to single shielded twisted-pair cable with RS-485 adapters and communicate with the MCC at 375 kilobits per second (kbps). The synchronous data link control (SDLC) protocol provides highly error-resistant data transmission.

These and other parameters are being molded into the proposed IEEE standard (P1073). Standardization will allow manufacturers to build their devices to these standards and eventually eliminate the need for a free-standing DCC.

20.5 Current Implementation

The bus is currently being used to gather data from, and send instructions to, infusion pumps in the ICUs and surgical suites at LDS Hospital. An anesthesia information system (AIS) has been designed that integrates data

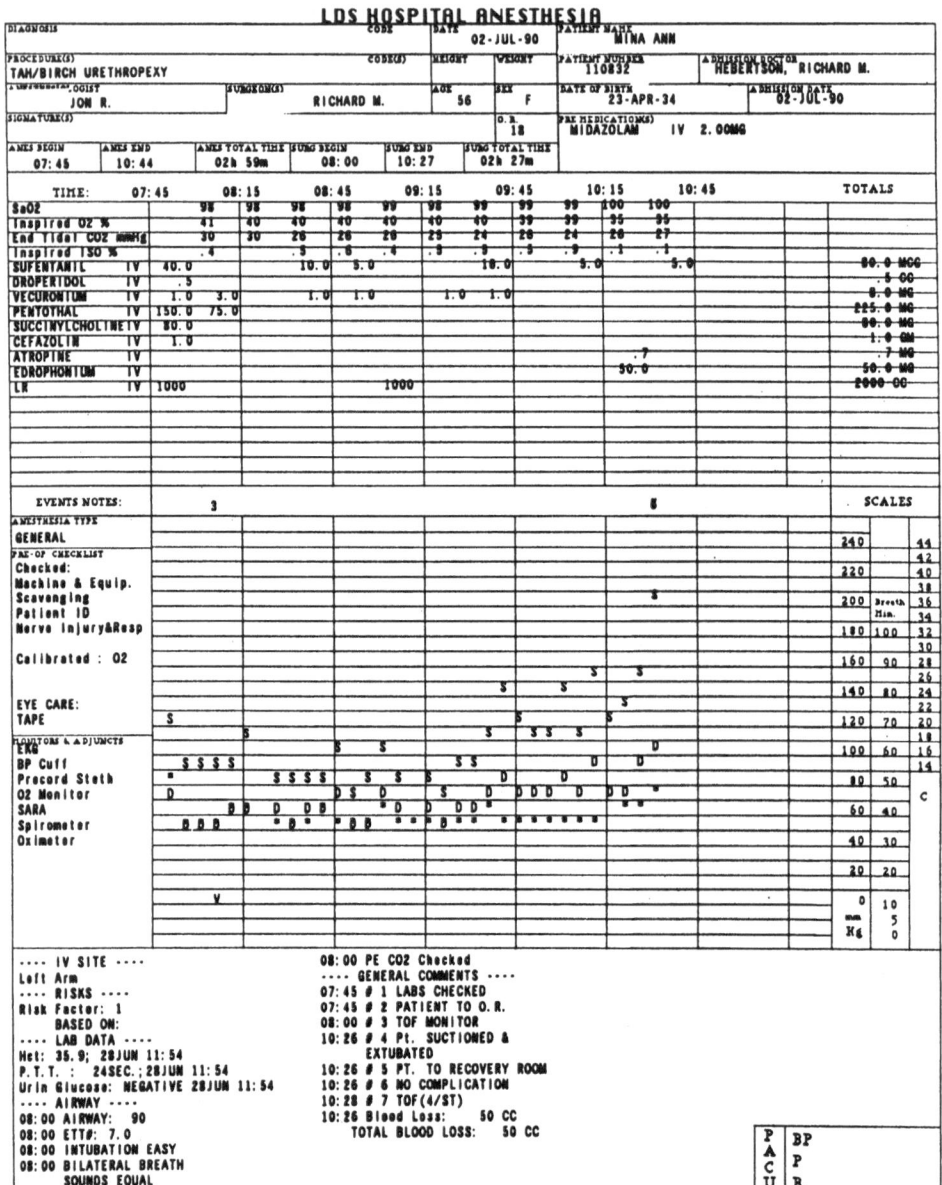

Figure 20.2. Automated anesthesia record (see text for description).

from blood pressure monitors, oximeters, and cardiac monitors for use by anesthesiologists during surgery. Data from the various devices come through an MIB installed in the surgery suite [7]. Anesthetic and other medication information come directly from the HELP computer and form an integrated anesthesia record (Figure 20.2).

Demographic and diagnostic information are shown at the top of Figure 20.2 along with premedication data and surgical times data. The next section shows medication data in a flow chart format with totals on the left hand side. The middle left hand side of the form shows preoperative procedures that have been performed by the anesthesiologist. These data have been entered from menu screens. Vital sign data (automatically captured from the MIB) are shown in the center of the form with comments data (also entered from menu-driven screens) shown at the bottom.

Research is under way to develop a knowledge base that will automatically screen for dangerous situations. The bus has also been used to gather data on various ventilator parameters. These data are in the process of being analyzed and compared to respiratory therapists charting data to see if recording of data from the bus is easier and/or more accurate. Physician and nurse response to the AIS has been strongly favorable.

References

[1] Hawley WL, Tariq H, Gardner RM. Clinical implementation of an automated medical information bus in an intensive care unit. *Symposium for computer applications in medical care (SCAMC)* 1988;12:621–624.

[2] Tariq H, Gardner RM, Hawley WL. Implementation of a medical information bus at LDS Hospital. Proceedings of the Annual International Conference of the IEEE Engineering in Medicine and Biology Society; 1988;10:1799–1800.

[3] Shabot MM. Standardized acquisition of bedside data: The IEEE P1073 medical information bus. *Intl J Clin Monit Comput* 1989;6:197–204.

[4] Gardner RM, Tariq H, Hawley WL, et al. Medical information bus: the key to future integrated monitoring. *Intl J Clin Monit Comput* 1989;6:205–209. Editorial.

[5] Gardner RM. Computerized management of intensive care patients. *MD Comput* 1986;3(1):36–51.

[6] Gardner RM, Hawley WL. Standardizing communications & networks in the ICU. Patient monitoring and data management conference. AAMI Technology Analysis and Review TAR 1986; No. 11–85:59–63.

[7] Wang X. *Design, development, implementation, and evaluation of a computerized anesthesia charting system.* Salt Lake City, Utah: University of Utah; 1990. Dissertation.

21
Hemodynamic Monitoring

21.1 Overview

The quest for many kinds of hemodynamic data motivated the computerization projects in the early 1960s that eventually grew into the HELP system. Whereas the role of actually capturing the signals and converting them to digital form has now been relegated to monitoring systems purchased from external vendors, hemodynamic measurements are still stored and manipulated by the HELP system [1–4]. For patients undergoing continuous blood pressure monitoring (either invasive or noninvasive), blood pressure readings are stored automatically by the HELP system every 5 to 15 minutes. Cardiac output measurements obtained via thermodilution techniques are routinely processed to yield a set of clinically useful cardiovascular parameters. Monitoring is performed most frequently in the ICUs and in the open-heart surgical suites.

21.2 Continuous Blood Pressure Monitoring

At our hospital blood pressure may be measured continuously by either an indwelling catheter (arterial line, which can give moment-to-moment blood pressure readings), or by a noninvasive pressure monitoring system (Dynamap™), which samples the blood pressure intermittently. In either case, the blood pressure values are transferred automatically to the HELP system.

As shown in Figure 21.1, the bedside monitors all feed their measurements in digital form to a central multiplexor. Each vendor provides a multiplexor with their system. The multiplexors manage the display of the signals on specialized (i.e., non-HELP system) displays at the nurses' stations. The multiplexors are also connected to a unix-based CRDS (Charles River Data Systems) minicomputer by an RS-422 serial link. Every 5 to 15 minutes the CRDS machine requests a copy of the data for each monitor

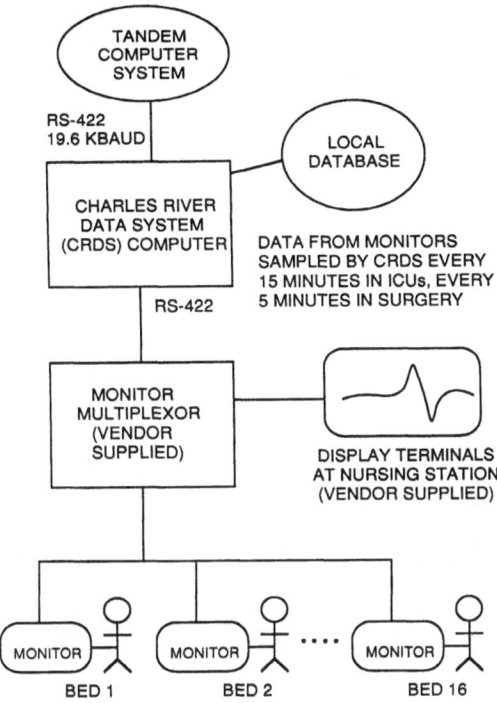

Figure 21.1. Connection of bedside hemodynamic monitors to HELP system (see text for further details).

through the multiplexor. For patients undergoing open heart surgery, the request for data occurs every 5 minutes. The data are stored locally on the CRDS machines and then rapidly sent in batch mode across high-speed serial lines to the Tandem for permanent storage in the computerized patient data file.

The CRDS machines also serve as the master communications controllers for the MIB (see Chapter 20) for each unit. Eventually, when the MIB standard has matured, bedside hemodynamic monitors may be connected directly to the MIB.

The hemodynamic measurements are printed and displayed graphically for easy reference in a variety of reports and are used for calculations of other cardiovascular parameters.

21.3 Cardiac Output Measurements

When a thermodilution cardiac output measurement is made, the nurse obtains five consecutive measurements. From these results the nurse chooses the three that are felt to be most representative of the patient's condition. These three results are automatically transmitted to the HELP computer where the result is averaged and stored in the patient's electronic file. These results, along with other measurements and calculations, appear in the cardiac output report (Figure 21.2).

The HELP system uses a knowledge base to interpret the parameters. The interpretations are printed below the numerical results along with cardiac medications the patient may be receiving.

```
                    C A R D I A C   O U T P U T   R E P O R T

              EDYTHE M.        NO.   089693 DR PEARL, JAMES E.        RM E601
HT    CM      WT104.80  KG      BSA  2.06 SQM          AGE 73    SEX F
TIME             CO    CI   HR  SV   SI  MP  MSP  PA  RA  PW  PVR  SVR  RWI  LWI
NORMAL HI      7.30  3.50  89 101   48 105 123  19 5.0 12 1.0   18 11.0   85
NORMAL LOW     2.90  2.80  49  47   38  70  80   9 1.0  4 0.5   12  8.0   48

JUL 02 03:59  12.20  5.91  95 128   62  85 101  18 4.0M 2 1.3    7 11.8   83
              HYPERDYNAMIC LV FUNCTION

JUL 01 04:00  14.00  6.79  97 144   70  74  87  19 8.0M 9  .7    5 10.5   74
              JUN 30 23:21  DOPAMINE (INTROPIN)    3 MCG/KG/MIN
              HYPERDYNAMIC LV FUNCTION

JUN 30 04:05  13.50  6.45 100 135   65  77  91  20 7.0M 8  .9    5 11.5   73
              JUN 30 02:32  DOPAMINE (INTROPIN)   10 MCG/KG/MIN
              HYPERDYNAMIC LV FUNCTION

JUN 29 04:15  12.00  5.73  92 130   62  81 102  19 7.0M 8  .9    6 10.1   79
              JUN 28 18:01  DOPAMINE (INTROPIN)    5 MCG/KG/MIN
              HYPERDYNAMIC LV FUNCTION

JUN 28 04:18  12.30  6.04 107 115   56  67  90  20 8.0M12  .7    5  9.1   59
              LV PARAMETERS ARE WITHIN NORMAL LIMITS

JUN 27 04:15  12.70  6.24 129  98   48  66  83  23 8.0M12  .9    5  9.8   46
              LV PARAMETERS ARE WITHIN NORMAL LIMITS

JUN 26 04:11   8.80  4.33 147  60   29  76  94  25 8.0M12 1.5    8  6.7   32
              MILD LV DYSFUNCTION

JUN 25 04:00  11.70  5.73 130  90   44  84 113  23 9.0M10 1.1    6  8.4   62
              LV PARAMETERS ARE WITHIN NORMAL LIMITS
(END)
```

Figure 21.2. Cardiac output report. Note interpretations and medications listed. See text for further details. CO = cardiac output, CI = cardiac index, HR = heart rate, SV = stroke volume, SI = stroke index, MP = mean pressure, MSP = mean systolic pressure, PA = pulmonary artery pressure, RA = right atrial pressure, PW = pulmonary wedge pressure, PVR = pulmonary vascular resistance, SVR = systemic vascular resistance, RWI = right ventricular work index, LWI = left ventricular work index.

HEMODYNAMIC PRESSURE REPORT   JUL 02 88 18:01   TO   JUL 03 88 06:00

SU KRISTINE   NO. *2210   RM E604

DATE	TIME	ARTERIAL** SP/DP	MP	HR	PA (MMHG) SP/DP	MEAN	WEDGE	LAP* (MMHG)	RAP* (MMHG)	CVP* (CMH2O)	AUG (MMHG)* SP/AP/DP
JUL 02 88	21:00	96/ 53	68	122	/						/
	20:45	99/ 56	70	120	/						/
	20:30	103/ 56	73	122	/						/
	20:15	97/ 52	68	122	/						/
	20:00	99/ 54		122	48/ 26	33	12				/
	20:00	98/ 53	68	120	/				10		/
	19:45	99/ 56	71	122	/						/
	19:30	97/ 55	68	122	/						/
	19:15	102/ 58	72	122	/						/
	19:00	104/ 60	76	124	/						/
	18:45	101/ 55	70	122							/
	18:30	/		124							/
	18:15	100/ 56	73	120	/						/

FINISHED: JUL 02 88 21:14

(END)

Figure 21.3. Hemodynamic pressure report including arterial and Swan-Ganz measurements. SP = systolic pressure, DP = diastolic pressure, MP = mean pressure, HR = heart rate, PA = pulmonary artery pressures, LAP = left atrial pressure, RAP = right atrial pressure, CVP = central venous pressure, AUG = augmented (used for balloon pump patients).

21.4 Reports

Hemodynamic measurements are displayed on the following routine HELP system reports:

1. *ICU Rounds Report* (Figure 4.4): This temporary report is produced daily for ICU housestaff rounds and contains, in addition to hemodynamic data, a significant amount of patient data organized by organ system.

2. *7-Day Report* (Figure 4.5): The 7-day report shows blood pressure for a period up to the past 7 days graphically. Medications and metabolic parameters are also displayed to assist in decision making.

3. *Computerized Nursing Shift Report* (Figure 13.10): This report is generated after each shift and becomes part of the permanent record. Blood pressure for the shift is shown graphically. Cardiac output parameters and other measurements and laboratory tests are listed.

4. *Cardiac Output Report* (Figure 21.2): Normal values for the parameters can be seen along with height, weight, and body surface area. Below this are the numerical results (most recent ones first), the relevant medications, and the HELP system's interpretation.

5. *Hemodynamic Pressure Report* (Figure 21.3): This report will print out hemodynamic pressure readings for the current or any other shift.

References

[1] Clemmer TP, Gardner RM. Data gathering, analysis, and display in critical care medicine. *Resp Care* 1985;30:586–598.
[2] Gardner RM. Computerized management of intensive care patients. *MD Comput* 1986;3(1):36–51.
[3] Gardner RM, Sittig DF, Budd MC. The computer in the ICU: match or mismatch? In: WC Shoemaker ed. *Textbook of Critical Care*. 2nd ed. Philadelphia: WB Saunders, 1989:248–258.
[4] Gardner RM, Monis SM, Oehler P. Monitoring direct blood pressure: algorithm enhancements. *IEEE Comput Cardiol* 1986;13:607–610.

22
Computerized Total Parenteral Nutrition Ordering and Nutritional Therapy Monitoring

22.1 Overview

Physicians at LDS Hospital order TPN through the computer. The computer prompts the physician to consider multiple possible regimens and the final order is evaluated by the HELP system's knowledge base to screen for certain inconsistencies. Concurrent review of nutritional therapy administration is done by the Nutritional Support Services (NSS) staff. The NSS makes use of several computer-generated reports to assist them in their duties.

22.2 Computer-Assisted Ordering

To order TPN for their patients, physicians enter their identification number at any one of the HELP system's terminals. After selecting the patient, the physician is presented with a series of screens that asks for the patient's height, weight, sex, and age (only if these data are not already present in the computerized database). From these data, the computer is able to calculate the patient's basal energy expenditure (BEE), which is printed on the final TPN order (Figure 22.1) and on the daily ICU Rounds Report (Figure 4.4). The physician enters whether he/she wants the patient to be fluid restricted or not, and whether he/she wishes the solution to be administered centrally or peripherally. The patient's most recent electrolyte values are then displayed for reference.

The physician indicates whether this prescription is for an adult or pediatric patient and the computer presents a list of solutions for the physician to choose from. The choices are standard solution, fluid restricted solution, renal failure solution, and modified (customized) solution.

The standard solution contains 500 ml of D_{50} and 500 ml of standard Aminosyn 8.5% (amino acid solution). The fluid-restricted solution con-

```
*************************    ****  CENTRAL   SOLUTION ****
TEST, ROOM ONE                           TEST    50000140 5/10/88  14:20
1-  65 YEARS OLD                   4- 150.00 cm  HEIGHT
2-        M                        5-  65.00 kg  WEIGHT
3- 1272.0 BEE                      6- NOT FLUID RESTRICTED
*****************************************************************
* PHARMACY ORDERS *                                             *
*                                                               *
*    500.0 ML   AMINOSYN      8.5%                              *
*    500.0 ML   DEXTROSE 50%                                    *
*     20.0 ML   LYPHOLYTE II                                    *
*      4.0 ML   SODIUM PHOSPHATE ( 4.0 mEq/ml )                 *
*      6.0 ML   MVI                                             *
*      0.7 ML   TRACE ELEMENTS                                  *
*     20.0 UN   REGULAR INSULIN                                 *
*   1030.7 ML   * TOTAL VOLUME *                                *
*                                                               *
*         MAKE 2 BOTTLES                                        *
*         SEND   500 ML OF LIPID EMULSION 10%                   *
*****************************************************************
*****************************************************************
* NURSING ORDERS *                                              *
*          - RUN TPN BOTTLES AT   72 ML PER HOUR                *
*          - RUN LIPID 10% EMULSION AT 125 ML PER HOUR          *
*               UNTIL   500 ML OF LIPID HAVE BEEN INFUSED       *
*          - ROUTINE MONITORING (WT., I&O, AND UR GLUC Q6H)     *
*****************************************************************
*****************************************************************
* LAB ORDERS - TO BE COMPLETED TOMORROW *                       *
*     - SMA-7                                                   *
*****************************************************************
EACH BOTTLE OF TPN (TOTAL VOLUME 1031 ML) SUPPLIES THE FOLLOWING:

(SCHEDULE THIS ORDER USING "MODIFED AMINOSYN      SOLUTION")

*    AMINOSYN       8.5%    500 CC  ( 41.0 GM PROTEIN)    *
*    DEXTROSE 50% 500cc  (end conc. 24.3%) 0.98 t.cal/ml TPN*
*    SODIUM       51.0 mEq     *  CHLORIDE   35.0 mEq     *
*    POTASSIUM    20.0 mEq     *  ACETATE    29.5 mEq     *
*    CALCIUM       4.5 mEq     *  PHOSPHATE  24.0 mEq     *
*    MAGNESIUM     5.0 mEq     *                          *
*    ZINC          3.5 mg      *  MVI-12      6.0 cc      *
*    COPPER        0.7 mg      *  INSULIN    20.0 units   *
*    MANGANESE     0.4 mg      *                          *
*    CHROMIUM      7.0 mcg     *                          *
*    *TOTAL VOL1030.7 cc *     *  (E-LYTE VOL 24.0 cc  )    *

* NUTRITIONAL SUMMARY FOR 24 HOURS:
    2224 ML TOTAL THERAPY
    1724 ML OF TPN GIVEN
     500 ML OF LIPID 10% EMULSION  (0.77 GM/KG/DAY)
     182 NON-PROTEIN CALORIES / GM NITROGEN
    1972 TOTAL NON-PROTEIN CALORIES ( 1.55 TIMES PATIENT'S BEE)
         1422 DEXTROSE CALORIES
          550 LIPID CALORIES
    2246 TOTAL CALORIES ( 1.77 TIMES PATIENT'S BEE)
      69 GM  PROTEIN  ( 1.1 GM / KG / DAY)

*****************************************************************
****  THIS IS NOT AN ORDER.  DO NOT SEND TO PHARMACY. ****
*****************************************************************

(END)
```

Figure 22.1. Computerized total parenteral nutrition (TPN) order.

tains D_{70} (so an isocaloric amount may be given with less volume). The renal failure solution contains Aminosyn RF (modified for renal failure patients) and D_{70}. Choosing the modified solution allows the physician to specify multiple parameters including total volume, lipid calories, protein calories, nonprotein calories, nitrogen/calorie ratio, g/kg/day of protein, and the type of amino acid solution.

Once the solution is chosen, the mixture of electrolytes can be determined. The pharmacy keeps two sets of electrolytes in stock that are suitable for most patients. They vary in their potassium, sodium, chloride, acetate, and magnesium concentrations. The physician is asked which of these he/she would like to use as a starting base. The electrolyte composition of the chosen mixture is displayed along with the patient's most recent laboratory electrolyte values. The physician may then accept the mixture as prepared or adjust the concentration of one or more of the components. (Choosing a prepared mixture is the most economical as modifications require manual preparation and incur a pharmacist's charge.)

Next, the patient's most recent protime (PT) is displayed and the physician may order vitamin K. If ordered, the vitamin K is administered via intramuscular injection, rather than added to the solution, due to compatibility problems. The computer then asks if the physician wishes to add insulin to the solution. Next, the physician can order blood tests to help monitor the patient's progress.

Finally, the TPN prescription is redisplayed, at which time it may be accepted or rejected by the physician. Free-text comments (that will appear on the printed order) may also be entered.

Once accepted, four copies of the order are printed. One is for the physician to sign. This is taken to the pharmacy and is the actual copy used in the preparation of the solution. Another copy is for the patient chart, and a third is a working copy for the nurse, clerk, or computer technician. The final copy is a charge copy and is printed in the pharmacy for billing purposes.

22.3 Alerts

HELP uses a knowledge base to screen incoming TPN orders for potentially adverse situations. The logic is invoked automatically and the alerts are displayed on the screen at the time the physician is entering the orders.

Examples include:

1. If the amount of potassium in the solution as ordered will deliver more than 8 meq/hr of potassium to the patient, HELP informs the physician of this fact and adds a nursing order for ECG telemetry monitoring to the TPN prescription.

2. Since calcium and phosphorus may precipitate in solution, if their respective concentrations are incompatible, this will be relayed to the physician, the order rejected, and a new order requested.

3. If the physician requests more than 150 cc/hr of TPN to be given through a peripheral IV line, or orders a solution with a final glucose concentration of greater than 10% to be given through a peripheral line, the computer will ask if this is actually what the physician wants to do.

The TPN-related alerts are an example of HELP's ability to assist with concurrent QA review.

22.4 Nutritional Support Services

The NSS at our hospital is a team of two dietitians and three nurses that is responsible for monitoring the quality of care related to the administration of parenteral and enteral nutrition.

The NSS daily reviews the charts of all TPN patients, including their TPN orders. If patients have TPN orders that are inconsistent with accepted practices, one of the NSS staff will interact with the physician. The NSS also receives, daily, on all patients receiving nutritional therapy, copies of laboratory results that may have an impact on, or be impacted by, the therapy. If they find laboratory values related to nutritional orders that are progressing toward an abnormal range, they will make sure the orders have been reassessed.

Other NSS functions include reviewing and charting the patient's nutritionally related statistics. They visit the patients daily, make sure that solutions are being administered properly, and make sure that nursing functions related to proper nutritional therapy are being performed.

These measures help to reduce the incidence of problems related to nutritional therapy in patients who may have multiple medical problems and are difficult to manage. These services are also cost effective as the staff works with physicians to help them use the therapy in the most efficient manner.

At our hospital NSS is also responsible for functions such as tapering of therapy and teaching and predischarge planning for patients who will be continuing their therapy at home.

22.5 Reports

The HELP computer prints a variety of reports related to nutritional therapy. These are primarily used by the NSS:

1. The TPN order (Figure 22.1) is printed in the pharmacy and other places as mentioned above.

2. Nutritionally related laboratory data (Figure 4.3) are printed for the NSS to review nutrition-related results. The reports include values for SMAC-20s, PT, PTT, platelet count, differential, lactic acid, magnesium, iron, urine values, thyroid studies, nutritional parameters, and others.

3. The NSS uses the 7-day summary reports (Figure 4.5) to follow the patient's intakes and outputs, weight, and complete nutritional data. The NSS also uses this report to review the patient's other medications to see if any of these may interact with the nutritional therapy.

4. The NSS maintains a computer list of active TPN patients, whis is updated manually. Patients taken off TPN are deleted at the end of each day, and new patients (determined by reviewing computerized pharmacy records) are added each morning.

5. The computer prints a daily list of patients receiving enteral therapy. This list must be updated manually by NSS.

6. All patients who have had a Miller-Frederick Tube (a radiographically placed intestinal feeding tube) inserted in the previous 72 hours are listed in a report each day (the 72-hr time selection is used to cover the weekends). The computer does this by searching the computerized radiology records for requests that read "GI tube placement." In this way the NSS can update the Enteral Patient List.

22.6 Data Integration

To generate its displays, alerts, and reports the computer uses information from the entire computerized patient file. Radiology orders are scanned to determine who had feeding tubes placed. Laboratory results are used for display and printed reports. Nursing and pharmacy data are printed on the 7-day summary report.

22.7 Statistical Analysis

Data on all patients who receive nutritional therapy are also entered into a separate program maintained on a PC by the Department of Critical Care Medicine. The data are collected manually by NSS and later entered by a data entry clerk. This allows data such as the reason for therapy, the length and type of therapy, and who the ordering physician was to be retrospectively analyzed. It is completely separate from the central computer's functions.

23
Computerized Protocols for Ventilator Management

23.1 Overview

The decision-making capabilities of the HELP system have been applied to the management of patients requiring mechanical ventilatory assistance. The computerized system uses computerized respiratory care charting data (see Chapter 14) and computerized blood gas data to make suggestions regarding the management of the inspired oxygen fraction (FiO_2) and positive end expiratory pressure settings (PEEP) in hypoxemic patients and respiratory rate, minute ventilation, and inspiratory:expiratory (I:E) ratio in patients with hypoventilation. The system is data-driven and new relevant clinical patient data automatically generate treatment suggestions. The decision logic employed in the system's computerized protocols was decided on by LDS Hospital pulmonary and intensive care physicians over an 18-month period. As of early 1990, the system had been used with more than 40 patients in the STRICU for a total of 12,000 hours and had generated more than 9200 decisions. Physicians had agreed with the computer's decisions logic (i.e., they had instituted the computer's recommendations) in more than 90% of the computer-generated decisions.

23.2 Motivation for the Protocols

The computerized ventilator management protocols were originally motivated by the need to study a new form of therapy for the adult respiratory distress syndrome (ARDS).[1] In the early 1980s, experiments suggested that

---

[1]ARDS can be precipitated by a variety of factors including pulmonary and nonpulmonary disease. It is characterized by bilateral alveolar chest x-ray infiltrates, decreased static chest compliance, low pulmonary capillary wedge pressure, and hypoxemia refractory to oxygen

extracorporeal CO_2 removal ($ECCO_2R$) could significantly decrease mortality in patients with the severe hypoxemic form of ARDS. However, the studies did not contain a control group of patients receiving traditional therapy against which the new therapy was measured, and thus the value of the new technique was questionable. Physicians at LDS Hospital who wanted to repeat the trial of ECCO2R in a controlled manner realized that to assure uniformity of treatment between the study and control arms of the new trial (as well as uniformity of treatment within each arm of the trial), protocol-based management of the ventilators would be necessary. Hand-written protocols for the management of hypoxemia and ventilation were designed to assure that care would be consistent for patients in each branch of the study. The protocols were designed to manage hypoxemia (the primary problem in ARDS) consistently. They were developed over 18 months by expert physicians at LDS Hospital. As the protocols became more and more complex, they became difficult to follow by hand and the likelihood of error increased. The decision was made to computerize the protocols and to use respiratory care and blood gas data already present in HELP to assist with the decision making.

Computerization of the protocol logic was shown to be feasible in a research project [1,2] and since then the project has undergone further enhancements to increase its utility and scope [3,4]. Whereas the protocols were initially designed to assist with the management of hypoxemia only, additional protocols have been written to assist with the management of ventilatory failure. Also, although the protocols were initially written to manage patients enrolled in the study and control arms of a trial of ECCO2R therapy for ARDS, the protocols are general enough so they have been able to be used for the management of other patients with hypoxemia or ventilatory failure.

23.3 Decision Logic

The paper-based versions of the hypoxemia protocols that served as the template for the computerized protocols were developed over an 18-month period by a group of LDS Hospital physicians expert in pulmonary disease [5]. The main module of the system is known as the CORE, for "continuous oxygenation and respiratory evaluator" (Figure 23.1). For patients enrolled in the protocols, the CORE collects the new clinical data and decides if

---

therapy. Traditional therapy is basically supportive ventilation and oxygenation while lung healing takes place. Overall mortality for ARDS is about 50% and for patients with severe hypoxemia mortality with traditional therapy has remained at 90% for the past two decades, as documented in multiple studies. ARDS affects 150,000 people in the United States each year.

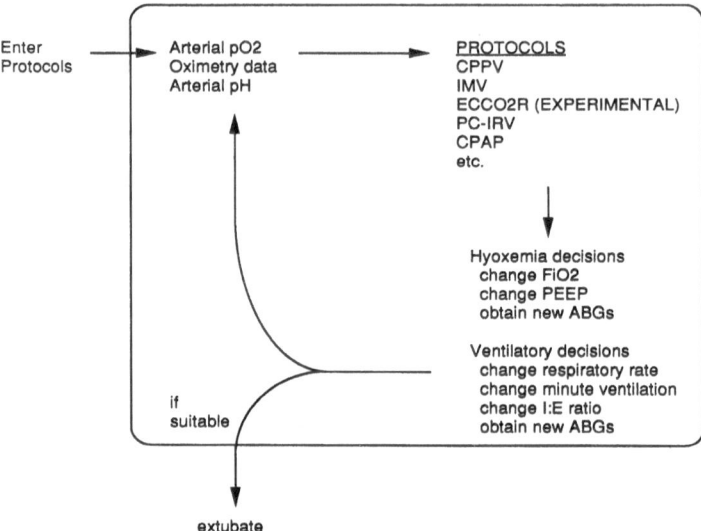

Figure 23.1. Continuous oxygenator and respiratory evaluator (CORE) module of ventilator management program.

hypoxemia is present, if hypoventilation is present, or if neither is present and the patient is suitable for a reduction in therapy. Depending on this initial decision by the CORE, other modules of the decision-making program are invoked. The results of blood gas measurements along with ventilator settings are used to determine what changes should be made in the FiO_2 and PEEP settings. The timing of follow-up blood gases is also specified. For hypoxemia, the storage of an arterial PO_2 or a classification based on pulse oximetry triggers further therapeutic suggestions until the protocol suggests the patient should be extubated. For patients receiving the experimental form of ARDS treatment (ECCO2R), specific protocols including this treatment are used. For the hypoventilation protocols, the program is triggered by pH data and suggests alterations in minute ventilation, respiratory rate, and I:E ratio.

The hypoxemia protocols were originally implemented using a blackboard control architecture (Figure 23.2) [2]. Incoming raw clinical patient data were first processed (i.e., calculated values were obtained) and then classified and summarized (i.e., the numerical values were converted to symbolic data). A therapy scheduler then determined exactly which ventilator management protocol should be employed (e.g., assist/control, intermittent mandatory ventilation, $ECCO_2R$, etc.). The management modules could use the intermediate decisions made by the program up to that point. The program's recommendations would be written to a journal file that could be examined by physicians, nurses, or respiratory therapists from bedside HELP terminals in the ICU. The program also kept track of the eventual therapeutic interventions and the extent to which they agreed with the program's suggestions.

The ventilator protocol program in current use has the same functionality as the initial program; however, the blackboard architecture has been abandoned and rather than creating intermediate decisions (i.e., the blackboard) en route to the final suggestion, decisions at every stage are now based on the raw data. The blackboard architecture was abandoned because [3] (1) data entry errors that were later edited could not affect the blackboard, (2) in the event of a system crash, there was no way to reconstruct the blackboard, and (3) the program based some decisions on prior recommendations and did not always know when previous suggestions had not been followed.

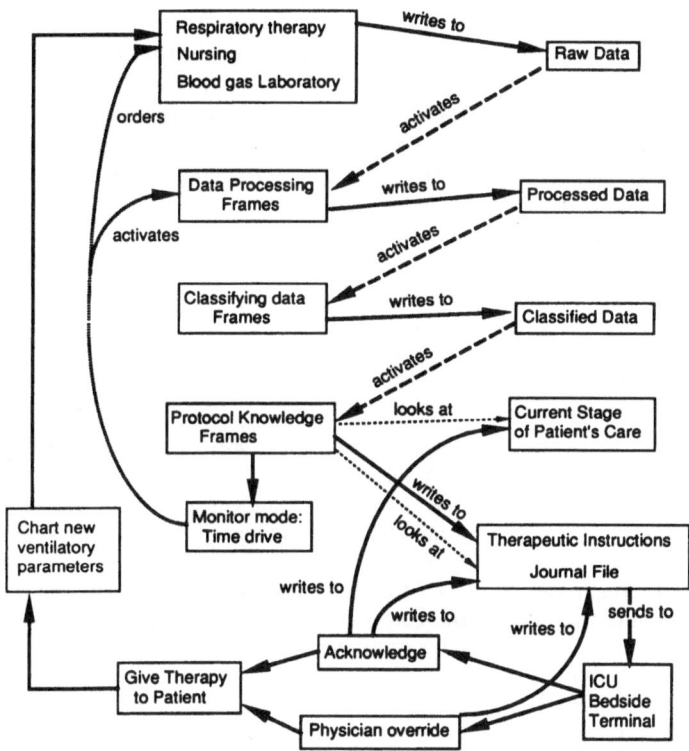

Figure 23.2. Blackboard control structure of original ventilator advice system. © 1989 Academic Press, New York. Reprinted with permission from "Implementation of a Computerized Patient Advice System Using the HELP Computerized Hospital Information System" by Sittig, Pace, Gardner et al., *Computers in Biomedical Research* 22:474–487, 1989.

23.4 Performance Evaluation

As of early 1990 the system had been used with more than 40 patients for more than 12,000 hours, and had generated 9200 decisions of which 90% had been followed by the physicians providing care to critically ill patients. Reasons for disagreement by the physicians with the computer's suggestions include [4] (1) disagreement of the staff with the underlying logic, (2) incorrect suggestions due to software bugs, (3) an incomplete database at the time of suggestion generation, (4) failures in the data drive, (5) incomplete protocol logic, (6) data entry errors, and (7) suggestions made based on data recorded while the patient was not in a steady state (e.g., transient hypoxemia due to suctioning). The frequency of disagreements is highest when major changes are made to the protocol logic or system architecture and decrease drastically when the logic is debugged over time and users become comfortable with the system. Also, since a complete database is necessary for accurate suggestions, procedures at LDS Hospital have been modified to assure that the relevant data (i.e., respiratory care charting and blood gas results) are entered into HELP in a timely manner.

The user interface is another aspect of the system that has been examined [3]. Since the decision-making system operates in the background in a multiuser, multitasking operating system, treatment suggestions must remain in the background or the process must seize control of the bedside terminal to deliver the suggestion. Since the terminals are used by multiple users for multiple purposes, the decisions remain in the background and those personnel interested in viewing the computer's suggestions must request the decisions to be displayed.

References

[1] Sittig DF. *ComPAS: A Computerized Patient Advice System*. Salt Lake City, Utah: University of Utah; 1988. Dissertation.

[2] Sittig DF, Pace NL, Gardner RM, et al. Implementation of a computerized patient advice system using the HELP computerized hospital information system. *Comput Biomed Res* 1989;22:474–487.

[3] East TD, Henderson S, Morris AH, et al. Implementation issues and challenges for computerized clinical protocols for management of mechanical ventilation in ARDS patients. *Symposium for computer applications in medical care (SCAMC)* 1989;13:583–587.

[4] Henderson S, East TD, Morris AH, et al. Performance evaluation of computerized clinical protocols for management of arterial hypoxemia in ARDS patients. *Symposium for computer applications in medical care (SCAMC)* 1989;13:588–592.

[5] Morris AH, Menlove RL, Rollins RJ, et al. A controlled clinical trial of a new 3-step therapy which includes extracorporeal CO_2 removal for ARDS. *Trans Am Soc Artif Intern Organs* 1988;34:48–54.

24
Surgery Monitoring on
the HELP System

24.1 Overview

Physiologic data from patients undergoing open-heart surgery at LDS Hospital are captured automatically by the HELP system from monitoring devices in the surgical suites (see Hemodynamic Monitoring, Chapter 21). Surgical milestones and clinical data (e.g., medications orders, times of key surgical events, urine output measurements, transfusions, etc.) are entered into the HELP system by computer technicians assigned to the thoracic surgical suites. The monitoring data and the other clinical data are combined to form a computerized surgical record.

The computerized surgery data are available for real-time review from any HELP system terminal. The surgical information is used by TICU nurses to anticipate the return of the patient to the nursing division and to apprise the family of the stage of surgery without phone calls to the Surgical Department. Physicians scheduled to operate later in the day follow the progress of earlier cases (either from a hospital terminal or from their offices via modem) to determine when a later case will begin. A printed copy of the surgical record is included in the patient's permanent medical record.

24.2 Complete Surgical Record

A complete computerized surgical record (Figure 24.1a–f, also known as the anesthesia record) is kept on patients who undergo procedures requiring use of cardiac bypass. In cases requiring bypass, a computer technician is assigned to the surgical case for the purpose of data entry. (At LDS Hospital, computer technicians are specially trained personnel who serve in the thoracic surgical suites and in the ICUs to enter clinical data into the HELP

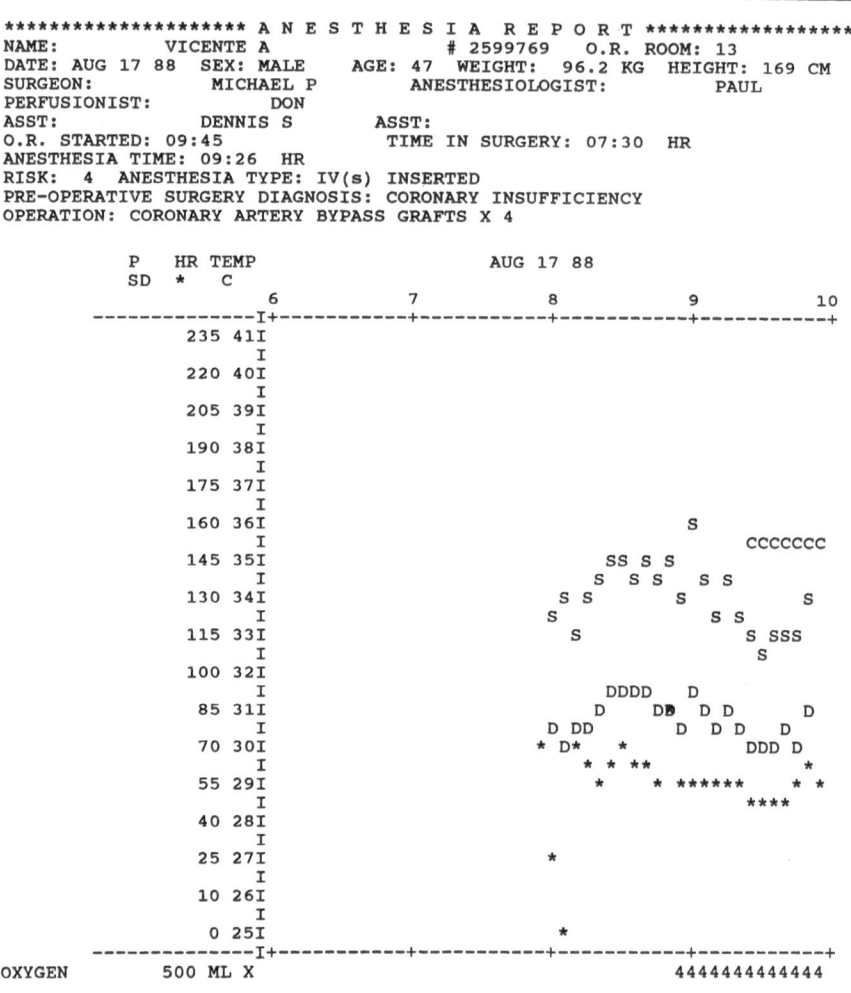

Figure 24.1.(a) Computerized anesthesia report (see text for description).

system, troubleshoot computer-related equipment problems, and perform clinical tasks such as arterial line insertions, catheter management, dressing changes, and other related hemodynamic functions.)

The computer technician enters coded comments related to progress of the surgery such as the timing of surgical events (e.g., anesthesia started, skin incision, etc.), urine outputs, transfusion administration, and medication charting into the HELP system. Data are entered in real time and are

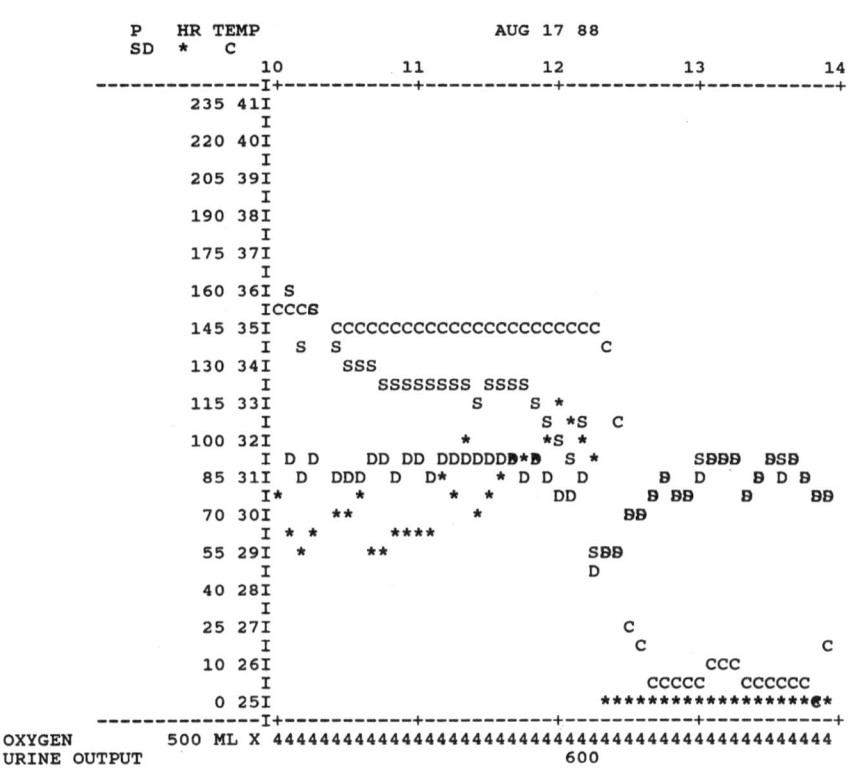

Figure 24.1.(b)

stored in the computer with a time stamp. All data are entered from a series of menu-driven screens and are thus entered as coded data rather than free-text entry.

The computer terminal is also used by nurses to order blood products and medications for patients during surgery. The processes for blood and medication ordering are the same as for other patients in the hospital (see Chapters 15 and 26 for further discussion).

The bypass patients are connected to computerized blood pressure, heart rate, and temperature monitoring devices from which the HELP system automatically collects data every 5 minutes (or other time interval, as desired). All the captured data are compiled into the Anesthesia Report, which becomes part of the patient's permanent medical record.

The anesthesia report invariably is a multipage document. Figure 24.1(a) shows patient identifying data along with the names of the surgeon, anesthesiologist, perfusionist, and assistants. The time the patient entered the operating room, total length of the surgery, time under anesthesia, patient's

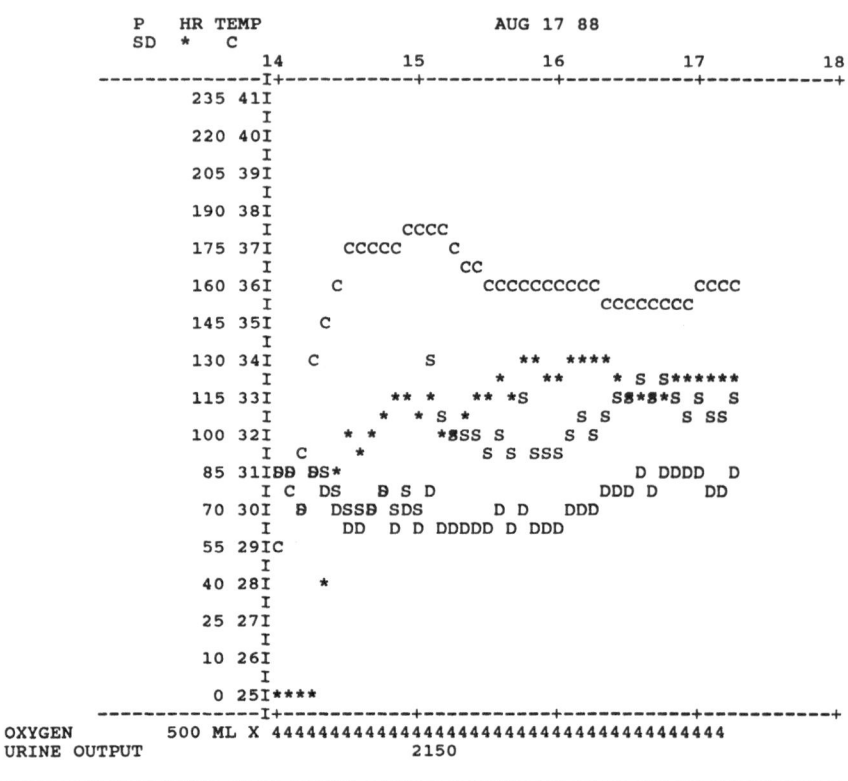

Figure 24.1.(c)

surgical risk (calculated by the anesthesiologist), type of anesthesia, preoperative diagnosis, and planned operation are shown next.

The remainder of Figure 24.1(a) and Figures 24.1(b) and (c) show the monitoring data. Systolic blood pressure (S), diastolic blood pressure (D), heart rate (*), and the patient's temperature (C) are shown. Oxygen flow (in multiples of 500 ml) is shown at the bottom. The date and time are shown at the top. The initiation of bypass can be seen in Figure 24.1(b) with the drop in heart rate to zero, the dropping of the temperature, and the equality of systolic and diastolic measurements. A urine output reading can also be seen in Figure 24.1(b). Figure 24.1(c) shows the reinstitution of the patient's cardiac function at just past 14:00 hours.

Figure 24.1(d) shows intake and output data and the beginning of a time-sequential record of the clinical events that occurred during the surgery. The record continues into Figures 24.1(e) and (f). Medications, personnel, diagnoses, planned procedures, the patient's position, surgical mile-

```
INTAKE (ML):  BLOOD                650   OUTPUT (ML):  PRE BY-PASS URINE    600
              COLLOID              165                 BY-PASS URINE       2150
              NON-BLOOD IV        3011                 EST O.R. BLOOD      1000
              BYPASS FLUID        2650
              CELL SAVER          1000
                 TOTAL            7476                    TOTAL            3750
```

```
AUG 17  06:29  O.R. MONITORING START
AUG 17  06:30  # MORPHINE     8.0 MGM, INJ, IM, NO.  10, DC'ED
               SCOPOLAMINE 0.3 MGM, INJ
AUG 17  07:53  ADMIT TO UNIT
AUG 17  07:53  ANESTHESIA STARTED
AUG 17  07:53  RISK  4
AUG 17  07:53  O.R.  ROOM    13
AUG 17  07:53  SURGEON   1432
AUG 17  07:53  ANESTHESIOLOGIST   1749
AUG 17  07:53  ASST. 1    209
AUG 17  07:53  PERFUSIONIST  1949
AUG 17  07:53  SUPINE
AUG 17  07:53  CORONARY INSUFFICIENCY
AUG 17  07:53  SURGERY TYPE, COMPUTER TECH, MONITORING CHRG START
AUG 17  07:53  WEIGHT  90.00 KG
               BODY SURFACE AREA  2.00 SQM
               HEIGHT  16.9 CM
AUG 17  07:54  CORONARY ARTERY BYPASS GRAFTS X 4
AUG 17  08:48  IV(s) INSERTED
AUG 17  08:48  LACTATED RINGERS 1000 ML, INJ, IVD, NO.  15, DC'ED
AUG 17  08:49  PANCURONIUM (PAVULON)   1.0 MGM, INJ, IV, NO.  18, DC'ED
AUG 17  08:49  # SUFENTANIL (SUFENTA)  0.5 CC, INJ, IV, NO.  19, DC'ED
AUG 17  08:50  # SUFENTANIL (SUFENTA)  1.0 CC, INJ, IV, NO.  20, DC'ED
AUG 17  08:52  # SUFENTANIL (SUFENTA)  2.0 CC, INJ, IV, NO.  21, DC'ED
AUG 17  08:52  SUCCINYLCHOLINE (ANECTINE)  120 MGM, INJ, IV, NO.  23, DC'ED
AUG 17  08:52  CEFUROXIME (ZINACEF)  1500 MGM, INJ, IV, NO.  26, DC'ED
AUG 17  08:52  ETIOMIDATE (AMIDATE)  14.0 MGM, INJ, IV, NO.  56, DC'ED
AUG 17  08:54  # SUFENTANIL (SUFENTA)  1.0 CC, INJ, IV, NO.  24, DC'ED
AUG 17  08:55  ORAL  INTUBATION
AUG 17  08:55  OXYGEN  2000 ML/MIN, INHAL, NO.  12, DC'ED
AUG 17  08:57  PANCURONIUM (PAVULON)   5.0 MGM, INJ, IV, NO.  25, DC'ED
AUG 17  09:09  FOLEY CATHETER INSERTED
AUG 17  09:10  # MIDAZOLAM (VERSED)  5.0 MGM, INJ, IV, NO.  27, DC'ED
AUG 17  09:11  SWAN GANZ INSERTION (SURGERY PROCEDURE)
AUG 17  09:11  SWAN GANZ INSERTION, TECH NAME
AUG 17  09:11  NORMAL SALINE   500 ML, INJ, IVD, NO.  16, DC'ED
AUG 17  09:12  # SUFENTANIL (SUFENTA)  2.0 CC, INJ, IV, NO.  28, DC'ED
AUG 17  09:15  # SUFENTANIL (SUFENTA) 15.0 CC, INJ, IVD, NO.  29, DC'ED
               D5W    150 ML, INJ
AUG 17  09:17  # SUFENTANIL (SUFENTA)  2.0 CC, INJ, IV, NO.  30, DC'ED
```

Figure 24.1.(d)

stones, transfusions, and so forth, have all been entered into the HELP
system by the computer technician.

24.3 Data Retrieval

Data from the computerized anesthesia record are immediately available
for review on hospital terminals. Nurses on the postoperative nursing units
can follow the patient's progress without having to call to the operating
rooms. This is helpful in anticipating the patient's return to the division

```
AUG 17 09:18 SURGICAL PREP
AUG 17 09:20 # SUFENTANIL (SUFENTA)   2.0 CC, INJ, IV, NO.   31, DC'ED
AUG 17 09:26 # SUFENTANIL (SUFENTA)   2.0 CC, INJ, IV, NO.   32, DC'ED
AUG 17 09:27 # SUFENTANIL (SUFENTA)   0.5 CC, INJ, IV, NO.   33, DC'ED
AUG 17 09:27 # MIDAZOLAM (VERSED)   2.5 MGM, INJ, IV, NO.   34, DC'ED
AUG 17 09:30 # SUFENTANIL (SUFENTA)   2.0 CC, INJ, IV, NO.   35, DC'ED
AUG 17 09:30 PANCURONIUM (PAVULON)   3.0 MGM, INJ, IV, NO.   36, DC'ED
AUG 17 09:39 TOWEL CLIPS
AUG 17 09:45 SKIN INCISION
AUG 17 09:45 O R START TIME
AUG 17 09:45 LEG CUTDOWN
AUG 17 09:45 # SUFENTANIL (SUFENTA)   1.0 CC, INJ, IV, NO.   37, DC'ED
AUG 17 09:49 STERNAL SPLITTING
AUG 17 09:54 STERNAL SPREAD
AUG 17 09:55 # MIDAZOLAM (VERSED)   2.5 MGM, INJ, IV, NO.   38, DC'ED
AUG 17 09:58 PERICARDIUM OPENED
AUG 17 10:00 # SUFENTANIL (SUFENTA)   1.0 CC, INJ, IV, NO.   39, DC'ED
AUG 17 10:00 FORANE (ISOFLURANE)   0.3 PRCT, INHAL, NO.   11, DC'ED
AUG 17 10:09 INTERNAL MAMMARY ARTERY DISSECTION COMPLETED
AUG 17 10:10 # SUFENTANIL (SUFENTA)   1.0 CC, INJ, IV, NO.   40, DC'ED
AUG 17 10:20 LACTATED RINGERS  975 ML, INJ, IVD, NO.   15, DC'ED
AUG 17 10:30 # SUFENTANIL (SUFENTA)   1.0 CC, INJ, IV, NO.   41, DC'ED
AUG 17 10:37 INTERNAL MAMMARY ARTERY DISSECTION BEGUN
AUG 17 10:40 VEIN REMOVED
AUG 17 11:40 # SUFENTANIL (SUFENTA)   3.0 CC, INJ, IV, NO.   42, DC'ED
AUG 17 11:45 # SUFENTANIL (SUFENTA)   2.0 CC, INJ, IV, NO.   43, DC'ED
AUG 17 11:50 # MIDAZOLAM (VERSED)   5.0 MGM, INJ, IV, NO.   44, DC'ED
AUG 17 11:52 HEPARIN  22000 UNITS, INJ, IV, NO.   45, DC'ED
AUG 17 11:55 PANCURONIUM (PAVULON)   5.0 MGM, INJ, IV, NO.   46, DC'ED
AUG 17 11:55 FORANE (ISOFLURANE)   0.0 PRCT, INHAL, NO.   11, DC'ED
AUG 17 12:02 PHENYLEPHRINE (NEOSYNEPHRINE)   0.05 MGM, INJ, IV, NO.   47, DC'ED
AUG 17 12:06 AORTIC ARCH CANNULATED
AUG 17 12:10 VENOUS UPTAKE CANNULA INSERTED
AUG 17 12:11 PARTIAL BYPASS
AUG 17 12:11 PRE BY-PASS URINE   600 ML
AUG 17 12:13 PANCURONIUM (PAVULON)   3.0 MGM, INJ, IV, NO.   48, DC'ED
AUG 17 12:14 COMPLETE BYPASS
AUG 17 12:16 VENTRICULAR SUMP INSERTED
AUG 17 12:18 AORTA CLAMPED
AUG 17 12:18 # SUFENTANIL (SUFENTA)   2.0 CC, INJ, IV, NO.   49, DC'ED
AUG 17 14:17 AORTA OPENED
AUG 17 14:17 TOTAL CROSS CLAMP TIME (HHMM)   159
AUG 17 14:27 PACER WIRES INSERTED
AUG 17 14:28 INSULIN REGULAR   10 UNITS, INJ, IV, NO.   50, DC'ED
AUG 17 14:55 # SUFENTANIL (SUFENTA)   1.5 CC, INJ, IV, NO.   51, DC'ED
AUG 17 14:55 # MIDAZOLAM (VERSED)   2.5 MGM, INJ, IV, NO.   52, DC'ED
AUG 17 15:00 PANCURONIUM (PAVULON)   3.0 MGM, INJ, IV, NO.   53, DC'ED
AUG 17 15:02 VENTRICULAR SUMP REMOVED
AUG 17 15:05 OFF BYPASS
AUG 17 15:05 TOTAL BYPASS TIME (HHMM)   253
AUG 17 15:05 BY-PASS URINE 2150 ML
AUG 17 15:05 CORONARY ARTERY BYPASS GRAFTS X 4
AUG 17 15:05 CARDIOPLEGIA SOLUTION INFUSED 1150 ML
AUG 17 15:05 BYPASS FLUID 2650 ML
AUG 17 15:07 VENOUS UPTAKE CANNULA REMOVED
AUG 17 15:10 PROTAMINE SULFATE 1%   30.0 ML, INJ, IV, NO.   54, DC'ED
AUG 17 15:17 AORTIC ARCH CANNULA REMOVED
AUG 17 15:30 LEG CLOSED
AUG 17 15:35 PACKED RBC   250 ML, IVD, NO.   55, DC'ED
```

Figure 24.1.(e)

```
                    O   POSITIVE
                    NO.        4039912
AUG 17 15:38 PROTAMINE SULFATE 1%   10.0 ML, INJ, IV, NO.   57, DC'ED
AUG 17 15:41 NORMAL SALINE   250 ML, INJ, IVD, NO.  16, DC'ED
AUG 17 15:42 DESMOPRESSIN (STIMATE)   27 MCG, INJ, IVD, NO.   58, DC'ED
                    D5W      50 ML, INJ
AUG 17 16:05 CALCIUM CHLORIDE   500.00 MEQ, INJ, IV, NO.   60, DC'ED
AUG 17 16:10 PANCURONIUM (PAVULON)    2.0 MGM, INJ, IV, NO.   61, DC'ED
AUG 17 16:10 # SUFENTANIL (SUFENTA)   1.0 CC, INJ, IV, NO.   62, DC'ED
AUG 17 16:26 # SUFENTANIL (SUFENTA)   0.5 CC, INJ, IV, NO.   63, DC'ED
AUG 17 16:33 PERICARDIUM CLOSED
AUG 17 16:35 CHEST TUBES INSERTED X 2
AUG 17 16:53 STERNUM CLOSED
AUG 17 16:58 PLATELETS   400 ML (RANDOM DONOR), IVD, NO.   64, DC'ED
                    O   POSITIVE
                    NO.       1302173
                    NO.       1302177
                    NO.       1302188
                    NO.       1302170
                    NO.         84729
                    NO.         84735
                    NO.         84726
                    O   NEGATIVE
                    NO.         84727
AUG 17 17:05 EST O.R. BLOOD LOSS   1000 ML
AUG 17 17:05 CELL SAVER   1000 ML
AUG 17 17:10 FRESH FROZEN PLASMA      165 ML, IVD, NO.   65, DC'ED
                    O   NEGATIVE
                    NO.       3084920
AUG 17 17:15 SKIN CLOSURE
AUG 17 17:15 OPERATION COMPLETED
AUG 17 17:19 ANESTHESIA STOPPED
AUG 17 17:19 O.R. MONITORING STOPPED
AUG 17 17:19 MONITORING CHRG STOP


     TECH  --------------------      ANEST  ----------------------
                    TIME OUT AUG 18 88  19:08
  (END)
```

Figure 24.1.(f)

as well as keeping the patient's family apprised of the progress of the surgery.

For bypass patients, the complete anesthesia record is available for review so the TICU nurses can be aware of the patient's status as well as the stage of surgery.

Physicians may follow the progress of preceding cases to help determine when a later case will begin. In addition to terminal review, physicians may use a PC and modem to review these data from their home or office (see Chapter 31 for further discussion).

25
Use of APACHE II
on the HELP System

25.1 Overview

The APACHE (for *a*cute *p*hysiological and *c*hronic *h*ealth *e*valuation) II system uses the values of clinical parameters measured in the first 24 hours of an ICU admission to estimate the likelihood of in-hospital mortality [1]. For patients in the LDS Hospital STRICU, the numerical APACHE II score is calculated automatically by the HELP system from clinical data already present in the computerized patient database and from hand-entered data. The APACHE II scores are printed in a report format each day for the STRICU medical director to review. The scores are also used as a measure of severity of illness when clinical studies from LDS Hospital are reported.

25.2 APACHE II

The APACHE II system, developed at George Washington University, yields a numerical value for each patient based on three components [2]: (1) an acute physiological score, (2) the patient's age, and (3) a chronic health evaluation. The acute physiological score (APS) is calculated from the 12 physiological variables shown in Table 25.1. Depending on the degree of abnormality, each variable yields a score of 0 to 4 (the larger numbers indicating greater abnormality). Up to six points are added for the patient's chronological age as shown in Table 25.2. The chronic health evaluation adds points if the patient has a history of severe organ system insufficiency or if the patient is immunocompromised. Two points are added if such a patient is postoperative an elective procedure and five points are added if such a patient is nonoperative or emergency postoperative.

APACHE II was validated in a study of more than 5000 patients [1] which showed a linear relationship between likelihood of hospital mortality

Table 25.1. Variables used to calculate APS component of APACHE II Score

1. Rectal temperature
2. Mean arterial blood pressure
3. Heart rate
4. Respiratory rate
5. PO_2 (or A–a gradient if $FiO_2 > 0.5$)
6. Arterial pH (or HCO_3-)
7. Serum sodium
8. Serum potassium
9. Serum creatinine
10. Hematocrit
11. White blood cell count
12. Glasgow coma score

and the numerical APACHE II score. Using a 50% decision rule (i.e., predicting death if the likelihood is greater than 50% and survival otherwise) the predictive value positive of APACHE II was 70% and the predictive value negative was 88% [2]. A refinement of APACHE II, APACHE III, was released in 1990.

25.3 Use of APACHE II on the HELP System

The HELP system is able to automatically calculate APACHE II scores because the patient's age and values for the physiological variables are contained in the computerized patient database from other programs (e.g, age from the ADT program, vital signs from computerized nurse charting, FiO_2 from computerized respiratory care charting, etc.). Chronic health evaluation data are not automatically present in the HELP system and must be manually entered by the nurses.

An APACHE II score is automatically calculated for each patient in the STRICU each day (even though the prognostic value of the score applies

Table 25.2. Points added to the APS based on age

Age	Points
< 44	0
45–54	2
55–64	3
65–74	5
> 75	6

```
*********************************************
*                                           *
*        REPORT FOR  DR. CLEMMER            *
*                                           *
*             JUN 19, 1988                  *
*                                           *
*********************************************
```

		APACHE / MOF	CHE / CV	AGE	HR / PULM	BP / REN	RR / HEP	AA / BOW	PH / CNS	NA / COG	K / SEP	CR	GCS	HCT	WBC
E601	CLARENCE* Apache	21	5	5	3	3	0	2	1	0	0	0	1	0	0
	MOF	4	0		3	0	0	0	0	0	1				
E602	HELEN S* Apache	22	0	6	2	4	0	2	3	0	0	3	1	0	1
	MOF	12	3		3	2	0	0	0	2	2				
E603	WILLEY* Apache	14	0	6	3	4	0	0	0	0	0	0	1	0	0
	MOF	5	1		3	0	0	0	0	0	1				
E604	KRISTINE* Apache	13	0	0	3	2	1	1	0	0	0	0	2	0	1
	MOF	6	0		3	0	0	0	0	0	3				
E605	DAVID* Apache	18	5	0	2	0	1	0	0	0	0	0	9	0	0
	MOF	6	0		3	0	0	0	2	0	1				
E606	MARK C.* Apache	18	5	0	3	3	0	0	1	0	0	0	1	2	0
	MOF	7	0		3	0	1	0	0	2	1				
E609	LAWRENCE C* Apache	9	0	3	2	3	0	0	0	0	0	0	1	0	0
	MOF	4	0		3	0	0	0	0	0	1				
E610	JANE* Apache	10	2	2	2	2	0	0	0	1	0	0	1	0	0
	MOF	6	0		3	1	1	0	0	0	1				
E611	COSGROVE* Apache	15	0	5	3	4	1	0	0	0	0	0	0	0	2
	MOF	6	0		3	0	0	0	0	0	3				
E612	BURTON* Apache	15	0	3	2	2	1	0	0	0	0	0	6	0	1
	MOF	4	0		3	0	0	0	1	0	0				

(END)

Figure 25.1. APACHE II report for medical director of Shock–Trauma–Respiratory Intensive Care Unit (STRICU). CHE = chronic health evaluation, HR = heart rate, BP = blood pressure, RR = respiratory rate, AA = arterial alveolar oxygen difference, PH = pH, NA = serum sodium, K = serum potassium, CR = serum creatinine, GCS = Glasgow Coma score, HCT = hematocrit, WBC = white blood cell count, MOF = multiorgan failure score, CV = cardiovascular system, PULM = pulmonary system, REN = renal, HEP = hepatic, BOW = nutrition/metabolic, CNS = neurological, COG = coagulation, SEP = infection.

only on the first hospital day). The scores are printed in a report for the director of the STRICU. Figure 25.1 shows the total APACHE II score for each of the STRICU patients along with the contribution to the score of each of the physiologic variables, the patient's age, and the chronic health evaluation. The APACHE II scores are used in an ad hoc manner in the STRICU and no decisions are made as a matter of routine based on the scores. APACHE II data have been used to stratify patients in clinical trials at LDS Hospital.

Some problems with the automatic calculation of APACHE II scores at LDS Hospital have resulted from nonrepresentative data present in the database. For example, if an erroneously high blood pressure is measured (e.g., a reading taken while the system is being flushed), an artificially high APACHE II score will be calculated. Also, patients receiving paralyzing medications will have high Glasgow scores not due to disease. The first problem could be solved by examining the patient data before calculating the APACHE II score and filtering out obvious nonphysiologic data. The second problem could be solved by modifying the nurse charting program to inquire if high Glasgow scores are due to medications. Also, at LDS Hospital the arterial-alveolar (A-a) gradient calculation has been corrected to account for pressure changes due to elevation (4600 feet) [3].

Figure 25.1 also shows the multiorgan failure (MOF) score for each patient. The MOF is a locally developed and as yet unvalidated severity of illness score that is based on computerized patient data. The MOF score appears on the ICU Rounds Report (Figure 4.4).

References

[1] Knaus WA, Draper EA, Wagner DP, Zimmerman JA. APACHE II: A severity of disease classification. *Crit Care Med* 1985;13:818–829.
[2] Seneff M, Knaus WA. Predicting patient outcome from intensive care: a guide to APACHE, MPM, SAPS, PRISM, and other prognostic scoring systems. *J Intens Care Med* 1990;5:33–52.
[3] Sittig DF, Gardner RM, Menlove RL, Clemmer TP. APACHE II—modifications for use at altitude. *Crit Care Med* 1987;15:890–891. Letter to the editor.

IV
Other Clinical Applications on the HELP System

26
Blood Ordering on
the HELP System

26.1 Overview

At LDS Hospital, orders for blood products and related tests (e.g., blood typing, Coombs tests, etc.) are entered directly into the HELP computer system by physicians and nurses through a set of menu-driven screens. To help comply with Joint Commission for the Accreditation of Healthcare Organizations (JCAHO) guidelines for blood usage, the computer seeks a reason for the request for blood products as part of the ordering process. When blood products are ordered, the computer automatically examines data present in the electronic patient file to determine if the order meets predetermined criteria. For example, if the reason given for ordering packed cells is anemia, the computer verifies that the patient's most recent hematocrit value is below a specified level. If a discrepancy is detected, the computer informs the physician who may then cancel the order, enter another reason, or override the computer. Each deviation from predetermined criteria is followed-up by quality assurance personnel.

A computerized log of all blood orders is kept. HELP may be used to display the log according to a variety of parameters (e.g., patient, physician, time period, etc.). These displays are used for review and QA purposes. More detailed reports are generated by downloading the log data to, and processing it with, a PC [1].

26.2 Quality Assurance in Blood Utilization

The motivation for the blood ordering program was to comply with the JCAHO requirements for monitoring of blood usage by the medical staff. In a 1987 document [2], the JCAHO outlined nine steps to be taken in blood usage review:

1. assign responsibility for blood usage review
2. delineate the scope of blood use in the hospital
3. identify the important aspects of blood use (e.g., aspects of use that have potentially serious consequences for patients and staff)
4. identify indicators of quality (i.e., measurable standards of quality to be monitored in order to determine whether the aspect of blood use conforms to current standards of appropriateness and quality)
5. establish criteria (i.e., specific standards to be used by the indicators)
6. collect and analyze data
7. take action to resolve identified problems
8. assess the actions and document improvement
9. communicate relevant information to the organization-wide quality assurance program

The JCAHO also recommends that all transfusion cases be thoroughly reviewed for a period of at least 6 months and only if blood usage is consistently justified may a sampling of high volume cases be substituted.

Traditional methods of blood usage review have been hampered by step 6, collection and analysis of data. Obtaining the needed data from paper-based medical records requires enormous investments of time and, for some parameters, such as whether or not the order meets criteria, may be impossible due to the incompleteness of the medical record.

At LDS Hospital, important aspects of blood use and indicators and criteria for the monitoring of blood usage were determined by a medical committee using national and local standards. After the approval of the medical staff, ordering of blood products through the HELP computer was instituted. Relevant data are now obtained at the time of order entry and data review is done with the computer rather than manually. The medical staff is being encouraged to use the program for all their blood ordering needs. If a physician uses the traditional order sheets or the phone for blood ordering, the order may then be entered into the computer system by a nurse.

26.3 Blood Ordering

The blood ordering program is entered from the ordering menu present on the main menu screen of most of the clinical terminals. The social security number of the physician or nurse acts as a security code/electronic signature and also indicates to the computer the type of provider entering the order. The computer then automatically displays blood orders in the past 24 hours, including the number of units dispensed and the number of units still available (Figure 26.1) and the patient's most recent relevant laboratory values (hematocrit, hemoglobin, platelet count, PT, PTT, and albumin) (Figure 26.2).

```
SMITH, JOHN                  NO.   2222222  AGE: 76  SEX: M  ROOM: E702
                          L D S   H O S P I T A L
                          BLOOD BANK REQUEST

                    BLOOD PRODUCTS ORDERED IN LAST 24 HOURS

1. 26-JUL 10:54  PRODUCT: PACKED CELLS    DOCTOR: JONES, JAMES E
                 UNITS:     2  TYPE: ROUTINE   REASON: ANEMIA
                 ******    2 UNITS DISPENSED      0 UNITS AVAILABLE *****
```

Figure 26.1. Blood availability data as displayed by blood ordering program. Name has been falsified.

```
      SMITH, JOHN      NO.   2222222  AGE: 76  SEX: M  ROOM: E702
                    L D S   H O S P I T A L
                    BLOOD BANK REQUEST

      MOST RECENT LAB DATA ENTERED OVER A 48 HOUR PERIOD:

          TEST              VALUE                TIME

          HCT               35.0            07/27 04:38
          HGB               12.2            07/27 04:38
          PLATELETS        237 K            07/27 04:38
          PT                10.9            07/26 04:00
          PTT                40             07/26 04:00
          ALBUMIN           NA
```

Figure 26.2. Patient laboratory data as displayed by blood ordering program. Name has been falsified.

After this, the computer displays a menu from which the provider may order a blood product or a related test:

Blood products
packed cells
fresh frozen plasma
platelets
leukocyte poor cells
whole blood
rhogam (appears for female patients only)
washed packed RBCs
leukocyte-poor whole blood
other blood products

Related tests
type and screen
hold clot only
blood type
antibody screening (indirect Coombs)
direct Coombs
Coombs packet (direct and indirect)
cord packet
Rh recheck

Choosing #9 will present another menu from which the provider may order granulocytes, cryoprecipitate, deglycerolized RBCs, factor XIII concentrate, factor IX concentrate, single donor plasma, or enter a free-text order. The free-text option is almost never used. If the order is for one of the related tests (column 2), the request is redisplayed for verification and then forwarded to the blood bank laboratory.

For blood products ordered, the user is prompted to enter the number of units requested. If the patient is a child (automatically determined from the patient's age as recorded in the patient identification file), the number of milliliters is requested. The type of order (i.e. stat, routine, preop, etc.) is then entered. After this, the provider is given the option to include directions for the administration of the product that will accompany the order. The computer presents the following screen:

1. see handwritten instructions
2. give the product now
3. give the product today
4. give for hematocrit less than_____.__
5. replace cc for cc from (free-text) (used in situations of continuing blood loss)
6. other (free-text)
7. give over__hours
8. give__units, hold__units

Next, a reason for the blood order is requested. A menu is provided and will vary depending on the product being ordered. A common menu (the one used for red cells and whole blood) is:

1. trauma
2. preop
3. blood loss
4. anemia
5. volume replacement
6. hypotension, tachycardia, hypoxemia
7. bypass (for use during surgery)
8. dialysis

If the reason given is anemia, the computer will check the most recent hematocrit in the electronic patient file. If preop is listed, the computer examines the patient file to ensure a procedure is listed and the number of units ordered for the procedure is appropriate.

For some reasons further information is sought. For example, for blood loss, volume replacement, and anemia the computer inquires if this is due to active bleeding, a loss of more than 500 cc, or third spacing. If the computer finds a discrepancy between the stated reason and the data in the electronic patient record, it will so inform the user. The blood order may still be entered; however, the provider must enter a textual reason from the keyboard (Figure 26.3).

The provider is then asked if he/she wishes to enter another order. When the order entry is complete, the computer redisplays the entire order with

```
                         L D S    H O S P I T A L
                          BLOOD BANK REQUEST
REASON FOR REQUEST: PACKED CELLS        2 UNITS

1 - PRE-OP
2 - BLOOD LOSS
3 - ANEMIA
4 - VOLUME REPLACEMENT
7 - HYPOTENSION, TACHYCARDIA, or HYPOXEMIA
8 - TRAUMA
11- BYPASS   <CARDIOPULMONARY>
12- DIALYSIS

REASON FOR THE REQUEST -> 3
ACCORDING TO LDSH STANDARDS, PATIENTS 76 YEARS OLD DO NOT NEED
BLOOD PRODUCTS UNLESS:    HCT < 35.0     OR      HGB < 11.5
PATIENT'S PARAMETERS ARE:
                HCT =      35.0    (27 04:38)
                HGB =      12.2    (27 04:38)

DO YOU STILL WISH TO MAKE THE REQUEST? (Y/N) Y
PLEASE GIVE A REASON FOR THE OVERRIDE (MAX 40 CHAR) ->DEMONSTRATION
            ----------------------------------------
                         L D S    H O S P I T A L
                          BLOOD BANK REQUEST
REASON FOR REQUEST: FRESH FROZEN PLASMA      2 UNITS

2 - BLOOD LOSS
4 - VOLUME REPLACEMENT
5 - COAGULATION DEFECT
7 - HYPOTENSION, TACHYCARDIA, or HYPOXEMIA
8 - TRAUMA
11- BYPASS   <CARDIOPULMONARY>

REASON FOR THE REQUEST -> 5
ACCORDING TO LDSH STANDARDS BLOOD PRODUCTS ARE NOT NEEDED FOR
COAGULATION DEFECTS UNLESS:  PT >= 17  OR  PTT >= 55  OR  FIBRINOGEN < 100
PATIENT'S PARAMETERS ARE:
                PT =   11   (26 04:00)
                PTT =   40   (26 04:00)
                FIBRINOGEN = NA

DO YOU STILL WISH TO MAKE THE REQUEST? (Y/N) Y
PLEASE GIVE A REASON FOR THE OVERRIDE (MAX 40 CHAR) ->ANOTHER DEMO
```

Figure 26.3. Examples of overriding computer logic in blood ordering program.

options to edit the order, cancel it, or send it to the laboratory. If the option to send it to the laboratory is chosen, the order is automatically printed in the blood bank.

In the blood bank the order is printed onto a paper copy and also onto a label that is placed directly onto an order form and the blood product. When the product becomes available the patient's floor is informed via telephone. These data (as well as a record of units dispensed) are also entered into the computer so the record of the number of units available/ dispensed will remain accurate. Data in the blood bank are entered into a PC LAN that communicates with the laboratory information system (LIS, Chapter 16), which in turn communicates with the Tandem.

26.4 Standing Orders

Physicians have the option of participating in a program that will automatically order blood for their scheduled operative cases. Data that indicate the procedures physicians commonly perform and the number and types of units of blood they routinely order for those procedures are stored in the computer. Every afternoon at 4 PM the computer checks the cases listed on the surgery schedule (stored in electronic form in the computer) for the following day. If the computer finds a procedure and a physician for which blood ordering data have been entered, it will automatically order the blood. This reduces the possibility of inadvertent omission of routine preoperative ordering of blood and is a time-saving device for surgeons and their office staff.

26.5 Data Integration

As mentioned above, laboratory, surgery scheduling, and other patient data are used in the program's data display and decision logic.

26.6 Evaluation

Computerized order entry has permitted thorough, detailed, and timely analysis of blood ordering practices at LDS Hospital despite the large volume of data (more than 4357 blood orders in the first quarter of 1988). Analyses of orders can be done by blood component, patient location, order method (direct entry, written, standing, etc.), and reason for the order. Subanalyses and cross analyses are easily available. Ordering practices of individual physicians and departments can be examined.

Some interesting results from analysis of data from 1988 and 1989 are that (1) orders for red cell or whole blood products accounted for half of all blood orders (in line with national standards), (2) 35% of all blood orders come from acute nursing units and 37% of the orders come from the ICUs (assumptions were that intensive care orders far outweighed acute care orders), (3) physicians are entering orders directly into the system 45.6% of the time, writing them 25.8% of the time, using standing orders 7.0% of the time, and using verbal or phone orders the rest of the time (direct physician entry or standing orders are preferred since the capture of the reason is ensured), and (4) red blood cells are ordered for preoperative reasons 41% of the time, for blood loss 16% of the time, and for anemia 15% of the time.

Since the reason for the blood order is captured and verified by the computer at the time of order entry, quality assurance functions are, for the most part, performed automatically. Orders that bypass the computer

system or where the physician has overridden the computer's preprogrammed criteria are manually reviewed. In the first quarter of 1988, these
represented 19% of all orders entered (871/4357). After more intensive
review of these cases by the Quality Assurance Committee, only 17
(0.375%) were found to be true exceptions to the accepted criteria for
appropriate usage of blood and were referred to the appropriate medical
committee. This represents a review of 100% of the hospital blood orders.

26.7 Conclusion

The blood ordering program was implemented in 1987 and has been shown
to be a successful QA tool. Current efforts are being directed at incorporat-
ing approaches that will increase direct physician utilization of the system.

References

[1] Gardner RM, Golubjatnikov OK, Laub RM, et al. Computer-critiqued blood
    ordering using the HELP system. *Comput Biomed Res* 1990;23:514–528.
[2] Van Schoonhoven P, Berkmann EM, Lehmann R. Fromberg R, ed. *Medical
    Staff Monitoring Functions—Blood Usage Review*. Joint Commission on Ac-
    creditation of Hospitals; 1987.

27
Computerized Gathering of Cardiac Catheterization Data

27.1 Overview

The desire to acquire hemodynamic data from patients undergoing cardiac catheterization at LDS Hospital in the early 1960s was the motivation for the project that eventually resulted in the current HELP system. In those early days, signals were acquired, analyzed, and reported by a CDC computer. Eventually, systems were developed to gather additional clinical data and the concept of a comprehensive computerized clinical database was born. Software expansion and hardware upgrades have led to today's HELP system.

Hemodynamic data are currently obtained by special-purpose pressure and flow monitors. Analog data from the special purpose monitors are transferred to a PC for analog–digital conversion, data display, and data management. The catheterization data are analyzed by the PC and are subsequently transferred to the HELP system. The catheterization results are printed by HELP in a comprehensive report that includes other relevant clinical data from the integrated patient database.

27.2 Cardiac Catheterization

Patients undergoing cardiac catheterization at LDS Hospital have hemodynamic variables measured through intravascular pressure catheters and dye dilution techniques. Pressure measurements may be taken from the right and left side of the heart and from the great vessels. Cardiac output (flow) is measured using the dye dilution techniques. From these data pressure gradients and cardiac valve areas are calculated.

Hemodynamic measurements in LDS Hospital's catheterization laboratories are obtained using Honeywell monitors. The data are transferred in analog form to a 80386-based PC. The PC contains an analog-digital converter and the signals are converted to digital form for further processing. All further manipulation of patient data is done using the PC. The data display and data management software for the personal computer software was written by a Department of Medical Informatics graduate student. The PC's main menu screen contains the following options:

1. patient admission
2. data review/edit
3. pressure measurements
4. pressure gradients
5. valve area calculations
6. cardiac output measurements
7. send data to Tandem (HELP computer)
8. run a Tandem program
9. left-to-right shunt calculations

Choice 1 allows patient identifying data to be entered into the system. Choice 2 allows data that have been entered (discussed shortly) to be reviewed and/or edited. Figure 27.1 displays the format in which the data are presented for review in printouts and on the computer screen. Patient identifying data and height, weight, and body surface area appear at the top. The code refers to the number of the measurement. Place refers to the anatomical location (e.g., aorta, right atrium, right ventricle, etc.). Heart rate, systolic pressure, diastolic pressure, mean pressure, and an end-

```
PATIENT ID# :25182
PATIENT NAME:NINA
HEIGHT:171.0                WEIGHT: 63.0                    BSA: 1.7
------------------------------------------------------------------------
CODE   PLACE   HR    SP       DP      MP   EDP    COND    TIME
------------------------------------------------------------------------
  2      AO    78   144       75     101          RAR    12-18-1987 13:49
  3      RA    78  A  8   V   8        6          RAR    12-18-1987 13:49
  4      RV    72    64        2             8    RAR    12-18-1987 13:50
  5      PA    96    47       23       33          RAR    12-18-1987 13:55
  6      PAW   82  A 27   V  21       17          RAR    12-18-1987 13:55
  7      LV    75   115        2             2    RAR    12-18-1987 13:56
  8      PAW   75  A 30   V  30       25          RAR    12-18-1987 13:56
  9      LV    77   145        1            12    RAR    12-18-1987 13:57
 10      PAW   77  A 16   V  21       13          RAR    12-18-1987 13:57
 13      PA    48    39       13       20          O2R    12-18-1987 14:14
 14      LV    53   138        4            10    RAR    12-18-1987 14:18
 15      AO    75   183       90      126          RAR    12-18-1987 14:28
------------------------------------------------------------------------
------------------------------------------------------------------------
CODE    CO    CI    HR    AT     BT   LOC    COND    TIME
------------------------------------------------------------------------
 11     2.47 1.42   80    8.3    7.8  PA     RAR    12-18-1987 14:08
 12     2.31 1.33   63   10.0    6.8  PA     RAR    12-18-1987 14:10
------------------------------------------------------------------------
------------------------------------------------------------------------
```

Figure 27.1. Cardiac catheterization data (see text for details).

diastolic pressure measurement are then displayed. The COND column refers to the patient's condition. RAR stands for room air—patient at rest, and O2R stands for (patient on) oxygen—patient at rest. The time of the measurement follows. Cardiac output measurements are displayed in the lower part of the report and include the cardiac output, cardiac index, appearance time (of the dye), buildup time, location, condition, and time.

If choice 3 is selected, waveforms from the pressure transducers and the ECG are shown. The PC can accept data from two pressure channels simultaneously.

To store a suitable waveform and the associated pressure readings, the physician or technician simply presses the space bar on the PC. The location of the pressure measurement is chosen from a list. The transducer's location, the time of the reading, and the average of the pressure readings for the previous five beats are the data used to generate the report shown in Figure 27.1. The pressure waveforms (averaged for five beats) may be displayed on the computer screen or printed. Figure 27.2 shows an ascending aorta pressure curve obtained from the catheterization computer. Patient

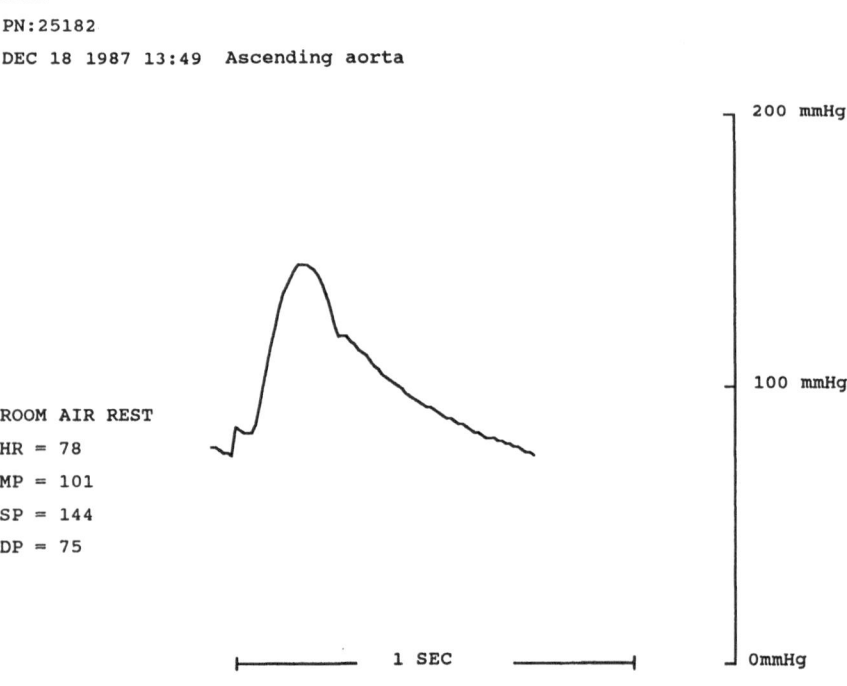

Figure 27.2. Ascending aorta pressure curve.

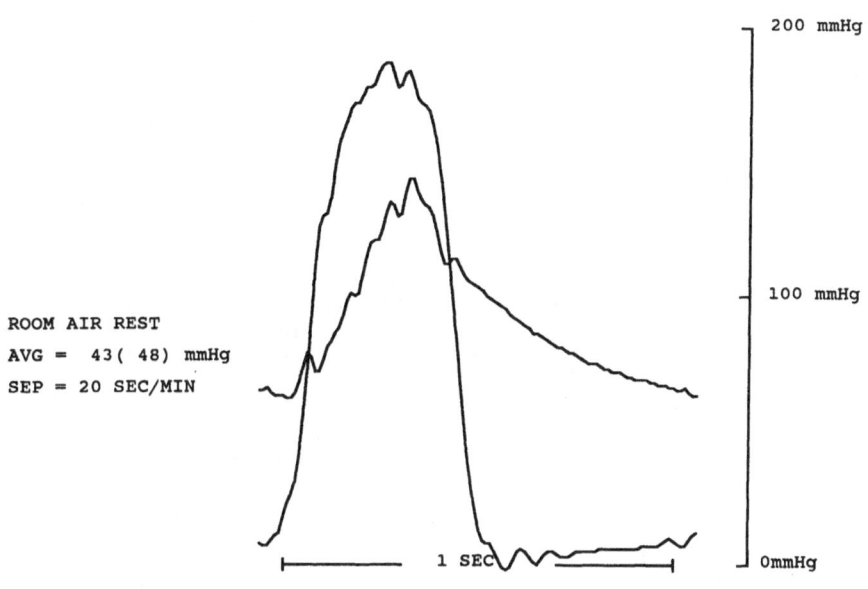

200 mmHg

100 mmHg

ROOM AIR REST

AVG = 43(48) mmHg

SEP = 20 SEC/MIN

1 SEC

0mmHg

Figure 27.3. Left ventricular and aortic pressure curves showing aortic stenosis (see text for details).

identifying data are displayed at the top left, pressure (in mm of mercury) and a time scale are displayed on the axes, and measurements for heart rate and mean, systolic, and diastolic blood pressures are displayed.

Choice 4 calculates pressure gradients across an orifice (e.g., the pressure gradient across the aortic valve will be calculated from pressure readings from the left ventricle and the aorta). Waveforms from two different locations may be superimposed for visual analysis of pressure gradients along with the numerical analysis. Figure 27.3 displays an aortic pressure curve overlying a left ventricular pressure curve. AVG refers to the aortic valve gradient with mean and peak values displayed. SEP stands for systolic ejection period. Choice 5 allows calculation of valve area from pressure gradient and cardiac output measurements.

The cardiac output (choice 6) is measured using dye dilution techniques. A measured amount of dye is injected into the pulmonary circulation and

```
                *** Cardiovascular Laboratory ** Catheterization Preliminary Report ***
         WESLEY EZRA                                LAB NO.: 18515     AGE: 79     SEX: M
PATIENT NO. 23308                      DR. FOWLES, ROBERT E                 DATE: 29JUN88
HEIGHT (CM): 186                       WEIGHT (KG): 64.40                 BSA(SQM): 1.86
```

TIME PRESSURE	LOCATION	STATE	SAT OR MMHG	HR
29 08:24	MID RIGHT ATRIUM	ROOM AIR-REST	9- 9(6)	73
29 08:25	MID RIGHT VENTRICLE	ROOM AIR-REST	60/ 0- 9	57
29 08:37	PULMONARY ARTERY TRUNK	ROOM AIR-REST	44/ 11(23)	57
29 08:38	PULMONARY ARTERY TRUNK	ROOM AIR-REST	42/ 11(22)	57
29 08:51	ASCENDING AORTA	ROOM AIR-REST	125/ 62(88)	72
29 08:54	MID LEFT VENTRICLE	ROOM AIR-REST	112/ 0- 17	70
29 08:55	MID LEFT VENTRICLE	ROOM AIR-REST	117/ 1- 38	71
29 08:55	MID LEFT VENTRICLE	ROOM AIR-REST	126/ 1- 21	57
29 08:55	MID LEFT VENTRICLE	ROOM AIR-REST	127/ 0- 30	57
29 08:55	PULMONARY ARTERY WEDGE	ROOM AIR-REST	9- 13(10)	57
29 08:56	MID LEFT VENTRICLE	ROOM AIR-REST	89/ 1- 26	81
29 08:56	PULMONARY ARTERY	ROOM AIR-REST	13- 13(10)	81
29 08:56	MID LEFT VENTRICLE	ROOM AIR-REST	126/ 2- 32	57
29 08:56	PULMONARY ARTERY TRUNK	ROOM AIR-REST	49/ 13(25)	57

```
INDICATOR DILUTION C
09:03    PULMONARY ARTERY TRUNK       ROOM AIR-REST
         CARDIAC OUTPUT            5.04      BUILDUP TIME                   12.00
         CARDIAC INDEX             2.92
         APPEARANCE TIME          12.50      HEART RATE DURING CATH          78
09:07    PULMONARY ARTERY TRUNK       ROOM AIR-REST
         CARDIAC OUTPUT            5.73      BUILDUP TIME                   10.00
         CARDIAC INDEX             3.32
         APPEARANCE TIME          12.50      HEART RATE DURING CATH          77
PRESSURE GRADIENT
              LEFT ATRIUM TO LEFT VENT              ROOM AIR-REST
         MEAN = 4     DIASTOLIC FILLING PERIOD = 34 S/MIN
              LEFT ATRIUM TO LEFT VENT              ROOM AIR-REST
         MEAN = 3     DIASTOLIC FILLING PERIOD = 22 S/MIN

** END OF DATA **
```

Figure 27.4. Preliminary catheterization report.

```
*** Cardiovascular Laboratory ** Catheterization Final Report ***
        WESLEY EZRA                          LAB NO.:18515    AGE: 79    SEX: M
PATIENT NO. 23308            DR. FOWLES, ROBERT E            DATE: 29JUN88
HEIGHT (CM): 186             WEIGHT (KG):  64.40             BSA(SQM): 1.86
```

		HR	REST	EXERCISE	HR
PH (UNITS)			7.42	.00	
PA CO2 (MM HG)			41	0	
PA 02 (MM HG)			55	0	
A 02 SAT (%)			90	0	
A 02 CONTENT (VOL%)			15.6	.0	
V 02 CONTENT (VOL%)			11.2	.0	
HOB: 12.1	02 CAPACITY (VOL%)		16.8		

PRESSURE DATA (MMHG):		HR	REST	EXERCISE	HR
MID RIGHT ATRIUM	RA	73	9- 9(6)	0- 0(0)	0
MID RIGHT VENTRICLE	RA	57	60- 0- 9	0- 0 0	0
PULMONARY ARTERY TRUNK	RA	57	45- 12(23)	0- 0(0)	0
PULMONARY ARTERY WEDGE	RA	69	11- 13(10)	0- 0(0)	0
MID LEFT VENTRICLE	RA	66	116- 1-27	0- 0- 0	0
ASCENDING AORTA	RA	72	125-62(88)	0- 0(0)	0

	AFTER DYE		AFTER NITRO	
FLOW DATA:	RA REST	RA EX	02 REST	02 EX
CARDIAC OUTPUT (L/MIN/)	5.4	.0	.0	.0
CARDIAC INDEX (L/MIN/M2)	3.1	.0	.0	.0
RATE (BEATS/MIN)	78	0	0	0
STROKE VOLUME (ML/BEAT)	69.5	.0	.0	.0

PRESSURE GRADIENTS: (MMHG)	RA-REST	O2-REST	02-EX	
DIAS FILL PERIOD (S/MIN)	28	0	0	0
MITRAL – MEAN	4	0	0	0
SYSTOLIC EJECT PERIOD (S/MIN)	0	0	0	0
AORTIC VALVE – MEAN	0	0	0	0

CALCULATED VARIABLES	RA-REST	RA-EX	02-REST	02-EX
PULM RESISTANCE	2.5	.0	.0	.0
SYST RESISTANCE	15.2	.0	.0	.0

** END OF DATA **

Figure 27.5. Final catheterization report.

then a blood sample withdrawn from the arterial side. The withdrawn sample is injected into a dye densitometer which uses dilution theory to calculate the cardiac output. The result is automatically transmitted to the PC. If oxygen saturations are measured at certain specific locations, the computer may be used to calculate the left-to-right shunt of the heart (choice 9).

Once the data have been entered, reviewed, and edited, they are sent to the Tandem (HELP) computer (choice 7) for report generation and long-term storage. The communication between the PC and the Tandem is achieved using a RS-232 serial communication link and communications software written locally in BASIC. The PC may use the same link to function as a standard HELP system terminal.

27.3 Connection with the Tandem Computer

Only numerical data from the PC are sent to the HELP computer. The actual waveform data are not sent. The individual pressure readings, the time they were taken, the location of the catheter tip, whether or not the patient was breathing supplemental oxygen, and if the patient was exercising or not are listed in a preliminary catheterization report (Figure 27.4) along with cardiac output and pressure gradient data. A final catheterization report is also produced (Figure 27.5). Pulmonary and systemic resistance are calculated and blood gas measurements are included. The final catheterization report becomes part of the patient's permanent medical record.

28
Electrocardiograms on the HELP System

28.1 Overview

At LDS Hospital ECGs are obtained using Marquette machines that print out tracings with a preliminary report. The data from the tracing are also stored on a floppy diskette and are subsequently transferred to MUSE (the *M*arquette *U*niversal *S*ystem for *E*lectrocardiography). The MUSE system is a minicomputer-based system that allows storage and retrieval of ECG data. After the preliminary interpretations have been reviewed and modified by a cardiologist, the final MUSE ECG interpretations are electronically transmitted to the HELP system where the findings are converted into HELP's standard PTXT codes. ECG findings and data are thus available to the HELP system for display, printout, and/or inclusion in decision logic.

28.2 Marquette Univeral System for Electrocardiography

In the past at LDS Hospital, ECG signals were captured and interpreted by the HELP system [1]. The Department of Cardiology recently converted to the MUSE system, which consists of microprocessor-based ECG machines (used to obtain the tracing and produce a preliminary interpretation) and a central minicomputer (for storage of the tracing data, editing of the interpretation, and other functions such as display of the tracing on a graphic display terminal and the ability to search for and retrieve stored ECGs based on various parameters) [2].

When an ECG is performed on a patient a copy of the tracing, with the machine's preliminary interpretation, is placed into the patient's record. The data are also stored on a diskette and are transferred to the MUSE minicomputer when the technician returns to the ECG department.

After the ECGs are entered into the MUSE system, the interpretations

255

are updated to include comparisons with previous ECGs. These interpretations are then reviewed by a cardiologist. The cardiologists agree with the computer's interpretation 80% of the time. If a modification is necessary, the correct interpretation is indicated on a copy of the tracing and is subsequently entered into the MUSE computer by a technician.

Cardiologists come to the ECG department to read ECGs at least twice and up to five times a day during the week. Electrocardiograms are read once daily on weekends. When the final ECG interpretation is entered into MUSE (Figure 28.1), a copy is electronically sent to the HELP computer, along with the numerical data. Because the ECGs are held in the department until they are read by a cardiologist, this is the rate-limiting step in transmission of the ECG data from the patient to the HELP system.

28.3 Integration with the HELP System

Electrocardiogram orders from nursing units are entered into the HELP system by ward clerks and nurses and are automatically printed out in the ECG department (see Order Entry, Chapter 11). The ECGs are usually performed within an hour. Electrocardiogram technicians are paged for stat orders.

After the ECG has been taken and the validated interpretation stored in the MUSE computer, electronic copies of the interpretation and associated numerical data are sent to the HELP computer. The results of the ECGs are transmitted to the HELP system as soon as they are validated.

A program to interface MUSE with HELP was written to convert the interpretation from terms that the MUSE system uses into terms that are defined in the HELP system database (Figure 28.2). In this manner, not only is the text of the interpretation available to the HELP system (for printout or display on any terminal), but the physiologic information contained therein is available for use by the HELP system's decision-making capabilities.

A data flow diagram is shown in Figure 28.3.

28.4 Conclusion

Like the laboratory system, the ECG system is an example of how a medical information system (in this case, MUSE) with specialized functionality has been integrated with HELP. Special-purpose programs have needed to be written to convert the data into a form understandable by HELP. The linkage allows the ECG data to be stored on HELP for hospital-wide data display, for the generation of comprehensive reports, and for computer-assisted decision making.

LDS HOSPITAL   NORMAL SINUS RHYTHM FIRST DEGREE AV BLOCK WITH

            WALLACE   OCCASIONAL PREMATURE SUPRAVENTRICULAR

ID: 4498732   LOC: EKG   COMPLEXES

21JUN38 IN LB CAU MALE ST & T WAVE ABNORMALITY, CONSIDER

MED: NONE   INFEROLATERAL ISCHEMIA OR DIGITALIS EFFECT

LOC: 0 ROOM: W470   CANNOT RULE OUT INFERIOR INFARCT, AGE UNDETERMINED

OPT: 10 BP: RM: W555   ABNORMAL ECG

ECG TAKEN: 03-MAY-88 14:20   WHEN COMPARED WITH ECG OF 01-MAY-88 20:14

VENT. RATE 96BPM   T WAVE INVERSION NOW EVIDENT IN INFERIOR LEADS

PR INTERVAL: 165 MS   MS

QRS DURATION 90 MS

QT/QTC 392/457 MS

P-R-T AXES 20 27 43   REFERRED BY: JOHN S REVIEWED BY: KEVIN COTE M. D.

Figure 28.1. A Marquette system (MUSE) ECG diagnostic report.

LDS HOSPITAL ECG REPORT

WALLACE AGE: 50 SEX: M NO. 432343   DR: TOWNER D. RM:

W555

ECG DATA

•• •••   5/3/88 14:20

                                VENT. RATE = 96 BPM

                                PR INTERVAL = 165 MS

                                QRS DURATION = 90 MS

                                QT/QTC = 392/457 MS

                                P AXIS = 20

                                R AXIS = 27

                                T AXIS = 43

        NORMAL SINUS RHYTHM

        FIRST DEGREE AV BLOCK WITH OCCASIONAL

        PREMATURE SUPRAVENTRICULAR COMPLEXES

        ST T WAVES ABNORMALITIES, INFEROLATERAL ISCHEMIA CONSIER

        DIGITALIS EFFECT SUSPECT

        INFERIOR INFARCTION CANNOT BE EXCLUDED AGE UNDETERMINED

        ABNORMAL ECG

        WHEN COMPARE WITH ECG OF 5/1/88 20:14

        T WAVE INVERSION PRESENT INFERIOR LEADS

    *** END OF REPORT ***

Figure 28.2. ECG findings from Fig. 28.1 converted into HELP format.

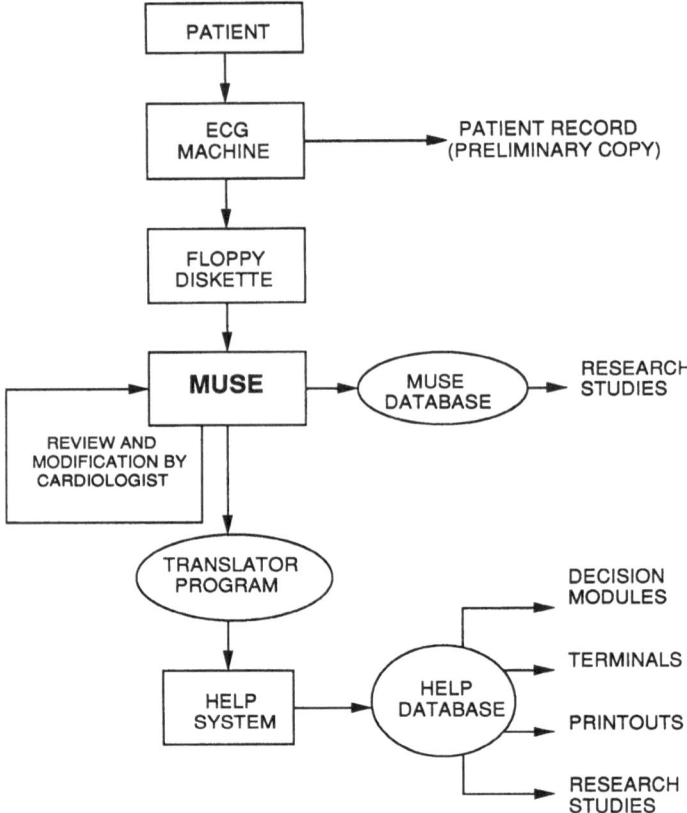

Figure 28.3. ECG data flow diagram at LDS Hospital.

References

[1] Pryor TA. *Automated Computer Analysis of the Electrocardiogram*. Salt Lake City, Utah: University of Utah; 1972. Dissertation.
[2] *ECG Systems Operator's/Training Manual*. Marquette Electronics Inc.

29
Use of the HELP System
by Radiology

29.1 Overview

The radiology module of the HELP system manages much of the workload in LDS Hospital's Radiology Department. Ordering of examinations, billing, film tracking, results reporting, and a large-scale mammography screening project are all managed by the computer.

The radiology module is also the focus of a great deal of research in computer-assisted decision making. Bayesean approaches to diagnostic reasoning (first used on the HELP system in the 1960s) have been exploited in the data-rich radiological setting in a variety of ways. Past clinical projects used the computer to infer the findings likely to be present on an x ray and a natural language processor currently extracts coded findings from free-text chest x-ray reports with the assistance of Bayesean reasoning. Current research is examining if the computer can determine which clinical data would be useful to a radiologist in accurately interpreting a patient's x ray. The clinical data (present in the HELP database) would be printed on the x-ray order form.

29.2 Order Entry and Billing

To order radiology and nuclear medicine studies, physicians complete a handwritten order form designed to elicit patient data relevant to the study. The unit clerk (or nurse) who picks up the order enters this information into the HELP computer (Figure 29.1). When the nurse or clerk chooses the patient on the HELP system, the patient's last five radiology studies are displayed. The reason for the study (as noted by the physician) is entered as free text followed by data related to any necessary preparation for the study the patient will receive. The method of transportation to be used (chair, cart, bed, portable, etc.) is then entered.

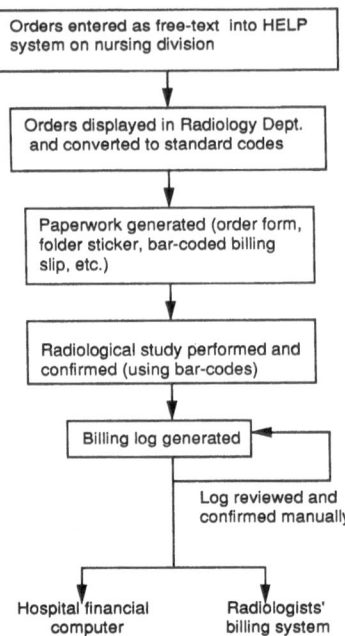

Figure 29.1. Radiology order entry and billing.

Next, information related to patient status is entered. Since this information will be transmitted to the Radiology Department, they will be aware if the patient is on oxygen, is pregnant, has fragile skin, is diabetic, has a fracture, is on isolation, or requires nurse assistance. The above choices are entered from a menu. Free text can also be entered. Next the type of examination is entered (as free text) followed by the priority (e.g., stat, ASAP, routine today, AM today, routine or AM tomorrow, future date, etc.). The entire order is then redisplayed for confirmation.

The computerized order is displayed on a HELP system terminal in the Radiology Department. A radiology clerk reviews the order and converts the free-text entries into standard, coded terms. In this way, "US Abd" and "ultrasound of the abdomen" receive the same test code. When the order has been completed, the departmental paperwork is generated, including a bar-coded order form, a bar-coded billing slip, a card for the orderly to use when picking up the patient, a file card, and a label for the film jacket. Bar code labels that identify the particular patient and study are printed by an interfaced Intermech bar-code printer.

After the study is performed, a bar-code reader is passed over the billing slip to indicate the study has been completed. A preliminary billing log is generated automatically from the bar-coded data. Any changes that need to be made to the billing log (e.g., as a result of modifications to the standard examinations) are made manually. Electronic copies of the completed billing log are sent to the LDS Hospital IBM financial computer and to the radiologists for their billing purposes.

29.3 Results Reporting

After the radiological examination is completed, the xrays are brought to the radiologist for review (Figure 29.2). Radiologists identify the patient and examination being reported on by the bar code on the order form and then dictate their findings into a Lanier dictation system.

The dictated reports are transcribed into PC network workstations using standard word processing programs. A Tandem networking product is used to transfer the text of the radiology report (as ASCII characters) from the PC network to the Tandem computer. Printouts of the radiology reports are generated by the Tandem for inclusion in the patient's chart and in the radiology folder. When chest x-ray reports are passed to the Tandem, they are examined by a natural language processing program so coded findings may be extracted. The natural language processor is described in detail below. The radiology reports may also be reviewed on HELP system terminals from anywhere in the hospital or remotely by modem. The Lanier dictation system also allows physicians to use touch-tone telephones to listen to a dictation in the Lanier system that has not yet been transcribed.

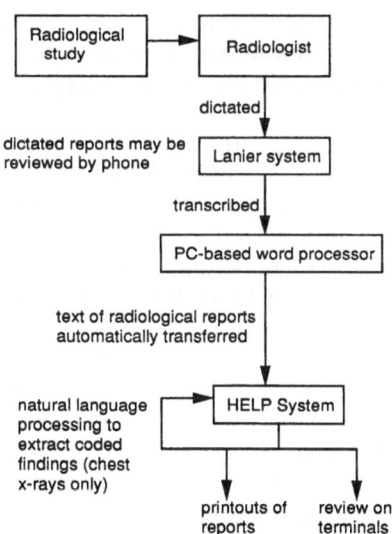

Figure 29.2. Radiology results reporting on the HELP system.

Voice recognition software has been explored as a means of reporting radiological results but as of this writing, no available system has been felt to be satisfactory.

29.4 Mammography Management Project

A system for managing mammography screening at LDS Hospital has been developed [1] and has been in use since 1985. The purpose of the project is to ensure correct and timely follow-up for women undergoing mammography. Data and decision making are managed by the HELP system.

When a patient arrives, demographic information is obtained. Next, the patient reviews an educational videotape and completes a computer-administered history questionnaire. The historical data (e.g., family history, pregnancy and menstrual history, etc.) are used later by decision-making programs. After the mammogram is performed, a radiologist uses a branching questionnaire format to enter mammographic findings into the HELP system. Data related to the technique (e.g., quality of the film, etc.) and technologists' observations (e.g., retraction of the nipple, etc.) are also entered.

The system automatically produces a letter to the patient describing the results and recommending appropriate follow-up. For negative results, timing of follow-up is determined by decision logic and depends on age and historical factors. When a patient is notified of suspicious findings, she is encouraged to follow up immediately with a physician. A sheet confirming receipt of the notice and a return envelope are included. Self-referred patients with positive findings are contacted directly to ensure they have arranged for follow-up.

A letter is also automatically sent to the referring physician (Figure 29.3). This letter is more detailed than the one sent to the patient and includes a detailed description of the findings with location, interpretation, clarifying comments, references, and the name of the interpreting radiologist. The contents of this letter (as well as the one sent to the patient) are constructed using the coded findings entered into HELP by the radiologist at the time the mammograms were interpreted.

The system is also designed to monitor follow-up of both abnormal and normal mammograms. The HELP computer automatically checks to see if patients with abnormal findings have follow-up studies performed within a specified time period. Hospital records are examined to see if any biopsy procedures were performed. If no follow-up procedure is found, a letter is sent to the physician asking what investigative steps have been taken. The system also sends letters to physicians whose patients require routine annual or biannual follow-up.

```
                          LDS HOSPITAL
                     DEPARTMENT OF RADIOLOGY
                   SALT LAKE CITY, UTAH  84143
                     TELEPHONE:  321-1791

PATIENT:         CONRA RAE - OPD                 RADIOLOGY #: 949382
ADDRESS:                                         PATIENT #:   606014
       SLC, UT   84103                           DR. NOYES, R. DIRK
       (   )364-

AGE:   34
Patient reported finding lump
THIS IS A DIAGNOSTIC OBSERVATION.

MAMMOGRAM: JUL 28 88
SUMMARY; Number of Observations:  1

   Observation # 1

          Findings: WELL DEFINED MASS MEASURING 5 MM, MAMMOPLASTY

          Location: LEFT  BREAST--UPPER OUTER QUADRANT, 2 CM DEEP

          Interpretation: NO CHANGE SINCE LAST MAMMOGRAM,FINDINGS ARE
                          INDETERMINANT, BUT CONSIDERING THE PATIENT'S AGE ARE
                          PROBABLY BENIGN,ULTRASOUND OF THE LESION IS
                          RECOMMENDED TO RULE OUT A SOLID VS A CYSTIC MASS

          Comment: THIS ABNORMALITY MAY BE A CYST. ULTRASOUND CAN DIFFERENTIATE
                   CYSTS FROM SOLID LESIONS. IF THE LESION IS SOLID, BIOPSY MAY
                   BE APPROPRIATE.

          Comment: IN PATIENTS WITH MAMMOPLASTY, A SIGNIFICANT AMOUNT OF BREAST
                   TISSUE IS OBSCURED BY THE BREAST IMPLANT.  A SMALL LESION
                   COULD BE MISSED BY MAMMOGRAPHY UNDER THESE CIRCUMSTANCES.
                   PHYSICIAN EXAMINATION IS AN IMPORTANT ADJUNCT TO MAMMOGRAPHY
                   IN THESE PATIENTS.

This mammogram was interpreted by Dr. Irena M. Tocino.
```

Figure 29.3. HELP-generated letter of mammography results to referring physician.

29.5 Extraction of Coded Findings from Reports

Since the early 1980s there has been a constant effort at LDS Hospital to use the results of radiological examinations in computerized decision logic. Radiological examinations contain a great deal of data pertinent to diagnosis; however, radiological results usually exist as free text and HELP requires coded findings to execute its decision logic. For example, the HELP CIDM (Chapter 18) is capable of diagnosing nosocomial pneumonias; however, the program must be aware if the patient's chest xrays are abnormal or not and if so, what type of findings are present. Two major projects have sought to incorporate radiological data into the HELP system for decision-making purposes.

29.5.1 Pre-Examination Prediction of Radiological Findings

In the early 1980s a system was developed that used clinical information available from the HELP database to generate a list of the most likely positive x-ray findings at the time the study was ordered [2–4]. Statistical studies were performed to determine which parameters from the computerized database (e.g., age, laboratory values, previous radiographic findings, etc.) and which data from the computerized radiology order (mode of transport, reason for the test, etc.) could be used to predict the likelihood of the presence of a particular radiographic finding. Further studies were done to determine which of these parameters were conditionally independent. A priori probabilities for the presence of particular findings were determined by medical experts and decision modules were written to calculate posterior probabilities based on relevant patient data. The calculations involved a sequential Bayesean analysis and probabilities were automatically determined for all potential findings whenever a radiological study was ordered.

The five findings with the highest probability of being present (in addition to "No significant abnormalities") were printed on the radiology reporting form for the radiologist to use in a checklist-like manner (Figure 29.4). Next to each positive finding the radiologist could indicate (1) if this was an initial finding, (2) if the finding had stayed the same, increased, or decreased (if the finding had appeared previously), (3) the magnitude of the finding (mild, moderate, severe), and (4) the likelihood of the presence of the finding (definite, likely, possible, indefinite). This method of reporting attempted to codify positive findings and complemented, rather than replaced, the dictated report.

The coded findings were later entered by secretaries into the HELP

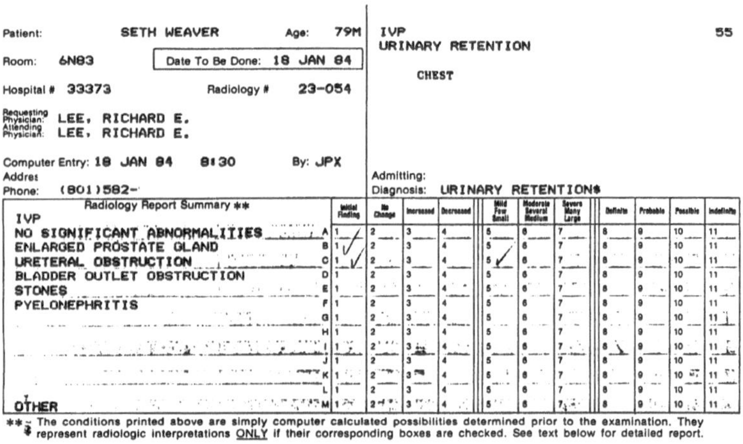

Patient: SETH WEAVER Age: 79M
Room: 6N83 Date To Be Done: 18 JAN 84
Hospital # 33373 Radiology # 23-054
Requesting Physician: LEE, RICHARD E.
Attending Physician: LEE, RICHARD E.
Computer Entry: 18 JAN 84 8:30 By: JPX
Addres
Phone: (801)582-

IVP 55
URINARY RETENTION
CHEST

Admitting:
Diagnosis: URINARY RETENTION*

Radiology Report Summary ** IVP		Initial Finding	No Change	Increased	Decreased	Mild Few Small	Moderate Several Medium	Severe Many Large	Definite	Probable	Possible	Indefinite
NO SIGNIFICANT ABNORMALITIES	A	1 ✓	2	3	4	5	6	7	8	9	10	11
ENLARGED PROSTATE GLAND	B	1 ✓	2	3	4	5	6	7	8	9	10	11
URETERAL OBSTRUCTION	C	1 ✓	2	3	4	5 ✓	6	7	8	9	10	11
BLADDER OUTLET OBSTRUCTION	D	1	2	3	4	5	6	7	8	9	10	11
STONES	E	1	2	3	4	5	6	7	8	9	10	11
PYELONEPHRITIS	F	1	2	3	4	5	6	7	8	9	10	11
	G	1	2	3	4	5	6	7	8	9	10	11
	H	1	2	3	4	5	6	7	8	9	10	11
	I	1	2	3	4	5	6	7	8	9	10	11
	J	1	2	3	4	5	6	7	8	9	10	11
	K	1	2	3	4	5	6	7	8	9	10	11
	L	1	2	3	4	5	6	7	8	9	10	11
OTHER	M	1	2	3	4	5	6	7	8	9	10	11

** The conditions printed above are simply computer calculated possibilities determined prior to the examination. They represent radiologic interpretations ONLY if their corresponding boxes are checked. See text below for detailed report.

Figure 29.4. List of most likely radiological findings for use by radiologist (see text for details).

computer system. The findings were shown to improve the accuracy of experimental HELP system programs that attempt to predict the patient's discharge diagnoses [5,6]. In one study, radiologists were reported to use the method 64% of the time [4]. However, this value fluctuated and a more consistent method of obtaining coded findings was sought.

29.5.2 Natural Language Processing of Radiology Text

In 1989, the checklist method of obtaining findings from radiologists was replaced by a HELP system program (SPRUS, for Special Purpose Radiology Understanding System) that automatically parses the free text of radiology reports for the purpose of extracting coded findings. The program has been piloted for chest x-ray reports [7] but research into extending the program into other areas is under way.

The program uses a semantic approach to find meaning in the text; that is, it has an expectation of encountering certain terms based on the context (in the case of the pilot study, chest radiographs). The program searches for findings, locations, and diseases or interpretations. When the program encounters a finding (e.g., localized alveolar infiltrate), it waits for a suitable anatomical location (e.g., left lower lobe) with which to link the finding. Finding–location pairs and diseases or interpretations (e.g., pneumonia) are stored in the patient database and are available to other HELP decision-making programs.

The expected finding–location pairs to be sought are determined by examination of the HELP diagnostic frames described in Chapter 5. A special purpose compiler (SPRUS-COM) examines the frames to determine all the disease settings in which chest radiological results (specifically, finding-location pairs or diagnoses) play a role. These terms are set aside and are used by SPRUS in the parsing of terms. The compilation occurs only once when the system is initialized and need not happen again unless the frames are changed. Extracting radiological knowledge from the frames is an example of using a general purpose knowledge frame (e.g., a description of a pulmonary disease) for a much narrower purpose (in this case, extraction of radiological findings). The set of terms used by SPRUS is expanded through the use of a thesaurus that defines synonyms (e.g., CHF is an acceptable equivalent for congestive heart failure).

The HELP knowledge frames and the hierarchical nature of HELP's data dictionary, PTXT (described in Chapter 3), are used in the process of disambiguation. As shown in Figure 29.5, a radiology report may refer merely to "infiltrates." As shown in Figure 29.6, in the HELP hierarchical dictionary, infiltrates is a more general term for (1) interstitial infiltrates, (2) diffuse alveolar infiltrates, or (3) localized alveolar infiltrates. SPRUS attempts to resolve ambiguities in language by establishing a context just as humans do. The context is established by examining the remainder of the text. In the example in Figure 29.5, the disease "adult respiratory distress

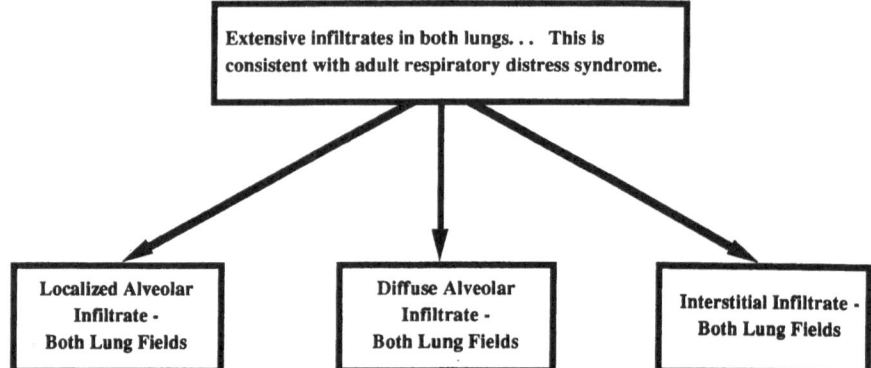

Figure 29.5. Example of ambiguity in radiological reporting. © 1990, RSNA. Reprinted with permission from "Computerized Extraction of Coded Findings from Free-Text Radiologic Reports," *Radiology* 544–545, February 1990.

syndrome" is mentioned. Since the HELP frame for adult respiratory distress syndrome contains probabilistic information regarding the presence of certain radiological findings in this disease, SPRUS is able to determine diffuse alveolar infiltrates are the most likely type of infiltrate in this disease. The ambiguity is thus resolved and the specific data can be entered into the patient database. Resolving ambiguities due to insufficient information in this manner has been found to be quite satisfactory thus far.

The results from SPRUS have been encouraging. When compared with

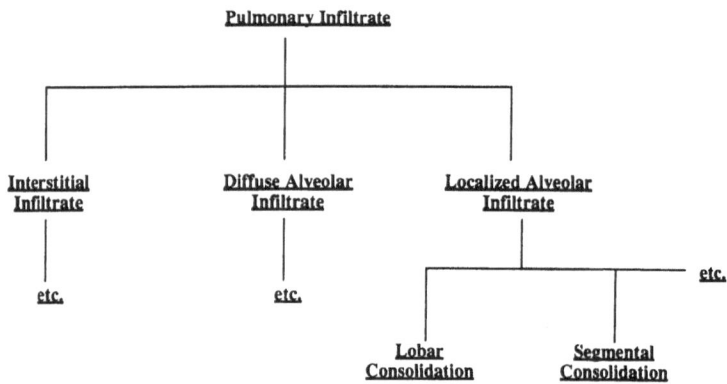

Figure 29.6. Example of hierarchical structure of medical terms in HELP system (used for disambiguation, see text for details). © 1990, RSNA. Reprinted with permission from "Computerized Extraction of Coded Findings from Free-Text Radiologic Reports," *Radiology* 544–545, February 1990.

a human extracting findings from the radiology reports, SPRUS had a sensitivity of 90% and a bad data rate (i.e., storing inaccurate data to the database) of 5% (of the total number of findings). In another set of data, disambiguation increased the sensitivity from approximately 70% to 90% while increasing the bad data rate from 5% to 6%, and thus disambiguation was felt to be a worthwhile process. Because of the semantic nature of the process, the accuracy of the program is very sensitive to the integrity of the thesaurus. For example, if a radiologist begins using a new term (e.g., LLL for left lower lobe) and the new term is not entered into the thesaurus, accuracy will suffer.

Semantic text processing requires a restricted domain and a specialized sublanguage. These restrictions pose difficulties as the project is expanded into areas such as general radiology and parsing of history and physical examination texts.

29.6 Future Uses of Decision Logic in Radiology

It has been demonstrated that radiologists are more accurate at reading films when presented with clinical data related to the patient's condition. Current research at LDS Hospital is the computer's ability to choose the most relevant data in the patient database to be included with the x-ray order form. The relevancy of the data will be determined by the impact of the data on the probability of the presence of each disease present in the HELP knowledge base, that is, the HELP frames. The frames will be evaluated with and without each clinical data element and the data elements found to alter most significantly the likelihood of disease will be presented to the radiologist. This is another example (similar to the semantic-based text processing described above) of a special purpose use of a general knowledge frame.

29.7 Conclusion

The radiology module of the HELP system provides workload management for the Radiology Department. Ordering, billing, and results reporting for radiographic studies are managed by the system. Computerized analysis of radiographic reports yields coded x-ray findings that are used by other HELP system modules. The text processing program (and other programs under development) make use of general purpose knowledge frames in the HELP knowledge base.

References

[1] Haug PJ, Tocino I, Clayton PD, Bair TL. Automated management of breast cancer screening. *Radiology* 1987;164:747–752.

[2] Clayton PD, Ostler DB, Gennaro JL, Beatty SS, Frederick PR. A radiology reporting system based on most likely diagnoses. *Comput Biomed Res* 1980;13: 258-270.

[3] Neeley JP, Ostler DB, Frederick PR, Clayton PD. Pre-examination prediction of radiological findings. *Invest Radiol* 1982;17:310-315.

[4] Clayton PD, Gray S, Frederick PR. A code-oriented radiology reporting system based upon pre-examination prediction of likely findings. Proceedings of the American College of Radiology 7th Conference on Computer Applications in Radiology; 1982:261-272.

[5] Clayton PD, Haug PJ, Gerard MJ, et al. The role of radiology findings in automated decision-making. Proceedings of the 8th ACR Conference on Computer Applications in Radiology; 1984:521-531; St. Louis, Mo.

[6] Clayton PD, Haug PJ, Shelton P, et al. Automated decision-making in radiology: advantages of an integrated system. *MEDINFO* 1986:624-628.

[7] Haug PJ, Ranum DL, Frederick PR. Computerized extraction of coded findings from free-text radiologic reports. *Radiology* 1990;174:543-548.

30
Computerized Pulmonary Function Testing

30.1 Overview

Pulmonary function testing at LDS Hospital is part of the integrated HELP computer system. Spirometric measurements are captured directly by the Tandem computer. Single-breath carbon monoxide diffusing capacity and body plethysmography data are entered into the computer by hand for further calculations and inclusion in reports. Interpretation of the data is done by the HELP system's knowledge base and decision logic. The results of pulmonary function tests may be reviewed on HELP system terminals or on printed reports. A copy of the printed pulmonary function test report is included in the patient's permanent medical record.

30.2 Data Stored

Standard spirometry values (e.g., FVC, $FEV_{0.5}$, FEV_1, $FEV_1/FVC\%$, FEV_3, $FEV_3/FVC\%$, $FEF_{25-75\%}$, IVC, IC, ERV, and expiration time) are measured using a volume sampling method [1] or calculated from the measured variables. Once obtained, the data are presented to the pulmonary technician for review. If the technician feels the data are valid they are transmitted automatically to the Tandem computer for interpretation and storage in the HELP patient file. If single-breath carbon monoxide diffusing capacity, lung volumes from tracer gas dilution, or lung plethysmography is ordered, these data are measured and entered into the computer system by hand. Instrumentation, techniques, measurements, and calculations in the pulmonary function laboratory at LDS Hospital conform to established regional and national standards [2].

30.3 Integration and Decision Logic

In addition to the spirometry values, the HELP system uses hemoglobin values from blood gas measurements when it interprets altered carbon monoxide diffusing capacity values. The HELP system uses logic defined in published guidelines [2] to interpret the pulmonary function test results.

Using published reference spirometric values, the computer can determine obstructive or restrictive patterns of flow, indicate the extent of the abnormality (mild, moderate, severe), and make qualifying statements (e.g., "Upper airway obstruction suggested" if $FEV_{0.5}/FEV_1 < 0.60$, or "Chest restriction may also be present; suggest lung volumes and DL_{CO}" if the measured FVC and IVC are both more than one 95% confidence interval below the predicted value). The computer also comments on the extent of improvement after bronchodilator therapy.

If no abnormalities are found and the effort is adequate, the program will interpret the spirogram as normal. Even in this situation it may have some comments such as, "Probably normal spirometry. Short exhalation may hide mild obstruction" if the expiratory time is less than 3 seconds. If maximal voluntary ventilation is low relative to the FEV_1 the program will so indicate, suggesting a poor effort or a neuromuscular problem.

Single-breath carbon monoxide diffusing capacity is similarly interpreted as normal or mildly, moderately, or severely reduced. If obstruction is present and the diffusing capacity:alveolar ventilation ratio is reduced, the computer will indicate that emphysema may be present. Alveolar ventilation (calculated) and hemoglobin values (from blood gas measurements) are considered as potential causes of altered DL_{CO} values. If both the DL_{CO} and the DL_{CO}:alveolar ventilation ratio are reduced, the computer will indicate that an abnormality of the gas exchanging surface may be present.

The computer compares the vital capacity (VC) from single-breath measurements and the forced vital capacity from the spirometry. If the VC is smaller, the computer will suggest the dilution test may have been a submaximal effort. If total lung capacity (TLC) measured by plethysmography exceeds TLC measured by tracer gas dilution methods, the HELP system will consider the difference between the two to be a measure of trapped air unavailable to gas dilution techniques.

30.4 Reports

In our hospital, review of the pulmonary function test data, with interpretation, is available from any terminal within minutes after the completion of the test. Hard-copy printouts are available from the printers at the nursing stations.

A sample printout of a pulmonary function test report is shown in Figure 30.1. Patient identifying data with age, sex, height, and weight are listed at

LDS Hospital Pulmonary Function Report

		NO.						06/10/88
SEX: M	AGE: 64	HT: 171	WT: 104.0	BP: 645		TEMP: 21.0		RM: PULM

SV4		PRE-BRONCHODILATOR				POST-BRONCHODILATOR		
		NORMAL	09:40			09:43		%
	PRED	LIMIT	DATA	%PRED		DATA	%PRED	Impr
FVC (L)	4.24	3.12	2.29	54		2.51	59	10
FEV .5 (L)	2.70	1.99	1.19	44		1.20	45	1
FEV 1 (L)	3.33	2.49	1.58	47		1.65	49	4
FEV1/FVC%	78.5	70.2	69.1	88		65.6	84	
FEV 3 (L)	3.90	2.88	2.03	52		2.16	55	6
FEV3/FVC%	92.1	87.5	88.9	97		86.3	94	
FEF 25-75%	3.19	1.52	.93	29		.85	27	−8
TIME (SEC)			7.51			8.15		

DLCO SINGLE BREATH DATA - VALUES FROM 1 TEST GIVEN 09:55

	PRED	NORMAL LIMITS	DATA	%PRED	CORRDATA	%PRED
DLCO SB	30.4	22.2	28.9	95	27.0	89
DL/VA SB	4.82	3.43	7.02	146	6.55	136
TLC SB (L)	6.47	8.08–4.86	4.12	64		
FRC SB (L)	3.36	4.82–1.90	1.79	53		
RV SB	2.18	2.94–1.42	1.48	68		
RV/TLC SB	.34		.36	107		
VC SB (L)	4.24		2.64	62		
IC SB (L)			2.33			
ERV SB (L)			.31	14		
Hb 9m/dl	15.7	17.7 −13.7	17.4			

PLETHYSMOGRAPH DATA—AVERAGE OF 2 TESTS

	PRED	NORMAL LIMITS	DATA	%PRED
TLC PL(L)	6.47	8.08–4.86	5.47	85

MILD AIRWAY OBSTRUCTION
MILDLY REDUCED VITAL CAPACITY
  Spirogram Not Clearly Improved Post Bronchodilator
NORMAL—TLC
MILD AIR TRAPPING PRESENT
  Estimated Trapped Air 1.35 L (PL—SB)
NORMAL DIFFUSING CAPACITY (DLCO)

Figure 30.1. Sample pulmonary function test report.

the top. Room temperature is also listed. Next the spirometric values are listed with predicted values, lower limits of normal, measured values (data), and percent of predicted value noted for each variable. The percent improvement after bronchodilator therapy is also listed.

Next are the results of the carbon monoxide diffusing capacity and lung volumes calculated from the single breath test. The measured value corrected for the hemoglobin concentration (from blood gas data) is also listed

(CORRDATA). The last measurement listed is the total lung capacity measured by plethysmography. The HELP system's interpretations of the above results follow the data. Comments with respect to airway obstruction, VC, TLC, air trapping, and diffusing capacity can be seen.

Inpatient reports are printed out in the Pulmonary Department, combined with a specialist's report, and sent to the nursing unit via interoffice mail for permanent inclusion in the medical record. Outpatient reports are mailed to the requesting physician.

References

[1] Ostler DV, Gardner RM, Crapo RO. A computer system for analysis and transmission of spirometry waveforms using volume sampling. *Comput Biomed Res* 1984;17:229–240.
[2] Morris AH, Kanner RE, Crapo RO, Gardner RM. *Clinical Pulmonary Function Testing; A Manual of Uniform Laboratory Procedures*, 2nd ed. Intermountain Thoracic Society, Salt Lake City, Utah. 1984.

31
Physicians' Remote Call-In Network

31.1 Overview

Physicians affiliated with LDS Hospital may use PCs to access patient information contained in the HELP system from their homes or offices [1]. Standard telecommunications software and modems installed on the physicians' PCs is used to connect (through telephone lines) to one of five modems that have been interfaced to the HELP computer. Physicians may perform data retrieval (e.g., laboratory, hemodynamics, etc.), blood ordering, ordering of TPN, and almost any function that can be performed on a HELP system terminal within the hospital. Physicians' office staff members may use the remote network to obtain demographic or financial information stored on the HELP system.

A questionnaire administered soon after the system was implemented revealed that users felt the laboratory and financial/demographic data availability was the system's most useful aspect [1]. Unauthorized access of data has not been a problem thus far. Eighty five to 90 participating users are currently accessing the system an average of 12 to 15 times per month per user.

31.2 System Components

To use the system, physicians must possess an IBM compatible PC with a 300-, 1200-, or 2400-baud Hayes-compatible modem. The telecommunications software is Procomm [2] that has been customized by the Department of Medical Informatics of LDS Hospital for use specifically in the remote call-in network. The software is installed in the physician's PC by the Department of Medical Informatics and the physician and his/her staff is trained in the use of the software.

Communications between the physician's PC and the HELP computer is accomplished through a batch file on the PC (i.e., the process is automatic after the user initiates the process with a single command). The phone number used to access the HELP computer, as well as the password identifying the particular physician, have been encrypted into the software as part of the security measures (see below).

Five modems attached to the HELP system permit communications with participating physicians' offices. If one modem is occupied, the next will be accessed automatically so that up to five physicians may use the system simultaneously. Once communications have been established, the physician's PC functions like a HELP system terminal.

There is currently no charge for physicians to use the system. To register to use the system, the physician must sign a confidentiality and indemnity statement (Figure 31.1) agreeing to treat computer-accessed information as confidentially as he/she would a paper-based record. Each "session" on the system is limited to 20 minutes.

31.3 Security

A six-level security system is used to guard against unauthorized data access using the system.

1. The first level is the legal agreement (Figure 31.1) signed by the physician stating they will be responsible for data accessed under their log-on code.

LDS Hospital Physician Information Network

Legal Agreement

I agree to maintain the diskette I am provided with in a secure place and will be responsible for any computer system access with the logon security code recorded on the diskette. I also agree to maintain the patient's privacy just as if I were on-site at LDS Hospital. I will notify the Department of Medical Informatics at LDS Hospital if my diskette is lost or stolen so that my security code can be removed from the system.

I further agree to hold LDS Hospital and Intermountain Health Care harmless from any claim in tort, contract, or other legal theory arising out of my use or use through my access code of the LDS Hospital information system (HELP), or of any information derived through such use.

_____   _____

Signature   Date

_____

Name (Please Print)

Figure 31.1. Physician confidentiality agreement for remote call-in function.

2. A copy protection scheme is used to prevent unauthorized reproduction of the diskette containing the communications software.

3. The phone numbers used to access the HELP system are unpublished and encrypted into the communications software.

4. Passwords necessary to gain entry to the system (even if one should be able to obtain the phone number) are also encrypted into the software.

5. Access is restricted to applications programs. System utilities are inaccessible.

6. When users register to use the system, they are informed that the use of the system will be logged and audited. The list of which physicians reviewed which patients' data, along with the computerized list of patients' attending and consulting physicians, can be analyzed to determine if data reviews are acceptable (i.e., done by patient's attending or consulting physician or an associate of one of these) or questionable (not acceptable). Review of this analysis is the final level of security.

31.4 Evaluation

Current utilization statistics have been quoted in the first section of this chapter. Utilization is heaviest between 10 AM and 2 PM. There is rarely more than one physician on the system at a time. Statistics have been kept on the utilization of the laboratory and financial/demographic data retrieval programs. Laboratory data retrieval is used slightly more often (56% vs. 44%).

A questionnaire has revealed that users feel the system generally is reliable, is well supported, improves productivity, is a valuable adjunct to the hospital-based HELP system, and is easy to use.

References

[1] Davidson MG. *A Physician Information Network*. Salt Lake City, Utah: University of Utah; 1987. Thesis.
[2] *Procomm Program Reference Manual, Version 2.4.2*. Columbia, Mo.: PIL Software Systems. 1986.

32
Computerized Urinary Catheter Culture Monitoring

32.1 Overview

The Infectious Disease Department at LDS Hospital has a long-standing interest in the problem of nosocomial bacteruria. As part of ongoing research, urinary culture specimens are collected daily from patients with indwelling urinary catheters. The computer assists the research by generating a daily list of patients who should have their urine collected. If certain data elements need to be collected manually (as occasionally happens when a particular therapeutic or prophylactic regimen is being studied), the computer prints the questions that need to be answered on the daily list, thus turning the list into a data collection tool. Special urinary catheter culture reports are generated by the computer and become part of the patient's permanent medical record.

32.2 Goals and Methods of the Program

The purpose of the program is early surveillance of catheterized patients and a framework in which to study factors involved in the etiology and prevention of catheter-related urinary infection. The program has been in place for more than 10 years. Past studies have examined the effect of drainage techniques [1] and epidemiologic factors [2] on infection rates.

When a patient has a Foley catheter or suprapubic tube inserted, an initial catheter specimen (ICS) slip is completed by the person inserting the tube. The ICS slip contains information on who inserted the tube, the date and time, and the patient's name. The slip (with the initial urine specimen) is then sent to the Infectious Disease Department and the data entered into the HELP system. Based on the ICS data, HELP generates a list each morning of all patients who need to have a specimen collected.

Using the computer-generated list (Figure 32.1), a laboratory technician collects the urine specimen from the patient each morning between 6 AM and 9 AM. Next to the patients' names are codes that are checked off by the technician:

```
                    Collection List for 06/04/1988
                          Printed at 06:50

    ******    E6    ******

    E602    111111111    SMITH, JOHN            COL   NF   CNC   CO

    E605    222222222    SMITH, JANE            COL   NF   CNC   CO

            *** Suprapubic patient ***

            JD?   Y | N     SO?  Y | N      NC?  Y | N      NB?  Y | N

            Site Culture Taken?          Y | N
            Local inflammation?          Y | N
            Pain?                        Y | N
            Drainage at site:  Serous?    Y | N
                               Pus?       Y | N
            Complications: Bleeding?         Y | N
                           Clogging?         Y | N
                           Catheter damage?  Y | N
                           Other (specify)   Y | N
            DC Status:     Normal
                           Catheter clogged
                           Catheter related bleeding
                           Traumatic
                           Bowel perforation
                           Damaged catheter
                           Pain
                           Doctor request
                           Patient request
                           Other (specify)

    E607    333333333    SMITH, ABE            COL   NF   CNC   CO

    E611    444444444    SMITH, FRED           COL   NF   CNC   CO

            *** New Suprapubic patient ***

            JD?   Y | N     SO?  Y | N      NC?  Y | N      NB?  Y | N

            Site Culture Taken?          Y | N
            Local inflammation?          Y | N
            Pain?                        Y | N
            Drainage at site:  Serous?    Y | N
                               Pus?       Y | N
            Complications: Bleeding?         Y | N
                           Clogging?         Y | N
                           Catheter damage?  Y | N
                           Other (specify)   Y | N
            DC Status:     Normal
                           Catheter clogged
                           Catheter related bleeding
                           Traumatic
                           Bowel perforation
                           Damaged catheter
                           Pain
                           Doctor request
                           Patient request
                           Other (specify)
            Black line drawn?   Y | N
            Urine meter?        Y | N
```

Figure 32.1. Specimen collection list for urinary catheter culture monitoring. Names have been falsified.

```
                            LDS HOSPITAL

                    DAILY CATHETER CULTURE RECORD

SMITH, JOHN JAMES  Hosp Number:   234432234  Phy:JONES, WILLIAM  Room: E711

*********************************************************************************
Type      Culture Date      Preliminary Report      Final Report

BLA       5/15/88                                    NO GROWTH

BLA       5/16/88                                    NO GROWTH

BLA       5/17/88                                    NO GROWTH

BLA       5/18/88                                    STREPTOCOCCUS SP
                                                     Less than 1K

BLA       5/19/88                                    NO GROWTH

BLA       5/20/88                                    ALBICANS
                                                     1K to 10K

BLA       5/21/88                                    ALBICANS
                                                     1K to 10K

BLA       5/22/88                                    ALBICANS
                                                     10K to 50K

BLA       5/23/88                                    ALBICANS
                                                     10K to 50K

BLA       5/24/88                                    ALBICANS
                                                     1K TO 50K

BLA       5/25/88          ALBICANS
                           1K to 10K

* -> Species identification and antibiotic susceptibilitytesting will
  be done upon request.  (Phone weekdays ext-1006, weekends ext-1858)



5/26/88.10:47   ***Please discard this report when the next is received.***
***Monitoring will be stopped after 10 days -- to continue call ext 1006.
```

Figure 32.2. Results of urinary catheter culture monitoring. Name has been falsified.

1. specimen collected (COL)
2. patient not found (NF)
3. no urine in bag (CNC, for could not collect)
4. catheter removed (CO, for catheter out)

If the patient is enrolled in one of the ongoing studies, additional questions (pertaining to the study) appear on the list for the technician to complete. Examples of study-related questions include patients' complaints about pain, the presence or absence of hematuria, status of the urinary drainage system, and deviations from accepted standards of care. (Note: the questions in Figure 32.1 were related to ongoing studies.)

Upon returning to the laboratory, the technician submits the specimens for culture and enters the data into the computer. Patients who have been marked "catheter removed" (#4, above) are automatically removed from the list for the following morning.

32.3 Reports

Cultures are read daily and then entered into the computer as either preliminary or final results. Every day at noon a report is generated by the computer for every patient with a catheter culture. These are taken to the nursing units for inclusion in the patient chart (Figure 32.2).

If the specimen fulfills the criteria for a nosocomial infection, the culture report will appear on the CIDM report for review by the Infectious Disease Department (see Chapter 18 for further discussion of CIDM). Follow-up of routine reports is left to the attending physician.

The computerized data are also used to generate a patient charge.

References

[1] Burke JP, Larsen RA, Stevens LE. Nosocomial bacteriuria: estimating the potential for prevention by closed sterile urinary drainage. *Infect Cont* 1986;7(2): 96–99.
[2] Larsen RA, Burke JP. The epidemiology and risk factors for nosocomial catheter associated bacteriuria caused by coagulase-negative staphylococci. *Infect Cont* 1986;7(4):212–215.

V
Inactive or Experimental HELP System Modules

This part of the book contains descriptions of HELP system projects that are experimental, under development, or currently inactive for one reason or another. In general, HELP system projects are born from research interests or to fulfill a practical need at LDS Hospital. Research projects may be suggested by Department of Medical Informatics faculty or students or by other personnel (clinical or administrative) at LDS Hospital. Practical projects attempt to solve clinical, administrative, or financial problems that arise at LDS Hospital.

Research projects are generally funded in the same manner as other kinds of medical research. Seed money often exists but continued development requires funding either from local or national agencies. Projects with a more practical goal tend to be funded by LDS Hospital and/or its managing corporation, IHC or other private sources.

Also, as major modifications are made to the HELP system (e.g., major hardware upgrades, new software platforms, etc.), existing programs need to be modified to remain functional. Programs that are little used and have few advocates may not attract the resources necessary for upgrading and may become inactive.

Most programs are evaluated as they are installed. An overview of evaluations of HELP system programs is contained in Chapter 6.

33
Computerized History Taking
on the HELP System:
Initial Efforts, Goals,
Methods, and Evaluations

33.1 Overview

A program has been developed at LDS Hospital that uses historical data gathered directly from the patient and a knowledge base developed by medical experts and information scientists to calculate the probability of the patient having a given disease. This program is a direct continuation of experiments performed by Dr. Homer R. Warner in the early 1960s to demonstrate the ability of the computer to determine diagnoses correctly from clinical data. The current program generates a list of the patient's most likely diagnoses using Bayesean inferencing strategies. Evaluation studies reveal that 75% of the time, the program includes the patient's actual disease (as determined retrospectively) in the list of the five most likely diseases.

Several strategies for determining which questions to ask the patient (in addition to an exhaustive list) have been attempted. It appears that when methods of logic are applied to the questioning process, the number of questions that needs to be asked is decreased with no loss of the program's predictive ability. Studies have shown that revision of the statistical parameters in the knowledge base based on review of the HELP database leads to an increase in the program's predictive ability.

Thus far the program has been limited to the domain of pulmonary diseases. Its use is hampered by intensive demands it places on the main computer's central processor. It is not in routine clinical use at this time but development is continuing.

33.2 Goals

The goal of this effort is for the computer to generate a list of the five most likely diagnoses based on the patient's history at the time of admission. If this can be done with a fairly high degree of certainty, the physician would have access to likely diagnoses. Also, historical data could be made a more routine part of the HELP system's decision logic.

33.3 General Considerations

Data are obtained from the patient in one of a variety of ways (to be discussed). Once gathered, the data are subjected to a knowledge base and a list of likely diagnoses is generated.

The knowledge base, constructed by medical specialists with the assistance of information scientists, contains probability tables (e.g., the probability of a disease being present given a certain symptom, or the probability of the absence of a disease given the absence of a symptom). The knowledge base uses a sequential application of Bayes' Theorem to calculate a probability for each diagnosis and then lists them, with the most likely at the top of the list.

Bayes' Theorem is one aspect of a mathematical approach to decision making (decision analysis) that uses the prior probability of a disease being present and the knowledge of the influence of given factors to calculate a final probability of a condition being present. One of Bayes' limitations is that it assumes conditional independence of symptoms; that is, that the presence of one symptom has no bearing on the probability of another being present. Certain forms of Bayes' Theorem assume that only one disease at a time is present. Despite these limitations it is one of the principal techniques in use today for inferring in medical situations [1,2].

The probabilities in the initial knowledge base were taken from the opinions of clinical experts. When a patient database was used to modify the probabilities, the program's ability to include the correct diagnosis increased (see below).

33.4 Data Acquisition

Efforts at LDS Hospital have concentrated on finding the most comfortable and efficient ways of obtaining the history from the patient as well as determining the value (in terms of diagnostic accuracy) of the gathered data [3–6]. Thus far programs to predict the presence or absence of pulmonary disease have been written. Patients admitted to the services of pulmonary specialists here at LDS Hospital have been the test group.

In a comparison of two techniques [5], one group of patients was given a 182-item, handwritten questionnaire. These data were subsequently entered into the computer for analysis. This group had to answer all the questions (i.e., all 182) in every instance. Another group responded to questions asked by the computer and entered answers directly into a computer terminal.

In the case of the computer-administered history, although the set of questions that could be asked was the same as with the handwritten questionnaire, the choice of which questions to ask was determined by the computer's logic. First, the patient was asked a select group of questions. From these initial questions, certain diagnostic possibilities were raised and the computer (based on the knowledge base) asked further questions to modify the probabilities. Before asking the patient a question, the computer checked the electronic patient file to see if the desired data are already present.

Answers to follow-up questions raise new diagnostic possibilities and new follow-up questions were required. Restriction and ordering of these questions is handled by a module of the program called the QUERY driver. The QUERY driver chooses the most appropriate question to ask next, based on the likelihood of obtaining a helpful answer. For this reason the entire process is called decision-driven data acquisition (DDA). In addition to setting the order of questions, the QUERY manager eliminates specific questions if more general questions are answered in the negative. It also truncates a line of questioning if the possibility of that disease being present becomes very small.

Another experiment [6] compared the DDA approach to two other methods. First, a computer-administered, branching questionnaire was developed that uses no logic other than symptom hierarchy to determine which questions to ask (i.e., if the answer to a general question is no it will not ask a more specific one). Second, since every patient may not have access to a terminal at the time the history needs to be taken, a two-stage paper questionnaire was developed. In this case the patient answers an initial set of questions. These would be entered into the computer which would then, based on its knowledge base, generate a second set of questions that have the greatest likelihood of giving positive results. The results of the two sets of questions are then evaluated by the knowledge base and a list of diagnostic probabilities generated.

33.5 Results

Each method in the above-mentioned experiments was evaluated for its ability to predict accurately the presence or absence of pulmonary disease. This was done by comparing the list of the most likely diagnoses generated by the computer to the patient's actual discharge diagnoses. The number of questions asked (or, patient effort) was also examined.

In the first experiment [5], the DDA method and the paper-based questionnaire were equally efficacious. If the patient actually had a pulmonary disease, it was included in the list of five most likely diagnoses 75% of the time. They were also similar in their ability to identify patients without pulmonary disease (DDA, 78% of the time; questionnaire, 68%).

The different approaches differed greatly, however, in the number of questions it took them to reach their conclusions. As mentioned, patients receiving the paper-based questionnaire had to answer 182 questions. With the DDA method, patients had to answer an average of only 51 questions. A subset, patients without pulmonary disease, were asked an average of only 25 questions (with an accuracy of 78%, as mentioned).

The second experiment, comparing three methods of data acquisition, showed the DDA method, the computer-administered branching questionnaire, and the two-stage paper questionnaire to include the correct diagnosis in the list of the five most likely 68%, 66%, and 82% of the time, respectively. Out of 209 possible questions that could be asked, the three methods used 48, 137, and 55, respectively.

33.6 Revision of Diagnostic Logic

Another study used analysis of a patient database to modify the probabilities contained in the knowledge base [7]. Data were collected on 536 patients and applied to 102 patients. In this study, data were collected from patients via a computer terminal. For the 10 most common diseases, the revised frames accurately included the actual disease in the five most likely possibilities 90% of the time, as opposed to 79% for the original frames. Research is currently underway to attempt to automate the revision process based on patient data contained in HELP.

33.7 Discussion

The results from one experiment [5] seem to show that diagnostic ability is not lost by using the computer's logic to restrict the number of questions. This would seem to suggest the computer picks the "right" questions not to ask. The second experiment shows that more logic is required than simple hierarchical structuring of questions. It also shows that a two-stage, paper-based questionnaire can be as efficacious as a computerized decision-driven method. This is encouraging for settings with limited computer resources. In general, the goal of being able to predict a patient's disease with high certainty based on computer analysis of admission history data seems to be feasible.

33.8 Future Directions

This program places heavy demands on the HELP computer's central processor. Overcoming this practical problem and expanding the knowledge base to include medical domains other than pulmonary disease are the next steps to be taken.

References

[1] Warner HR. *Computer-assisted medical decision-making*. New York: Academic Press Inc; 1979.

[2] Weinstein MC, Fineberg HV. *Clinical Decision Analysis*. Philadelphia: WB Saunders; 1980.

[3] Gerard MJ, Haug PJ, Morrison WJ, et al. A computer system for diagnosing pulmonary disease. Proceedings of the American Association of Medical System Informatics; 1984:119–123.

[4] Haug PJ, Warner HR, Clayton PD, Schmidt CD, Pearl JE, Farney RJ. A computer-directed history: functional overview and initial experience. *MEDINFO*. 1986:849–852.

[5] Haug PJ, Warner HR, Clayton PD, et al. A decision-driven system to collect the patient history. *Comput Biomed Res* 1987;20:193–207.

[6] Haug PJ, Rowe KG, Rich T, et al. A comparison of computer-administered histories. Proceedings of the American Association of Medical System Informatics; 1988:21–25.

[7] Haug PJ, Clayton PD, Shelton P, et al. Revision of diagnostic logic using a clinical database. Proceedings of the American Association of Medical System Informatics; 1987:238–242.

34
Computer-Assisted Pregnancy Management

34.1 Overview

A computerized outpatient obstetrical management system is under development at LDS Hospital [1]. The system uses computer-administered questionnaires to obtain pertinent history, physical examination, and laboratory data from the patient, nurse, and physician. The gathered data are reviewed by an expert system that uses HELP's frame-based medical logic [2]. The expert system determines if any suggestions (e.g., diagnoses, treatment suggestions, further testing, etc.) are warranted. An early evaluation of the computer's recommendations showed an 80% sensitivity and 88% specificity when compared with clinicians' eventual decisions to order a fetal ultrasound examination.

The system is ready for an initial clinical implementation and further evaluation. For this purpose, HELP system terminals have been installed in a local obstetrical group's office and testing is due to begin in 1991.

34.2 Data Collection

A woman coming to the obstetrician's office uses a computer terminal to answer questions concerning fetal and maternal status. The system uses a logical, DDA [3] approach to gather data from the patient (see Chapter 33 for discussion of DDA).

The interview begins with screening questions. Answers to the initial questions trigger computerized logic, which generates further questions to confirm or reject considered hypotheses. Answers to follow-up questions may likewise generate further questions. Occasionally, information from the nurse and/or physician (e.g., specific physical findings or laboratory results) may be required to confirm or reject a considered hypothesis. In these instances, the computer will store the questions until the nurse or physician is logged onto the system. A negative screening history results in a brief questionnaire whereas positive results generate additional complaint-specific questions.

After the patient has entered the historical data and has undergone a physical examination, the physician and nurse use the computer to enter screening physical examination and laboratory data. Similarly to when the patient is entering history data, the HELP system's decision logic continually reviews the data entered and seeks further data to confirm or reject considered hypotheses.

34.3 Frame-Based Medical Knowledge Representation

The data elements (e.g., history, physical, laboratory, previous decisions) are represented in the HELP system in coded form. The coded items are manipulated logically, algebraically, or statistically for use in medical decision making.

HELP uses a frame-based system [2] to represent medical knowledge. An example of a HELP system obstetrical frame is shown in Figure 34.1. The structure and functioning of HELP frames is described in Chapter 3.

Table 34.1 shows a partial listing of the decisions made by the current set of obstetrical frames. A total of 110 frames are currently planned. The logic for these frames was developed by three obstetricians, an information scientist, and a graduate student meeting weekly for 15 months. Development of the frames included identification of the logic for a particular

---

OB Frame

    Title: Pregnancy Management - Trophoblastic Disease (7.42.13).

    Type: Diagnostic

    Author: Peter Haug

    Message: "Consider Gestational Trophoblastic Disease".

    Variables: duration_of_pregnancy **as** (DURATION OF PREGNANCY FRAME),

                 systolic_BP **as** (SYSTOLIC BLOOD PRESSURE),

                 diastolic_BP **as** (DIASTOLIC BLOOD PRESSURE).

    Logic: **If** (systolic_BP GE 140 **or** diastolic_BP GE 90) **and** duration_of_pregnancy LT 20 weeks **then conclude.**

    Evoke: If systolic_BP GE 140 **or** diastolic_BP GE 90

    Ask: (systolic_BP,diastolic_BP) physician/nurse

Figure 34.1. Example of HELP obstetrical frame.

Table 34.1. Partial list of decisions made by obstetrical frames

1. History of fetal losses suggest patient may be subject to habitual abortion.
2. Possible reasons for patient's history of habitual abortion are: ⟨positive＿history⟩.
3. Consider workup for habitual abortion: 1) ultrasound to confirm pregnancy 2) cervical cultures for mycoplasma & chlamydia 3) chromosmal studies on parents 4) urinalysis 5) quant HCG 6) creatinine and BUN 7) One hr post prandial glucose, 8) T3 and T4, 9) ANA, LE prep, and anti-SS antibody for Lupus, 10) Antiphospholipid antibody studies—specifically anticardiolipin and lupus anticoagulant.
4. Excessive bleeding in early pregnancy. Possible Hydatidiform mole. Suggest ultrasound and quantitative HCG (beta unit).
5. Possible Hydatidiform mole. Suggest Ultrasound and quantitative HCG (beta unit). Reasons: ⟨explanations⟩.
6. Consider Septic Abortion.
7. Consider ultrasound to assess vaginal bleeding beyond 20 weeks.
8. Consider ectopic pregnancy or spontaneous abortion. If these are absent, workup other causes of pelvic pain such as PID, appendicitis, ruptured ovarian cyst, etc.
9. Suggest follow-up with quantitative HCG and possibly ultrasound for suspicion of ectopic pregnancy or spontaneous abortion.
10. Possible pregnancy with IUD in place. Suggest vaginal exam to check for string and/or ultrasound to determine presence and location of IUD.
11. Patient is at risk for incompetent cervix.
12. Consider Ultrasound to evaluate internal cervical OS at 16 weeks gestation. Patient is at risk for incompetent cervix.
13. Patient has premature painless cervical dilation at ⟨number⟩ weeks gestation. Reasons: ⟨explanations⟩.
14. Mother is RH ⟨negative/positive⟩ ⟨recommendations⟩
15. Father is RH ⟨negative/positive⟩ ⟨recommendations⟩
16. Mother's Repeat Indirect Coomb's Test ⟨results⟩ ⟨recommendations⟩
17. Antibodies are ⟨messages⟩.
18. ⟨message1⟩ amniocentesis ⟨message2⟩
19. First Delta O.D. 450 value ⟨results⟩
20. Repeat Delta O.D. 450 value in zone ⟨results⟩.
21. Patient has had ⟨number⟩ previous cesarean section(s).
22. Suggest trial of labor for previous c-section only if personnel are available to handle an emergency c-section in 30 minutes or less. Reasons: Patient has only had one previous c-section, incision type was ⟨incision⟩, dilation was at least 3 cm, and indications for the c-section was ⟨indication⟩.
23. Patient is a diabetic—Treated with ⟨drugs⟩, ⟨dietary⟩.
24. Known diabetic: Suggest eye exam for diabetic retinopathy.
25. Known diabetic: Order a SMAC-20, Hemoglobin A1C, and 24-Hour urine for creatinine clearance, total protein. Repeat as indicated by lab and clinical course.
26. Consider hospitalization for diabetic patient ⟨reasons⟩.
27. Known diabetic: Suggest fetal ultrasound in the first trimester to assess gestational age.
28. Known insulin dependent diabetic: Suggest fetal echocardiogram at 20 weeks to assess the cardiovascular system.
29. Known insulin dependent diabetic: Suggest Ultrasound at 32 weeks to assess fetal growth and look for evidence of macrosomia.

decision, writing of the frame, revision of the frame, testing of the frame on sample patient data, and final modification of the frame based on the results of testing.

34.4 Reports

The system is being designed to be able to produce reports of the patient history, data input by the nurse and physician, and a list of decisions generated by the HELP system.

34.5 Evaluation

A prototype version of the program (using paper-based rather than computer-administered questionnaires) was evaluated on 185 patients to test the logic contained in the decision frames. The computer suggested one or more management actions in 55% (102/185) of the patients. Using the gathered data, the system was able to determine the duration of pregnancy, the expected date of confinement (EDC), and the Rh compatibility for all 185 patients.

The frame for suggestion of fetal ultrasound was evaluated for 142 patients. The computer correctly identified the 17 patients for whom the test was inappropriate for reasons of gestational age (< 21 weeks, > 36 weeks). Of the remaining 125 patients, the computer correctly identified 80% (36/45) who eventually went on to have fetal ultrasound, and 88% (70/80) of those who did not, thus showing good correlation with expert clinicians' eventual decisions.

34.6 Future Directions

The system is ready to be tested and evaluated clinically. Terminals have been installed in the office of a local two-person obstetrical group. Data collection and further evaluations are scheduled to begin shortly.

References

[1] Haug PJ, Hebertson RM, Heywood RE, et al. Computer-assisted pregnancy management. *Symposium on computer applications in medical care (SCAMC)* 1987;11:158–161.
[2] Pryor TA, Clayton PD, Haug PJ, Wigertz OB. Design of a knowledge driven HIS. *Symposium on computer applications in medical care (SCAMC)* 1987;11:289–293.
[3] Haug PJ, Warner HR, Clayton PD, et al. A decision-driven system to collect the patient history. *Comput Biomed Res* 1987;20:193–207.

35
Currently Inactive HELP Programs

35.1 Computerized Digoxin Intoxication Alerting

35.1.1 Overview

A program was written [1] that uses laboratory data, ECG findings, pharmacy and drug interaction data, history, physical findings, and a knowledge base to monitor patients automatically for existing signs of, and predisposing factors to, digoxin intoxication. The program was evaluated in a randomized, double-blind study. Patients receiving a computer-generated alert were more likely to receive an appropriate physician action.

The program was written at a time when ECG analysis was performed directly HELP system programs. When the MUSE ECG system was installed (see Chapter 28), a new format was chosen for storing ECG findings on HELP. Alteration of the digoxin intoxication alerting program to recognize ECG findings in the new format has not yet been performed and the program is currently not operational.

35.1.2 System Design

Conditions that may contribute to a state of digoxin intoxication were determined by cardiologists and incorporated into a series of logic modules (Table 35.1). These modules use the patient's age, vital signs, and laboratory and medication data in their determinations. Because these data already exist in the HELP system from other sources, no additional input of data is necessary for the system to function.

The alert modules were incorporated into the HELP system and were designed to be automatically activated at a fixed time each night (i.e., to be "time-driven"). When alerting conditions were satisfied, a digoxin alert report (Figure 35.1) was automatically printed on the patient's nursing unit. This was placed in the chart for review by the physician.

35.1.3 Evaluation

Over a 3-month period, the computer recognized 396 patients who satisfied the alerting conditions. Randomly, the computer transmitted and printed the alerts for 211 (53%) patients. The charts of these patients and the remaining 185 (47%), for which the alerts were withheld, were later reviewed to identify physician actions with possible relation to digoxin administration. The actions performed, along with statistical analysis of the variation between the alert and nonalert groups, is shown in Table 35.2.

As can be seen, physicians ordered digoxin levels, withheld digoxin, changed beta-blocker orders, ordered potassium supplements, and ordered potassium levels significantly more often in the alerted patients.

35.1.4 Current Status

Because the developers of the program were not available when crucial changes were made to the HELP operating system, the module did not receive the modifications to keep it operational. These modifications could be performed at some point; however, currently the program is not functional.

35.2 Computerized Arrhythmia Management

35.2.1 Overview

A program has been developed within the HELP system [2] that generates treatment suggestions for patients with atrial or ventricular arrhythmias. The program makes use of laboratory, hemodynamic, and pharmacy data already stored in the HELP system and refines its suggestions if optional data such as physical exam data or past history are entered.

When evaluated by cardiologists the program was felt to make appropriate recommendations. The program has not found routine clinical use because, to be worthwhile in complicated cases, a large amount of patient data not routinely present in the HELP database must be entered by hand. Physicians have rejected routine use of the program because of the manual data entry and the program is not currently being used clinically.

35.2.2 Data Stored

The program formulates treatment suggestions for atrial flutter, atrial fibrillation, multifocal atrial tachycardia, paroxysmal supraventricular tachycardia, premature ventricular complexes, various forms of ventricular tachycardia (sustained, nonsustained, hypotensive, nonhypotensive), and ventricular fibrillation. The physician first enters the type of arrhythmia to be treated and then enters any clinical data (e.g., history, physical findings,

Table 35.1. Characteristics of digoxin intoxication alert frames

Alert module	Alert text	Decision criteria
Low weight	Low body weight of (* kg) may require reduced digoxin dosage	Digoxin dosage ≥ 0.25 mg/day and (male weight < 55 kg, female weight < 40 kg)
Old age	Age of patient (* yr) may require reduced digoxin dosage	Digoxin dosage ≥ 0.25 mg/day and age ≥ 75 years
High serum digoxin level	Last serum digoxin level (* ng/ml) indicates that digoxin therapy [ ] = must be closely monitored = is currently contraindicated	Serum digoxin level > 2 ng/ml but < 3 ng/ml Serum digoxin level > 2 ng/ml but < 3 ng/ml Serum digoxin level ≥ 3 ng/ml
Low serum potassium level	Last serum K − (* mEq.l): suggest [ ] = that digoxin be withheld until corrected = that serum K − be monitored	Serum potassium ≤ 3.0 mEq/liter Serum potassium > 3.0 mEq/liter but ≤ 3.6 mEq/liter
Renal insufficiency	Last serum creatine (* mg/100 ml) suggest renal impairment: the dose of digoxin may need to be reduced	Serum creatine level ≥ 1.6 mg/100 ml
No serum potassium	The serum potassium has not been done: suggest that K − be monitored	No serum potassium level found
Concurrent beta-blocking agent	Concurrent use of digoxin and beta-blocking agents can result in bradycardia: suggest monitoring ventricular rate	Current beta-blocking prescription
Concurrent quinidine	Concurrent use of digoxin and quinidine can result in increased serum digoxin levels: suggest monitoring serum digoxin level	Current quinidine prescription

		Current calcium channel blocking agent prescription
Concurrent calcium channel blocking agent	Concurrent use of digoxin and calcium channel blocking agents can induce bradycardia: suggest monitoring ventricular rate	
Acid–base disorder	Blood pH (*) out of normal limits: suggest caution in use of digoxin	pH ≤ 7.3 or pH > 7.5
Hypoxemia	Blood PO$_2$(* mm Hg): suggest care in use of digoxin	Arterial PO$_2$ ≤ 60 mm Hg
Atrial tachycardia with block	ECG finding of atrial tachycardia with block suggests possible digoxin toxicity	ECG finding exists
Junctional arrhythmia	ECG indicates [ ]; use digoxin with caution	
	= junctional tachycardia	ECG finding exists
	= junctional rhythm	ECG finding exists
Ventricular arrhythmia	Ventricular [ ] present: suggest caution in use of digoxin	
	= premature complexes	ECG finding exists
	= tachycardia	ECG finding exists
Sinoatrial block	ECG finding of second degree SA block: consider modification of digoxin therapy	ECG finding exists
Atrioventricular block	[ ] AV block suggests caution in use of digoxin	
	= first degree	ECG finding exists
	= second degree	ECG finding exists
	= complete	ECG finding exists
Acute infarction	Acute infarction: consider reduction in digoxin dosage	ECG finding exists

Digoxin Alert Report — 06/15/83 15:20

Patient: XXXXXXXXXXXX Number 4981213
Dr: XXXXXXXXXXXXXXX Room: 4W09
Digoxin dosage: 0.250 MG, TABS

ALERT 18: The serum potassium has not been done. Suggest that serum K+ be
   monitored.
ALERT 22: Concurrent use of digoxin and quinidine can result in increased serum
   digoxin levels. Suggest monitoring serum digoxin levels.
ALERT 39: First degree AV block. Suggest caution in use of digoxin.

Figure 35.1. Sample digoxin intoxication alert report. © 1984 Elsevier, New York.
Reprinted with permission from "Application of a Computerized Medical Decision-
Making Process to the Problem of Digoxin Intoxication" by White, Lindsay, Pryor
et al., *Journal of the American College of Cardiology*, 573–575, September 1984.

Table 35.2. Physician action in response to digoxin intoxication alerts

	Frequency for alert group	Frequency for nonalert group	Weighted ratio Al/Nal	Statisical p value
Alert days	260	246		
Serum digoxin determination ordered	48	17	2.67	< 0.0001 S
Digoxin withheld	27	9	2.84	< 0.002 S
Digoxin discontinued	5	2	2.37	< 0.14 NS
Digoxin dose reduced	5	1	4.73	< 0.06 NS
Quinidine changed	2	1	1.89	< 0.30 NS
Beta-blocking agent changed	4	0	—	< 0.03 S
Potassium supplement ordered	69	48	1.33	< 0.04 S
Serum potassium determination ordered	117	89	1.24	< 0.02 S
Oxygen delivery increased	42	32	1.24	< 0.16 NS
Concern of toxicity in note	5	1	4.73	< 0.06 NS
Electrocardiogram ordered	36	29	1.17	< 0.25 NS
Any action taken	175	136	1.22	< 0.003 S

© 1984 Elsevier, New York. Reprinted with permission from "Application of a Computerized
Medical Decision-Making Process to the Problem of Digoxin Intoxication" by White, Lindsay,
Pryor et al., *Journal of the American College of Cardiology*, 573–575, September 1984.
Al = alerted, Nal = nonalerted, NS = not statistically significant, S = statistically signifi-
cant.

test results, or other data not already stored by the HELP system) he/she feels could be relevant to the therapeutic choice. The data items that may be entered are listed in Table 35.3.

If only limited data are available from the above sources, the computer's recommendations will have marginal value. Therefore an "ask" mode was developed for the computer to obtain information it needs to make suitable therapeutic suggestions.

35.2.3 Data Integration, Decision Logic, and Reports

In formulating its logic, the computer uses laboratory data, hemodynamic data from indwelling catheters, and pharmacy data regarding side effects and dosages of drugs, all already present in the HELP system. The program can also use previous decisions derived by the HELP system.

The computer's logic base tries to emulate the thinking of a cardiologist. In forming its therapeutic decisions it considers parameters such as the patient's clinical condition, possible etiological factors for the arrhythmia, contraindications to potential treatments, and goals of therapy. The program proceeds stepwise through its logic until it has reached a therapeutic choice that is a reasonable approach to the intended goal.

The suggested therapy (or therapies) are printed in a short report giving treatments, dosages, and which parameters should be followed. Examples are shown in Figure 35.2.

35.2.4 Evaluation

The program was run on 27 patients, each having one of the above abnormal rhythms. These were evenly divided between atrial and ventricular arrhythmias. Appropriate data were entered from the patient's chart and data in the computerized database were reviewed for accuracy before running the computerized arrhythmia management (CAM) program. The results were reviewed by cardiologists who felt the recommended therapy to be appropriate in every case.

35.2.5 Future Directions

At present the program is not in routine use. When active, the program was used primarily for difficult cases. At that point, entry of all the relevant data (e.g., past therapeutic attempts, clinical condition) was cumbersome and hindered comfortable use of the program. An approach that will make all relevant data more conveniently available to the program needs to be developed.

Table 35.3. Patient data that may be entered into the CAM system

A ventricular arrhythmia
An atrial arrhythmia
Medical History
  myocardial infarction   systemic lupus
  unstable angia   rheumatic arthritis
  hyperthyroidism   hypertension
  hypothyroidism   asthma
  alcoholism   diabetes
  prostatic hypertrophy
  chronic obstructive pulmonary disease
  organic heart disease other than coronary heart
Treadmill and echo tests
  a treadmill test was done today
  abnormal ST depression was found from treadmill test
  leaflet prolapse from echo test
  mitral stenosis from echo test
Inefficacy of drugs (new patient) [a]
Contraindications to drugs (new patient) [a]
Congestive heart failure from physical examination
  audible S3
  rales, more than 50% lung fields and moist inspiratory
  interstitial/alveolar edema
  appears dyspneic
  cold periphery skin
  appears fatigued
  abnormal level or consciousness
  low urine output
  cyanotic nails
Associated findings:
  chest pain, rule out myocardial infarction
  poor perfusion
  anxious (as a sign for excess of catecholeamines)
  hepatic dysfunction
  diarrhea within the last 24 hours
Vital signs:
  current arterial systolic pressure
  current arterial diastolic pressure
  current heart rate
  heart rate at onset of the arrhythmia

[a]Drugs included are: lidocaine, procainamide, bretylium, verapamil diltiazem, nifedipine, quinidine, disopyramide, phenytoin, digoxin, propranolol, bronchodilators, alpha sympathomimetics, beta sympathomimetics, tricyclic antidepressants, beta sympatholytic agents, phenothiazines, and lithium.

```
EMERGENT TREATMENT
ATRIAL FLUTTER: SUGGEST CARDIOVERSION. CONSIDER RAPID ATRIAL PACING FOR FOLLOWING CONTRAINDICATORS
DRUG THERAPY: GIVE DIGOXIN IV BECAUSE OF HEART RATE GREATER THAN 100
-- ELECTRICAL CARDIOVERSION: FOR FLUTTER 50 WATT-SEC, INITIALLY USING STANDARD
        TECHNIQUE
-- START CLASS I ANTIARRHYTHMICS:
            QUINIDINE SULFATE 300-400 Q6H 4-6 DOSES
        OR PROCAINAMIDE 375-500 Q4H 4-6 DOSES
        OR DISOPYRAMIDE 150 Q6H 4-6 DOSES
-- START ANTICOAGULATION: GIVE HEPARIN IV OR SQ. (TO PREVENT THROMBOEMBOLISM ON CONVERSION)
RISK OF REVERSION IS LOW.   RECOMMEND THAT IN 2 WEEKS D/C:
    -- ANTICOAGULATION
    -- DIGITALIS
    -- CLASS I ANTIARRHYTHMICS

            Recommended treatment for patient with atrial flutter.


        DIAGNOSIS:   HYPOTENSIVE SUSTAINED VT
        ACUTE MI

SUGGESTED TREATMENT:
GIVE ELECTRICAL CARDIOVERSION AT 100-300 WATT-SEC.  ON SYNCHRONIZED MODE:  TO PREVENT RECURRENCE FOLLOW UP WITH LIDOCAINE
GIVE LIDOCAINE:
LOADING DOSE: 3 MG/KG TOTAL (BOLUS METHOD) INITIALLY 1 MG/KG.  THEN 0.5 MG/KG EVERY 2-5 MIN UP TO 3 MG/KG TOTAL OVER 10-20 MIN.
SIMULTANEOUSLY, A 2-4 MG/MIN INFUSION (AVERAGE 3 MG/MIN) IS BEGUN AND CONTINUED FOR THE PERIOD OF THERAPY, MINIMUM OF 24 HOURS SUGGESTE
CHECK FOR TOXICITY.
ORAL CONTINUATION IF NEEDED: RECOMMEND QUINIDINE BECAUSE DRUG OF FIRST CHOICE.
GIVE QUINIDINE SULFATE:
INITIALLY: FOR AGE GREATER THAN 70 OF WEIGHT LESS THAN 50 KG: 200 MG Q6H
OTHERWISE:  300-600 MG.
INCREASE AS NEEDED UP TO 400-600 MG Q6H FOLLOWING RESPONSE TO ECG (QRS<QT) AND BLOOD LEVELS
ALTERNATIVE SALTS INCLUDE GLUCONATE AND POLYGA LACTURONATE
COMMENTS:

            Recommended treatment for patient with sustained hypotensive ventricular tachycardia
```

Figure 35.2. Examples of suggestions from computerized arrhythmia management system. © 1984 IEEE, New York. Reprinted with permission from "Computerized Arrythmia Management in Patients in a Coronary Care Unit" by Pryor, Goldberg, Brown et al., *Computers in Cardiology, 26, 29,* 1984.

35.3 Use of Treatment Protocols Within the HELP System

35.3.1 Overview

Because of the logical nature of many medical treatment situations, there is great opportunity to computerize them. Protocols for ventilator management (and their implementation in HELP) are discussed in detail in Chapter 23. Other projects involving the use of protocols on the HELP system include a graphics-based knowledge editor for protocol knowledge and an evaluation study demonstrating that computer protocols can help to direct medical care in an outpatient setting.

35.3.2 PROTOKOL: A Program to Assist in Protocol Development

Protocols can be used to characterize programmed approaches to medical problems. PROTOKOL [3], a program that allows medical personnel to design flow charts (protocols), was developed to help medical experts' represent their knowledge.

The protocol is designed by a physician working alone or working with a knowledge engineer on a PC in a graphics-based environment. This knowledge editor permits the physician to choose the objects commonly found in a flow diagram and incorporate them into the diagram. Once the protocol is completed, a module of PROTOKOL converts the knowledge contained in the graphical image into the syntax of the HELP frame language. (The newly created frames may be used by other HELP system programs too.) The PROTOKOL-generated frames may then be executed just like any other HELP frames.

PROTOKOL was used at LDS Hospital to automate a protocol to locate patients for, and to manage the patients' participation in, the Cardiac Arrhythmia Suppression Trial (CAST), a nationwide study to evaluate the efficacy of various medications' ability to decrease ventricular ectopy and sudden death in postmyocardial infarction patients. The program was run for 8 weeks and screened patients from three nursing units for eligibility for the study. The computer screened 484 patients and made more than 4000 decisions. Four eligible patients were found and managed. When the decisions were reviewed manually and compared to the original written protocol, less than 1% were found to be in error. All the errors stemmed from incorrect interpretations of the written protocol into a logical scheme.

The pilot use of the program in the CAST study demonstrated its potential. Further work is necessary to complete the development of this program into a routine clinical tool.

35.3.3 Computerized Interactive Protocol System

Protocols were developed [4] to assist physician's assistants treating minor medical conditions at a rural clinic that had only intermittent physician support. Criteria for diagnostic, therapeutic, and referral recommendations were adopted and coded into HELP decision logic in four areas: (1) upper respiratory and ear, nose, and throat disorders, (2) urinary tract infections, (3) chronic hypertension, and (4) chronic diabetes. The system was implemented on a minicomputer at the clinic.

Patients entered history data directly into a computer terminal. After reviewing the history, the physician's assistants performed a physical examination and entered the results into the computer.

The computer would apply the data to its logic base. If it had enough data to generate suggestions, it would do so. The logic was designed to obtain additional data from the physician's assistant when required. New data would prompt consideration of other possibilities and further questioning. When the data set was complete, the computer would display and print a report of the data that it used and the decisions that it reached (Figure 35.3).

The system was evaluated for patient and provider satisfaction and its

---

```
                  *** PROTOCOL SYSTEM REPORT ***
                              ENTP

12/ 2/77      3000040

'PATIENT, DEMO A.
*** CURRENT HISTORY QUESTIONS WITH YES ANSWERS:
    HAVE YOU FELT PAIN OR ITCHING IN EITHER EAR?
    DO YOU HAVE A FEELING OF FULLNESS OR PRESSURE IN YOUR EAR?
    DO YOU NOW HAVE A RUNNY, STUFFY, OR CONGESTED NOSE?
    HAVE YOU HAD A FEVER OR CHILLS WITH THIS ILLNESS?
    DO YOU HAVE A COUGH?
*** CURRENT PHYSICAL EXAM DATA:
    EAR, NOSE, AND THROAT, TYMPANIC MEMBRANE, INFLAMED
    EAR, NOSE, AND THROAT, TYMPANIC MEMBRANE, FLUID IN MIDDLE EAR
    VITAL SIGNS, TEMPERATURE (C X10)      375
    EAR, NOSE, AND THROAT, EXTERNAL CANAL, NORMAL BILATERAL
    EAR, NOSE, AND THROAT, PARANASAL SINUSES, NORMAL BILATERAL
    EAR, NOSE, AND THROAT, POSTERIOR PHARYNGEAL WALL, NORMAL
    EYE EXAMINATION, CONJUNCTIVA AND SCLERAE, NORMAL
    EAR, NOSE, AND THROAT, TONSILS, NORMAL
    EAR, NOSE, AND THROAT, MASTOIDS, NORMAL
    CHEST AND LUNG EXAM, PULMONARY AUSCULTATION, NORMAL
*** OTHER CURRENT DATA USED IN DECISIONS:
*** CURRENT DECISION LIST:
    DIAGNOSIS:   OTITIS MEDIA
    DIAGNOSIS:   SEROUS OTITIS
    DIAGNOSIS:   VIRAL UPPER RESPIRATORY INFECTION
    TREATMENT: PENICILLIN V 500 MG Q.I.D. X 10
    WARNING:     ALLERGIC HISTORY SUGGESTS THAT THE PATIENT SHOULD BE
                 WATCHED AFTER PENICILLIN ADMIN.
```

Figure 35.3. Printout of protocol-driven patient encounter. See text for details. © 1980 Academic Press, New York. Reprinted with permission from "Experience with a Computerized Interactive Protocol System Using HELP" by Cannon and Gardner, *Computers and Biomedical Research* 13:399–409, 1980.

impact on diagnostic tests, treatments, and referrals. Out of 60 patients, 55 (92%) felt the computer was an acceptable means of accepting data and the same amount felt the self-history adequately expressed their problem to the provider. Only two patients reported having problems using the terminal for data entry.

The physician's assistants had a data entry error rate of 1.1% (22/2010 items). Most of these were with the first few patients, probably reflecting an early unfamiliarity with the system. Interviews revealed the physician's assistants were generally comfortable using the system. Assistants spent an average of less than 2 minutes per patient at the terminal. (Total computer time, i.e., patient history, analysis, prompting, and report generation, was 10.3 minutes per patient.)

Analysis of the clinical data showed the computerized protocol recommended antibiotics unnecessarily in cases of sore throat far fewer times (8 vs. 30, out of 175) than the assistants without the use of the computer. Four times the computer recommended referral to a physician unnecessarily. This was due to incorrect computer interpretation of long-standing symptoms that did not reflect complications of the patients acute problem. The assistants had none of these "false-positive" referrals. The computer and the assistants had similar rates of accurately deciding when to culture a sore throat.

The system was installed for purposes of evaluation and is not currently operating.

References

[1] White KS, Lindsay A, Pryor TA, Brown WF, Walsh K. Application of a computerized medical decision-making process to the problem of digoxin intoxication. *J Am Coll Cardiol* 1984;4:571–576.

[2] Pryor TA, Goldberg RD, Brown WF, Anderson JL. Computerized arrythmia management in patients in a coronary care unit. *Comput Cardiol* 1984;11:25–30.

[3] Smith JC. *Protokol: A System-Building Aid for Developing Protocol-Type Knowledge Bases.* Salt Lake City, Utah: University of Utah; 1988. Dissertation.

[4] Cannon SR, Gardner RM. Experience with a computerized interactive protocol system using HELP. *Comput Biomed Res* 1980;13:399–409.

Appendix A

Acknowledgments

The following personnel have been involved in the development of each of the modules listed. The authors would also like to acknowledge the support of the Information Systems Division at Intermountain Health Care, Inc., and the administration of LDS Hospital in the ongoing development of the HELP system.

Admit-Discharge-Transfer:
  Pam Shelton, primary programmer, Intermountain Health Care, Information Systems Division
  Tammy Smith, admitting supervisor

Medical Records:
  Bette Maack, Director, LDS Hospital Medical Records
  Mark Bradford, primary programmer, Clinical Services
  Reed Gardner, Ph.D., Department of Medical Informatics

Quality Assurance:
  Julie Jacobsen, Director of Quality Assurance
  Reed Gardner, Ph.D., Department of Medical Informatics
  Joe Hales, graduate student, programmer

Surgery Scheduling:
  Reed Gardner, Ph.D., Department of Medical Informatics
  Ronald Jensen, Department of Surgery
  Jacquie Augason, Department of Surgery
  Kurtis Scherting, primary programmer
  Kathy Delaplaine, programmer

Order Entry:
   T. Allan Pryor, Ph.D., Department of Medical Informatics
   Dennis Stansfield, programmer, Intermountain Health Care, Information Systems Division

Dietary:
   Josh Wisham, Intermountain Health Care, Information Systems Division, design and original implementation
   Stephanie Weems, primary programmer
   Elizabeth Reams, Chief clinical dietitian

Emergency Room:
   Judy Prince, R.N., Head Nurse, Emergency Room
   Terry Clemmer, M.D., Director of Critical Care Medicine
   Reed Gardner, Ph.D., Department of Medical Informatics
   Joel Russell, graduate student, programmer

Nursing:
   Connie Klingle, R.N., Department of Medical Informatics
   Diana Willson, R.N., Department of Medical Informatics
   Nancy Nelson, R.N., Department of Nursing
   Brenda Rosebrock, R.N., Department of Medical Informatics
   Dickey Johnson, R.N., 3M Health Information Systems
   Ann Tinker, R.N., 3M Health Information Systems
   Marijo Burkes, R.N., Department of Medical Informatics, Evaluations
   Marge Budd, R.N., Assistant Director of Nursing
   Judy Blaufuss, R.N., Assistant Director of Nursing
   T. Allan Pryor, Ph.D., Department of Medical Informatics
   Kathy Delaplaine, programmer
   Li Song, programmer

Respiratory Care:
   C. Gregory Elliot, M.D., Medical Director, Pulmonary Department
   Deon Simmons, R.R.T., Department of Respiratory Care, Quality Assurance Coordinator
   Kip Enger, primary programmer

Pharmacy:
   Russell Hulse, R.Ph., Associate Director, LDS Hospital Pharmacy
   Keith Larsen, R. Ph., Intermountain Health Care, Information Systems Division, primary programmer

Laboratory:
   Stan Huff, M.D., Department of Medical Informatics
   Scott Evans, Ph.D., primary analyst/programmer, Department of Infectious Diseases
   Fred Miya, M.D., Medical Director, Clinical Laboratory

Laboratory Alerts:
 Karen Bradshaw Tate, Ph.D., original design, implementation, and evaluation
 Reed M. Gardner, Ph.D., Department of Medical Informatics

Infectious Disease:
 John Burke, M.D., Chief, Department of Infectious Diseases
 Scott Evans, Ph.D., Department of Infectious Diseases, primary programmer
 Lane Stevens, B.S., graduate student
 Dave Classen, M.D., Department of Infectious Diseases
 Stan Pestotnick, Department of Infectious Diseases

Blood Gas:
 Robert Crapo, M.D., Department of Pulmonary Medicine
 Olaf Golubjatnikov, Chief Medical Technologist, Blood Gas Laboratory
 Steven Howe, graduate student

Medical Information Bus:
 William Hawley, Director, Bioinstrumentation
 Hasan Tariq, M.S., primary programmer
 Yang Weiqun, graduate student, ventilators
 Xiaodong Wang, graduate student, surgery
 Reed Gardner, Ph.D., Department of Medical Informatics

Hemodynamic Monitoring:
 Reed Gardner, Ph.D., Department of Medical Informatics
 Julie Parker, R.N., Cardiovascular Monitoring
 William Hawley, Director, Bioinstrumentation

Total Parenteral Nutrition:
 Terry Clemmer, M.D., Director of Critical Care Medicine
 James Orme Jr., M.D., Director of Critical Care Medicine
 Lorraine Horne, R.N., Head Nurse, Nutritional Support Staff
 Keith Larsen, R.Ph., primary programmer, Intermountain Health Care, Information Systems Division
 Stephanie Weems, programmer
 Randy Holt, programmer, Intermountain Health Care, Information Systems Division

Ventilator Protocols:
 Alan Morris, M.D. Department of Pulmonary Medicine
 Tom East, Ph.D., Department of Anaesthesia
 Susan Henderson, programmer
 Dean Sittig, Ph.D., Doctoral thesis on ComPAS; Currently at Yale University, Department of Anaesthesia

Surgery Monitoring:
   Reed Gardner, Ph.D., Department of Medical Informatics
   Julie Parker, R.N., Cardiovascular Monitoring
   Ronald Jensen, Department of Surgery
   Jacquie Augason, Department of Surgery
   Kurtis Scherting, programmer

Blood Bank:
   Reed Gardner, Ph.D., Project Director
   Myron Laub, M.D., Director, Blood Bank
   Sonia Deford, primary programmer
   Olaf Golubjatnikov, data analyst, graduate student

Cardiac Catheterization:
   Xiaodong Wang, primary programmer, graduate student
   Reed Gardner, Ph.D., Project Director
   Hyram Marshall, M.D., Director of Heart Catheterization Laboratory,
      Department of Cardiology

Electrocardiography:
   Ween-Ping Tarng, primary programmer, graduate student
   Allan Pryor, Ph.D., Department of Medical Informatics
   Frank Yanowitz, M.D., Department of Cardiology

Radiology:
   Peter Haug, M.D., Department of Medical Informatics
   Philip Frederick, M.D., Department of Radiology
   Tupper Kinder, programmer
   Allison Meier, programmer
   Dave Ranum, graduate student, natural language processing

Pulmonary Function Testing:
   Robert Crapo, M.D., Department of Pulmonary Medicine
   Steve Berlin, Director, Pulmonary Laboratory

Urinary Catheter Culture Monitoring:
   John Burke, M.D., Director, Department of Infectious Diseases
   Lane Stevens, B.S., primary programmer

History:
   Peter Haug, M.D., Department of Medical Informatics

Obstetrics:
   Peter Haug, M.D., Department of Medical Informatics, Project director
   Richard Hebertson, M.D., Department of Obstetrics
   Reed Heywood, M.D., Department of Obstetrics
   Joyce Sager, graduate student
   Craig Swapp, programmer
   Mike Twede, M.D., Department of Obstetrics

Protocols:
  Carr Smith, Ph.D., Doctoral thesis on PROTOKOL; Currently at EMTEK Corporation, Phoenix, Arizona

Tools:
  Dennis Stansfield, Intermountain Health Care, Information Systems Division
  Sara Grover, Intermountain Health Care, Information Systems Division
  Robert Greely, 3M Corporation
  Tupper Kinder, LDS Hospital

Appendix B

HELP System Dissertations and Theses

The following doctoral dissertations and masters theses have been submitted to the Department of Medical Informatics, University of Utah.

[1] Wang P. *The Frame-Based Model of Nursing Care Plans on the HELP System*. December, 1989. Thesis.

[2] Al-Hashimi MAS. *A Pre-Formatted Report Generating System: Optimized Real-Time Clinical Information Retrieval for the HELP System*. December, 1989. Thesis.

[3] Smith V. *Time Study of the Tandem Structured Query Language*. September, 1989. Thesis.

[4] Ranum DL. *SPRUS: A Knowledge Based Understanding System for Radiology Text*. August, 1989. Dissertation.

[5] Crawford BD. *Preliminary Analysis of Protocols for a Home Health Care Expert System*. August, 1989. Thesis.

[6] Albright FS. *Use of Fuzzy Logic in the Interpretation of Electrocardiograms*. June, 1989. Thesis.

[7] Wong TW. *A Medication Ordering Knowledge Acquisition and Management Tool*. June, 1989. Thesis.

[8] Bennett D. *A Relational Query Interface to the LDS Hospital Patient Database*. March, 1989. Thesis.

[9] Rich T. *Two Aspects of Computerized Patient History that Improve Bedside Pulmonary Diagnosis*. March, 1989. Thesis.

[10] Russell JE. *A Computerized Emergency Department Log*. March, 1989. Thesis.

[11] Tarng WP. *Interface of a Commercial ECG System to the HELP System*. December, 1988. Thesis.

[12] Sittig DF. *COMPAS: A Computerized Patient Advice System to Direct Ventilatory Care*. July, 1988. Dissertation.

[13] Bradshaw KE. *A Computerized Laboratory Alerting System to Warn of Life Threatening Events.* June, 1988. Dissertation.

[14] Smith JC. *PROTOKOL: A System-Building Aid for Developing Protocol-Type Knowledge Bases.* May, 1988. Dissertation. *ILIAD: A Computer Program for Managing Personal Bibliographic Knowledge.* 1986. Thesis.

[15] Zheng Z. *Physician's Computerized Charting in a Newborn Intensive Care Unit.* June, 1988. Thesis.

[16] Shelton P. *Analysis of the Information Content of Medical Data Using a Frame-Based Medical Diagnostic System.* August, 1987. Thesis.

[17] Davidson MG. *A Physician Information Network.* June, 1987. Thesis.

[18] Thomas CH. *Injury Severity Scores: A Review of Current Designs and the Development of a New Objectively Based Score.* March, 1987. Thesis.

[19] Smith DZ. *Design and Evaluation of Time-Alignment Methods for Averaging Low Level Electrocardiographic Data.* 1987. Thesis.

[20] Harada S. *The Accuracy of Radiologic Interpretations with Medical Decision Logic.* 1987. Thesis.

[21] Rossi JA. *The Utilization of Patient Acuity Data for Predicting Staffing Needs.* December, 1986. Thesis.

[22] Sawdey R. *Computer Networking Issues for an Integrated Hospital Information System.* November, 1986. Thesis.

[23] Meldrum KC. *An Evaluation of Database Software in a Medical Environment.* August, 1986. Thesis.

[24] Nyugen L. *Transferability of Medical Knowledge: A Case Study Between INTERNIST-I and HELP.* March, 1986. Dissertation.

[25] Anderson CL. *Techniques for Physician Review of a Patient History Data in the HELP Computer System.* December, 1985. Thesis.

[26] Gould G. *The Diagnosis and Treatment of Coronary Artery Disease.* June, 1985. Dissertation.

[27] Hicken RR. *Computerization of Infection Control and Culture Data Use in Surveillance.* March, 1985. Thesis.

[28] Andrews RD. *A Computerized Respiratory Therapy Information System and the Use of Portable Computers for Data Entry.* March, 1985. Thesis.

[29] Gerrard MJ. *Computerized Automated Pulmonary Disease.* December, 1984. Thesis.

[30] Evans RS. *Computer-Assisted Reporting and Monitoring of Microbiology/Infectious Disease Data and Antibiotic Therapy.* August, 1984. Dissertation.

[31] Ostler DV. *A Computer System for Analysis and Transmission of Spirometry Waveforms Using Volume Sampling.* March, 1984. Thesis.

[32] Beus ML. *Computer Categorization of Spirometry Data Using Statistical Data.* December, 1982. Thesis.

[33] Barlow GK. *An Evaluation of Different Methods Used in Applying Bayes' Rule to Medical Decision Making.* August, 1981. Thesis.

[34] Smith CM. *Design and Implementation of an X-Ray File Tracking System.* June, 1980. Thesis.

[35] Beatty SS. *An Application of Automated Decision Logic in Diagnostic Radiology.* June, 1980. Thesis.

[36] Gennaro JL. *A Computerized Radiology Reporting/Ordering System Based on Most Likely Diagnosis.* March, 1980. Thesis.

[37] Yeh MK. *Computerized Decision Making Applied to the Diagnosis of Lactic Acidosis*. August, 1979. Thesis.

[38] Rothert SW. *A Computerized Approach to Dietary Analysis*. December, 1978. Thesis.

[39] Dick RS. *A Computer-Assisted Instruction System for Teaching Decision Making in Clinical Medicine*. August, 1978. Dissertation.

[40] Ricord LC. *Computer Assisted Decision Making Applied to Anemia*. March, 1978. Dissertation.

[41] Cannon SR. *A System for Computerized Interactive Protocols Using HELP*. June, 1977. Dissertation.

[42] Clark SJ. *An Automated System for Prediction and Prevention of Adverse Drug Reactions*. June, 1977. Dissertation.

Appendix C

HELP System Publications

[1] Warner HR, Olmsted CM, Rutherford BD. HELP — a program for medical decision making. *Comput Biomed Res* 1972;5:65–74.

[2] Warner HR, Rutherford BD, Houtchens B. A sequential bayesean approach to history taking and diagnosis. *Comput Biomed Res* 1972;5:256–262.

[3] Pryor TA, Warner HR. Some approaches to computerized medical diagnosis. In: Jaquez JA, ed. *Computer Diagnosis and Diagnostic Methods*. Springfield, Ill: Charles C. Thomas; 1972:241–254.

[4] Warner HR. A computer-based information system for patient care. In: Bekey GA, Schwartz, eds. *Hospital Information Systems*. New York: Marcel Dekker; 1972: 293–332.

[5] Pryor TA, Lindsay AE, England RW. Computer analysis of serial electrocardiograms. *Comput Biomed Res* 1972;5:709–714.

[6] Johnson JD, Warner HR. Role of the computer in coronary care. In: Meltzer LE, Dunning AJ, eds. *Textbook of Coronary Care*. Publisher: Amsterdam: Excerpta Medica; 1973.

[7] Warner HR. Health Evaluation through Logical Processing — HELP. Proceedings of Medis '73. Osaka International Symposium on Medical Information Systems. Kanasi Institute for Information Systems; October 4–6, 1973.

[8] Gardner RM. Computerized intensive care monitoring at LDS Hospital — progress and development. *Comput Cardiol* 1974;1:97–105.

[9] Warner HR. Can automation make interactive medical history taking feasible and acceptable? Proceedings of Health Service Industry Workshop; February, 1974;No.4:11.

[10] Warner HR, Woolley FR, Kane RL. Computer assisted instruction for teaching clinical decision-making. *Comput Biomed Res* 1974;7:564–574.

[11] Warner HR, Olmsted CM, Rutherford BD. HELP — A program for medical decision making. Proceedings of Technical Workshops. The Automation Research Council; February, 1974;No.5:247.

[12] Yanowitz FG, Pryor TA, Morgan JD, Warner HR. The HELP system for

medical decision making. Proceedings of the 4th Annual Conference of the Society for Computer Medicine; November 7-10, 1974; New Orleans, La.

[13] Pryor TA, Ridges JD. A computer program for stress test data processing. *Comput Biomed Res* 1974;7:360-369.

[14] Pryor TA, Morgan JD, Clark SJ, England W, Warner HR. HELP—A computer system for medical decision making. Proceedings of the National Computer Conference & Exposition; May, 1974.

[15] Gardner RM, Cannon GH, Morris AH, Olsen KR, Price GA. Computerized blood gas interpretation and reporting system. *IEEE Comput* 1975;8(1):39-45.

[16] Gardner RM. The place of computers in medicine. *IEEE Comput* 1975;8(1): 19. Guest Editor's Introduction.

[17] Pryor TA, Morgan JD, Clark SJ, Miller WA, Warner HR. HELP: A computerized system for medical decision making. *IEEE Comput* 1975;8(1):34-38.

[18] Hulse RK, Clark SJ, Jackson JC, Warner HR, Gardner RM. Computerized medication monitoring system. *Am J Hosp Pharm* 1976;33:1061-1064.

[19] Warner HR, Pryor TA, Clark SJ, Morgan JD. Integration of computer support for institutional practice: The HELP system. Computer applications in health care delivery. 7th Annual Meeting of Society Advanced Medical Systems (SAMS); 1976; Houston, Tex.

[20] Yanowitz FG, Pryor TA, Frost DA. A computerized ECG alarm system for the coronary care unit. *Comput Cardiol* 1976;3:431-433.

[21] Gardner RM, Scoville DP, West BJ, Bateman B, Cundick RM Jr, Clemmer TP. Integrated computer systems for monitoring of the critically ill. *Symposium on computer applications in medical care (SCAMC)* 1977;1:301-307.

[22] Gardner RM, Clemmer TP. Computerized protocols applied to acute patient care. In: *Advances in Automated Analysis*. Mediad Inc. Technicon International Congress; 1977;1:158-163.

[23] Frost DA, Yanowitz FG, Pryor TA. Evaluation of a computerized arrhythmia alarm system. *Am J Cardiol* 1977;39:583-587.

[24] Warner HR. First the electrocardiogram—then what? *Am J Cardiol* 1978;41: 115-118.

[25] Gardner RM, Clemmer TP. Computerized protocols applied to emergency and acute care. *Emerg Med Serv* 1978;7:90-93, 180.

[26] Pryor TA, Gardner RM, Clayton PD, Warner HR. A distributed processing system for patient management. *Comput Cardiol* 1978;5:325-328.

[27] Warner HR. Progress, problems and promises. 8th Annual Conference of the Society for Computer Medicine; October 12-14, 1978; Minneapolis, Minn.

[28] Warner HR. Knowledge sectors for logical processing of patient data in the HELP system. *Symposium on computer applications in medical care (SCAMC)* 1978;2:401-404.

[29] Warner HR. *Computer-Assisted Medical Decision-Making*. New York: Academic Press; 1979.

[30] Beatty SS, Gennaro JL, Frederick PR, Clayton PD. A radiology reporting system which is based upon an individualized list of most likely diagnoses. Proceedings of the 7th New England Bioengineering Conference; 1979:259.

[31] Johnson DS, Ranzenberger J, Herbert RD, Gardner RM, Clemmer TP. A computerized alert program for acutely ill patients. *J Nurse Admin* 1980;10:26-35.

[32] Cannon SR, Gardner RM. Experience with a computerized interactive protocol system using HELP. *Comput Biomed Res* 1980;13:399–409.

[33] Clemmer TP, Gardner RM, Orme JF Jr. Computer support in critical care medicine. *Symposium on computer applications in medical care (SCAMC)* 1980; 4:1557–1561.

[34] Pryor TA, Warner HR, Gardner RM. HELP—a total hospital information system. *Symposium on computer applications in medical care (SCAMC)* 1980;4: 3–7.

[35] Warner HR, Clark SJ, Larsen KG, McNeil F. The HELP system as a tool of monitoring physician prescribing patterns. Proceedings of the 13th Hawaii International Conference on System Sciences; Shriver, Sprague, eds. Western Periodicals Co; January, 1980; 3:254–255.

[36] Warner HR. HELP—An approach to hospital-wide artificial intelligence. In: Computer-Assisted Decision Making Using Clinical and Paraclinical (Laboratory) Data. Statland BE, Bauer S, eds. Tarrytown, N.Y.: Mediad Inc; 1980.

[37] Clayton PD, Ostler DB, Gennaro JL, Beatty SS, Frederick PR. A radiology reporting system which is based on the most likely diagnosis. *Comput Biomed Res* 1980;13:258–270.

[38] Clayton PD, Ostler DB, Gennaro JL, Beatty SS, Frederick PR. A radiology system based on most likely diagnoses. Proceedings of the 13th Hawaii International Conference on System Sciences; Schriver, Sprague, eds. Western Periodicals Co; 1980; 3:71.

[39] Pryor TA, Clayton PD, Gardner RM, Waki R, Warner HR. HELP—a hospital-wide system for computer-based support of decision-making. Proceedings of the 14th Annual Hawaii International Conference on Systems Sciences; Jan. 8, 1981.

[40] Waki R, Clayton PD, Jensen RL, Yanowitz FG, Liddle HV. Patient specific decision analysis using the HELP system. *Symposium on computer applications in medical care (SCAMC)* 1981;5:237–242.

[41] Gardner RM, Clemmer TP, Morris AH. Computerized medical decision-making—an evaluation in acute care. In: Prakash O, ed. *Computers in Critical Care and Pulmonary Medicine*. New York: Plenum Press; 1982;2:147–150.

[42] Gardner RM, Crapo RO, Morris AH, Beus ML. Computerized decision-making in the pulmonary function laboratory. *Resp Care* 1982;27:799–808.

[43] Gardner RM, West BJ, Pryor TA, et al. Computer-based ICU data acquisition as an aid to clinical decision-making. *Crit Care Med* 1982;10:823–830.

[44] Pryor TA, Gardner RM, Clayton PD, Warner HR. The HELP system. *Symposium on computer applications in medical care (SCAMC)* 1982;6:19–27.

[45] Blum BI, Lindberg DAB, Barnett GO, Warner HR, Lenhard RE, McDonald CJ. Information systems and patient care. *Symposium on computer applications in medical care (SCAMC)* 1982;6:3–7.

[46] Neeley JP, Ostler DB, Frederick PR, Clayton PD. Preexamination prediction of radiographic findings. *Invest Radiol* 1982;17:310–315.

[47] Waki R, Clayton PD, Jensen RL, Yanowitz FG, Liddle HV. HELP based decision analysis applied to coronary artery disease. *Comput Biomed Res* 1982; 15:188–202.

[48] Clayton PD, Gray S, Frederick PR. A code-oriented radiology reporting system based upon pre-examination prediction of likely findings. American Col-

lege of Radiology 7th Conference on Computer Applications in Radiology. 1982: 261–272.

[49] Gould BL, Clayton PD, Jensen RL, Liddle HV. Effects of early graft patency on long term morbidity. *Comput Cardiol* 1982;9:457–460.

[50] Pryor TA, Clayton PD, Larsen KG. Test ordering and medical decision making: a synergistic relationship. *Comput Healthcare* 1982;3(6):36–40.

[51] Pryor TA, Gardner RM, Clayton PD, Warner HR. The HELP system. *J Med Syst* 1983;7:87–102.

[52] Gardner RM. Information management—hemodynamic monitoring. *Semin Anesth* 1983;2:287–299.

[53] Warner HR. The tradeoffs between timesharing and dedicated computers in a medical setting. Proceedings of the AAMSI Conference '83; May, 1983: 305–307; San Francisco, Calif.

[54] Warner HR, Haug P. Medical data acquisition using an intelligent machine. Proceedings of MEDINFO '83 Seminar; July 1983: 582–584; Amsterdam, Holland.

[55] Cengiz M, Ranzenberger J, Johnson DS, Killpack AK, Lumpkin RW, Pryor TA. Design and implementation of computerized nursing care plans. *Symposium on computer applications in medical care (SCAMC)* 1983;7:561–564.

[56] Morris AH, Kanner RE, Crapo RO, Gardner RM. *Clinical Pulmonary Function Testing; A Manual of Uniform Laboratory Procedures*, 2nd ed. Salt Lake City: Intermountain Thoracic Society; 1984.

[57] Pryor TA, Warner HR, Gardner RM. HELP—A Total Hospital Information System. In: Blum BI, ed. *Information Systems for Patient Care*. New York: Springer-Verlag; 1984;109–128.

[58] Bradshaw KE, Gardner RM, Clemmer TP, Orme JF Jr, Thomas F, West BJ. Physician decision-making—evaluation of data used in a computerized iCU. *Intl J Clin Monit Comput* 1984;1:81–91.

[59] Gardner RM. Tomorrow's electronic hospital is here today. *IEEE Spectrum* 1984;21:101–103.

[60] Gardner RM. Computers in the ICU. *Med Electr* 1984;15:129–135.

[61] Gardner RM, Pryor TA, Clayton PD, Evans RS. Integrated computer network for acute patient care. *Symposium on computer applications in medical care (SCAMC)* 1984;8:185–188.

[62] Pryor DB, Barnett GO, Gardner RM, McDonald CJ, Stead WW. Measuring the value of information systems. *Symposium on computer applications in medical care (SCAMC)* 1984;8:26–28.

[63] Gardner RM, West BJ, Pryor TA. Distributed data base and network for ICU monitoring. *Comput Cardiol* 1984;11:305–307.

[64] Pryor TA. The role of clinical data on the computerized interpretation of ECGs. Computerized interpretation of the electrocardiogram. Proceedings of the 1983 Engineering Foundation conference; 1984:277–284.

[65] White KS, Lindsay A, Pryor TA, Brown WF, Walsh K. Application of a computerized medical decision-making process to the problem of Digoxin intoxication. *J Am Coll Cardiol* 1984;4:571–576.

[66] Preston K Jr, Fagan LM, Huang HK, Pryor TA. Computing in medicine. *IEEE Comput* 1984;17(10):294–313.

[67] Chapman R, Ranzenberger J, Killpack AK, Pryor TA. Computerized chart-

ing at the bedside: promoting the nursing process. *Symposium on computer applications in medical care (SCAMC)* 1984;8:700–702.

[68] Johnson DS, Ranzenberger J, Pryor TA. Nursing applications on the HELP system. *Symposium on computer applications in medical care (SCAMC)* 1984;8: 703–708.

[69] Killpack AK, Budd MC, Chapman RH, Ranzenberger J, Johnson DS, Pryor TA. Automating patient acuity in critical care units from nursing documentation. *Symposium on computer applications in medical care (SCAMC)* 1984;8: 709–711.

[70] Grams S, Dvorak RM, Pryor TA, Childs BW. Panel: trends in health care information systems. *Symposium on computer applications in medical care (SCAMC)* 1984;8:139–142.

[71] Pryor TA, Goldberg RD, Brown WF, Anderson JL. Computerized arrhythmia management of patients in a coronary care unit. *Comput Cardiol* 1984;11: 25–30.

[72] Kwok WY, Pryor TA, Hagan AD, Yanowitz FG. The effect of computer reported clinical information on the cardiologist's behavior in the interpretation of ECG's. *Comput Cardiol* 1984;11:39–44.

[73] Giles DJ, Thomas RJ, Osborn AG, et al. Lumbar spine: pretest predictability of CT findings. *Radiology* 1984;150:719–722.

[74] Liddle HV, Gould BL, Jones PD, Clayton PD. Conditional probability of multiple coronary graft failure. *J Thorac Cardiovasc Surg* 1984;87:526–531.

[75] Gould BL, Clayton PD, Jensen RL, Liddle HV. Association between early graft patency and late outcome for patients undergoing artery bypass graft surgery. *Circulation* 1984;69:569–576.

[76] Gerard MJ, Haug PJ, Morrison WJ, et al. A computer system for diagnosing pulmonary disease. Proceedings of the American Association for Medical System Informatics; May, 1984:119–123; San Francisco, Calif.

[77] Clayton PD, Haug PJ, Gerard MJ, et al. The role of radiology findings in automated medical decision-making. Proceedings of the 8th ACR Conference on Computer Applications in Radiology; May, 1984:521–531; St. Louis, Mo.

[78] Ostler DV, Gardner RM, Crapo RO. A computer system for analysis and transmission of spirometry waveforms using volume sampling. *Comput Biomed Res* 1984;17:229–240.

[79] Evans RS, Gardner RM, Bush AR, et al. Development of a computerized infectious disease monitor (CIDM). *Comput Biomed Research* 1985;18:103–113.

[80] Gardner RM. Computerized data management and decision making in critical care. *Surg Clin N Am* 1985;65(4):1041–1051.

[81] Andrews RD, Gardner RM, Metcalf SM, Simmons D. Computer charting: an evaluation of a respiratory care computer system. *Resp Care* 1985;30:695–707.

[82] Clemmer TP, Gardner RM. Data gathering, analysis, and display in critical care medicine. *Resp Care* 1985;30:586–598.

[83] Clayton PD, Pryor TA, Gardner RM, Haug PJ, Warner HR. HELP: A medical information system with decision making capability. In: Roger FH, Gronroos P, Tervo-Pellikka R, O'Moore R, eds. *6th International Congress of Medical Informatics*. Heidelberg: Springer-Verlag; 1985:127–131.

[84] Clayton PD, Gardner RM, Larsen KG, Pryor TA, Warner HR. HELP/ PATHLAB integration—A decade of experiences using an expert system interfaced to a clinical laboratory system. Fifth International Conference on Computing in Clinical Laboratories; June, 1985: 98–103; Stuttgart, Federal Republic of Germany.

[85] Clayton PD, Pryor TA, Gardner RM, Warner HR. HELP—a medical information system which combines automated medical decision-making with clinical data review and administrative support. In: Koller S, Reichutz PL, Ueberla K, eds. *Medizinische Informatik und Statistik*. Heidelberg: Springer-Verlag; 1985: 62:266–272.

[86] Ostler MR, Stansfield JD, Pryor TA. A new, efficient version of HELP. *Symposium on computer applications in medical care (SCAMC)* 1985;9:296–297.

[87] Bekemeyer WB, Crapo RO, Calhoon S, Cannon CY, Clayton PD. Efficacy of chest radiography in a respiratory intensive care unit—a prospective study. *Chest* 1985;88:691–696.

[88] Gould BL, Clayton PD. The effect of quality of life estimates on the treatment decisions for coronary artery disease. *IEEE Compu Cardiol* 1985;12:127–130.

[89] Gardner RM, Evans RS, Andrews BS. Impact of a clinical information system on hospital costs. In Kuhn RL, ed. *Frontiers of Medical Information Science*. New York: Praeger; 1985:81–89.

[90] Gardner RM. Computerized management of intensive care patients. *MD Comput* 1986;3(1):36–51.

[91] Gardner RM. Artificial intelligence in medicine—is it ready? *Intl J Clin Monit Comput* 1986;2:133–134. Editorial.

[92] Gardner RM, Monis SM, Oehler P. Monitoring direct blood pressure: algorithm enhancements. *IEEE Comput Cardiol* 1986;13:607–610.

[93] Albright FS, Pryor TA. The effect of fuzzy logic on the stability of serial computerized ECG interpretations. *Comput Cardiol* 1986;13:687–690.

[94] Gardner RM, Clausen JL, Cotton DJ, et al. Computer guidelines for pulmonary laboratories. *Am Rev Resp Dis* 1986;134:628–629.

[95] Evans RS, Larsen RA, Burke JP, et al. Computer surveillance of hospital-acquired infections and antibiotic use. *JAMA* 1986;256:1007–1011.

[96] Crapo RO, Gardner RM, Berlin SL, Morris AH. Automation of pulmonary function equipment—user beware! *Chest* 1986;90:1–2. Editorial.

[97] Blaufuss J. Promoting the nursing process through computerization. *MEDINFO* 1986;585–586.

[98] Clayton PD, Haug PJ, Shelton P, et al. Automated decision-making in radiology: advantages of an integrated clinical information system. *MEDINFO* 1986; 624–628.

[99] Haug PJ, Warner HR, Clayton PD, Schmidt CD, Pearl JE, Farney RJ. A computer-directed patient history: functional overview and initial experience. *MEDINFO* 1986;849–852.

[100] Warner HR, Detmer DE, Peay WJ. IAIMS implementation and administration at the University of Utah. *MEDINFO* 1986;945–946.

[101] Gardner RM, Pryor TA, Clayton PD, Evans RS, Warner HR. The HELP system: A system for clinical decision-making. *MEDINFO* 1986;1143.

[102] Gardner RM, Hawley WL. Standardizing communications and networks in

the ICU. Patient Monitoring and Data Management Conference. AAMI Technology Analysis and Review; 1986; TAR No. 11–85:59–63.

[103] Dudeck J, Clayton PD, Sebald P, et al. Erfarung bei der Anpassung des Krankenhaus — Informations Systems HELP an die Deutsche Umgeburg. In: *Medizinishe Informatik in der Schweiz*. Hagenbuch RE, ed. Basel: Schwabe; 1986; 127–146.

[104] Jacobson JT, Johnson DS, Ross CA, Conti MT, Evans RS, Burke JP. Adapting disease-specific isolation guidelines to a hospital information system. *Infect Cont* 1986;7(8):411–418.

[105] Haug PJ, Hebertson RM, Heywood RE, et al. Computer-assisted pregnancy management. *Symposium on computer applications in medical care (SCAMC)* 1987;11:158–161.

[106] Johnson D, Wigertz G, Pryor TA. Nurse Charting on the HELP system. MIE Conference; 1987.

[107] Pryor TA, Warner HR, Gardner RM, Clayton PD, Haug PJ. The HELP system development tools. In: Orthner H, Blum BI, eds. *Implementing Health Care Information Systems*. New York: Springer-Verlag; 1989:363–383.

[108] Clayton PD, Haug PJ, Pryor TA, Wigertz OB. Representing a medical knowledge base for multiple uses. *AAMSI* 1987;289–293.

[109] Haug PJ, Warner HR, Clayton PD, et al. A decision-driven system to collect the patient history. *Comput Biomed Res* 1987;20:193–207.

[110] Haug PJ, Tocino I, Clayton PD, Bair TL. Automated management of breast cancer screening. *Radiology* 1987;164:747–752.

[111] Johnson D. Decisions and dilemmas in the development of a nursing information system. *Comput Nurs* 1987;5:94–98.

[112] Gardner RM, Ostler DV, Duffy OH. Computers in the emergency room. *Int Med Specialist* 1987;8(3):105–114.

[113] Gardner RM. Computer technology and its application to cardiovascular nursing. *J Cardiovasc Nurs* 1987;1:69–71.

[114] Pryor TA, Clayton PD, Haug PJ, Wigertz OB. Design of a knowledge driven HIS. *Symposium on computer applications in medical care (SCAMC)* 1987;11: 289–293.

[115] Clayton PD, Delaplaine KH, Jensen RD, Bird B, Evans RS, Cannon CY. Integration of surgery management and clinical information systems. *Symposium on computer applications in medical care (SCAMC)* 1987;11:393–395.

[116] Clayton PD, Evans RS, Pryor TA, et al. Bringing HELP to the clinical laboratory — use of an expert system to provide automatic interpretation of laboratory data. *Ann Clin Biochem* 1987;24:S1-5 – S1-11.

[117] Bouhaddou O, Haug PJ, Warner HR. Use of the HELP clinical database to build and test medical knowldege. *Symposium on computer applications in medical care (SCAMC)* 1987;11:64–67.

[118] Evans RS, Gardner RM, Burke JP, et al. A computerized approach to monitor prophylactic antibiotics. *Symposium on computer applications in medical care (SCAMC)* 1987;11:241–245.

[119] Halford G, Pryor TA, Burkes M. Measuring the impact of bedside terminals. *Symposium on computer applications in medical care (SCAMC)* 1987;11:359–362.

[120] Johnson DS, Burkes M, Sittig D, Hinson D, Pryor TA. Evaluation of the

effects of computerized nurse charting. *Symposium on computer applications in medical care (SCAMC)* 1987;11:363-367.

[121] Sittig DF. Computerized management of patient care in a complex, controlled clinical trial in the intensive care unit. *Symposium on computer applications in medical care (SCAMC)* 1987;11:225-232.

[122] Haug PJ, Clayton PD, Shelton P, et al. A revision of diagnostic logic using a clinical database. Proceedings of the AAMSI Congress; 1987:238-242.

[123] Gardner RM, Pryor TA, Clayton PD, Evans RS, Johnson DS. Use and evaluation of the HELP clinical decision-making system. *Symposium on computer applications in medical care (SCAMC)* Demonstration, 1987.

[124] Andrews BS, Gardner RM. Portable computers used for respiratory care charting. *Intl J Clin Monit Comput* 1987;5:45-51.

[125] Haug PJ, Rowe KG, Rich T, et al. A comparison of computer-administered histories. Proceedings of the AAMSI Congress; 1988:21-25.

[126] Gardner RM, Sittig DF, Budd MC. The computer in the ICU: match or mismatch? In: Shoemaker WC, ed. *Textbook of Critical Care.* 2nd ed. Philadelphia: WB Saunders; 1989; 248-258.

[127] Gardner RM. Patient monitoring applications. In: Shortliffe EH, Perrault LE, Fagan LH, Wiederhold GCM, eds. *Medical Informatics.* Reading, Mass: Addison-Wesley; 1990:366-399.

[128] Bradshaw KE, Sittig DF, Gardner RM, Pryor TA, Budd M. Improving efficiency and quality in a computerized ICU. *Symposium on computer applications in medical care (SCAMC)* 1988;12:763-767.

[129] Sittig DF, Elliott CG, Wallace CJ, Bailey P, Gardner RM. Computerized screening for identification of adult respiratory distress syndrome (ARDS) patients. *Symposium on computer applications in medical care (SCAMC)* 1988;12: 698-702.

[130] Hawley WL, Tariq H, Gardner RM. Clinical implementation of an automated medical information bus in an intensive care unit. *Symposium on computer applications in medical care (SCAMC)* 1988;12:621-624.

[131] Ostler DC, Gardner RM, Logan JS. A medical decision support system for the space station health maintenance facility. *Symposium on computer applications in medical care (SCAMC)* 1988;12:43-47.

[132] Shabot MM, Chamberlin WH, Gardner RM, Leyerle BJ. Extracting severity of illness: Tapping the power of the electronic medical record. *Symposium on computer applications in medical care (SCAMC)* Panel; 1988.

[133] Huff SM, Evans RS, Gandhi S, Jensen B. Development of a regional laboratory healthcare network. *Symposium on computer applications in medical care (SCAMC)* 1988;12:643-647.

[134] Prokosch HU, Pryor TA. Intelligent data acquisition in order entry programs. *Symposium on computer applications in medical care (SCAMC)* 1988; 12:454-458.

[135] Ranum DL. Knowledge based understanding of radiology text. *Symposium on computer applications in medical care (SCAMC)* 1988;12:141-145.

[136] Budd M, Blaufuss J, Propotnik T, Maynard J, Klingle C, Pryor TA. An automated patient classification system for staffing, billing and productivity measurement. *Symposium on computer applications in medical care (SCAMC)* 1988;12:785-789.

[137] Blaufuss JA, Tinker AK. Computerized falls alert — a new solution to an old

problem. *Symposium on computer applications in medical care (SCAMC)* 1988; 12:69–72.

[138] Pryor TA. The HELP medical record system. *MD Comput* 1988;5:22–33.

[139] Prokosch HU, Hulse RK, Wall M, Wong TW, Pryor TA. New decision support concepts for the pharmacy application in the HELP system. Proceedings of the EFMI Special Topic Congress "Expert Systems and Decision Support in Medicine"; 1988; Hannover, West-Germany.

[140] Sittig DF, Pace NL, Gardner RM, Beck E, Morris AH. Implementation of a computerized patient advice system using the HELP clinical information system. *Comput Biomed Res* 1989;22:474–487.

[141] Bradshaw K, Gardner RM, Pryor TA. Development of a computerized laboratory alerting system. *Comput Biomed Res* 1989;22:575–587.

[142] Prokosch HU, Wong TW, Pryor TA. Medication ordering based on a predictive knowledge base. *Artificial Intelligence in Medicine*, 1989;1:41–48.

[143] Bradshaw D, Sittig DF, Gardner RM, Pryor TA, Budd M. Computer-based data entry for nurses in the ICU. *MD Comput* 1989;6(5):274–280.

[144] Pryor TA. Computerized Nurse Charting. *Intl J Clin Monit Comput* 1989;6: 173–179.

[145] Pryor TA. Computers in medicine. *Utah Business Magazine*; November, 1989.

[146] Nelson BD, Gardner RM, Ostler DV, Schulz JM. Medical Impact Analysis for the Space Station. *Aviat Space Environ Med* 1990;61:169–175.

[147] Elliott CG, Simmons D, Schmidt CD, Enger K, Greenway L, Gardner RM. Computer assisted medical direction of respiratory care. *Resp Manage* 1989;19: 31–35.

[148] Gardner RM, Ostler DV, Nelson BD, Logon BS. The role of smart medical systems in the space station. *Intl J Clin Monit Comput* 1989;6:91–98.

[149] Sittig DF, Gardner RM. International symposium on computer-assisted decision support and database management in anesthesia, intensive care, and cardiopulmonary medicine. *Intl J Clin Monit Comput* 1989;6:1–6. Summary.

[150] Gardner RM. Pulmonary function laboratory standards. *Resp Care* 1989;34: 651–660.

[151] Gardner RM. HELP: Integrating Clinical Data Collection and Decision Making. New Horizons in Health Information. New York State Department of Health; January 1989.

[152] East TD, Henderson S, Morris AH, Gardner RM. Implementation issues and challenges for computerized clinical protocols for management of mechanical ventilation in ARDS patients. *Symposium for computer applications in medical care (SCAMC)* 1989;13:583–587.

[153] Henderson S, East TD, Morris AH, Gardner RM. Performance evaluation of computerized clinical protocols for management of arterial hypoxemia in ARDS patients. *Symposium for computer applications in medical care (SCAMC)* 1989;13:588–592.

[154] Lundsgaarde HP, Gardner RM, Menlove RL. Using attitudinal questionnaires to achieve optimization. *Symposium for computer applications in medical care (SCAMC)* 1989;13:703–708.

[155] Sittig DF, Gardner RM, Elliott CG. Screening for adult respiratory distress syndrome patients: use of the HELP hospital information system. *J Clin Engin* 1989;14:237–243.

[156] Gardner RM, Bradshaw KE, Hollingsworth KW. Computerizing the intensive care unit: current status and future directions. *J Cardiovasc Nurs* 1989;4: 68–78.

[157] Larsen RA, Evans RS, Burke JP, Pestotnik SL, Gardner RM, Classen DC. Improved peri-operative antibiotic use and reduced surgical wound infections through use of computer decision analysis. *Infect Cont Hosp Epidemiol* 1989; 10:316–320.

[158] Haug PJ, Clayton PD, Tocino I, et al. A clinical audit in the radiology department based on information theory. *Radiology* 1989;173;352.

[159] Haug PH, Clayton PD, Shelton P, et al. Revision of Diagnostic Logic Using a Clinical Database. *Med Dec Mak* 1989;9:84–90.

[160] Haug PJ, Ranum DL. Computerized extraction of coded findings from free-text radiology reports. *Radiology* 1990;174:543–548.

[161] Huff SM, Evans RS, Gandhi S. An LIS to HIS Interface Using CCITT Standard Protocols. *Symposium for computer applications in medical care (SCAMC)* 1989;13:692–695.

[162] Haug PJ, Frederick PR. A Proposal for Computerized Medical Audit in Radiology. Medical Informatics and Education International Symposium. 1989: 613–617; The University of Victoria, Victoria, B.C., Canada.

[163] Pryor TA, Dupont R, Clay JA. MLM-based order entry system: the use of knowledge in a traditional HIS application. *Symposium for computer applications in medical care (SCAMC)* 1990;14:579–583.

[164] Haug PJ, Hoak S. Veristat: a support tool for knowledge development. *Symposium for computer applications in medical care (SCAMC)* 1990. In press.

[165] Huff S, Warner HR. Comparison of Meta-1 and HELP terms: implications for clinical data. *Symposium for computer applications in medical care (SCAMC)* 1990. In press.

[166] Tate KE, Gardner RM, Weaver LK. Impact of a computerized laboratory alerting system on patient care and outcome. *MD Comput* 1990;7:296–301.

[167] Pestotnik SL, Evans RS, Burke JP, Gardner RM, Classen DC. Therapeutic antibiotic monitoring: surveillance using a computerized expert system. *Am J Med* 1990;88:43–48.

[168] Gardner RM, Tariq H, Hawley WL, East TD. Medical information bus: the key to future integrated monitoring. *Intl J Clin Monit & Comput* 1989;6:205–209. Editorial.

[169] Sittig DF, Gardner RM, Morris AH, Wallace CJ. Clinical evaluation of computer-based respiratory care algorithms. *Intl J Clin Monit & Comput* 1990; 7:177–185.

[170] Gardner RM, Laub RM, Golubjatnikov OK, Evans RS, Jacobson JA. Computer critiqued blood ordering using the HELP system. *Comput Biomed Res* 1990;23:514–528.

[171] Lepage E, Gardner RM, Laub RM, Golubjatnikov OK. Use of blood components in a hospital with a computerized information system. Submitted for publication.

[172] Sittig DF, Gardner RM, Pace NL, Morris AH, Beck E. Computerized management of patient care in a complex, controlled clinical trial in the intensive care unit. *Comput Meth Prog Biomedicine* 1989;30:77–84.

[173] Gardner RM, Shabot MM. Computerized ICU data management: pitfalls and promises. *Intl J Clin Monit Comput* 1990;7:99–105.

[174] Gardner RM, Hulse RK, Larsen KG. Assessing the effectiveness of a computerized pharmacy system. *Symposium for computer applications in medical care
(SCAMC)* 1990;14:99–105.

[175] Kuperman GJ, Gardner RM. Impact of the HELP system on the LDS Hospital Medical Record. *Symposium for computer applications in medical care
(SCAMC)* 1990;14:673–677.

[176] Henderson S, Crapo RO, East TD, Morris AH, Gardner, RM. Computerized
clinical protocols in an intensive care unit: how well are they followed? *Symposium for computer applications in medical care (SCAMC)* 1990;14:284–288.

[177] Evans RS, Burke JS, Pestotnick SL, Classen DC, Menlove RL, Gardner RM.
Prediction of hospital infections and selection of antibiotics using an automated
hospital data base. *Symposium for computer applications in medical care
(SCAMC)* 1990;14:663–667.

[178] East TD, Morris AH, Clemmer TP, et al. Development of computerized
critical care protocols – a strategy that really works! *Symposium for computer
applications in medical care (SCAMC)* 1990;14:564–568.

[179] Grandia LD, Gardner RM. Quality and productivity results of expert system-
based clinical hospital information systems (HELP). American Hospital Association Meeting; March 1990.

[180] Evans RS, Burke JP, Classen DC, et al. Computerized identification of
hospital-acquired infections and high risk patients. Submitted for publication.

Appendix D

LDS Hospital statistics for 1990

Total acute admissions	20,563
Acute care days	107,035
Average length of stay	5.21 days
Number of beds	475
Average occupancy	293
Percent occupancy	62%
Hospital budget	$200 million
Medical informatics budget	$1.2 million
Charity care	$3.2 million
Total emergency cases	22,277
Total ICU admits	2,335
Total laboratory procedures	926,707
Radiation therapy cases	28,272
Maternity admissions	3,345
Total deliveries	3,600
Total surgical procedures	14,549
Open heart surgeries	837
Total hip replacements	360
Transplant procedures	127
Kidney	75
Heart	24
Liver	16
Pancreas	12
Total radiology procedures	83,400
X-ray	57,998
Ultrasound	6,051
CT scans	6,330
Mammography	6,076
Nuclear medicine	3,159
Magnetic resonance imaging	2,538
Angiography	1,248

Index